PREHISTORIC COPPER MINING IN EUROPE

Prehistoric Copper Mining in Europe

5500–500 BC

WILLIAM O'BRIEN

OXFORD
UNIVERSITY PRESS

OXFORD
UNIVERSITY PRESS

Great Clarendon Street, Oxford, OX2 6DP,
United Kingdom

Oxford University Press is a department of the University of Oxford.
It furthers the University's objective of excellence in research, scholarship,
and education by publishing worldwide. Oxford is a registered trade mark of
Oxford University Press in the UK and in certain other countries

First Edition published in 2015

Published in the United States of America by Oxford University Press
198 Madison Avenue, New York, NY 10016, United States of America

British Library Cataloguing in Publication Data
Data available

Library of Congress Control Number: 2014949006

ISBN 978–0–19–960565–1

This book is dedicated to
Paul Ambert and Miguel Ángel de Blas Cortina:
from Ross Island to Cabrières and El Aramo.

Preface

From its earliest use in Europe some 8,000 years ago, the ability to produce copper marked a new stage in the human relationship with the natural world. This book is the first comprehensive review of geo-archaeological research conducted on some five millennia of prehistoric copper mining in Europe. There is a broad-ranging look at field data collected over the past 150 years or so by archaeologists, mining engineers, and geologists. These regional studies come together for the first time to present a remarkable story of early human achievement.

The study commences with the oldest known copper mines in Europe, dating to the sixth and fifth millennia BC in the Balkans. It examines the record of Chalcolithic (Copper Age) and Bronze Age mining for many parts of Europe, from the eastern and central Mediterranean to Iberia, southern France, Britain, and Ireland, the Eastern Alps and central Europe, as far as the Russian Urals. The focus is on research findings from key sites, including such famous mining centres as the Mitterberg in Austria, Kargaly in Russia, the Great Orme in Wales, and those examples in Cyprus from where the name copper derives. The history of research in each region is examined, as is the survival of early mine archaeology in different geological settings. Survey and archaeological excavation provides a wealth of data on technological processes of mineral prospecting, ore extraction, and smelting. An analysis of this information offers various insights into the organization of early copper production as part of a *chaîne opératoire* that operated within a wider economic and social context. The impact of early copper mining and smelting on the environment is also considered.

The study of these mines has progressed from antiquarian speculation to the first systematic research conducted in Austria in the late nineteenth century. The early work at the Mitterberg mines near Salzburg by Matthäus Much, Johann Pirchl, Georg Kyrle, and Oliver Klose, followed in 1932 by a seminal publication by Karl Zschocke and Ernst Preuschen, laid a foundation for research in the modern era. The pace of work steadily increased from the 1960s with fieldwork programmes in different parts of Europe led by such researchers as Evgenij Chernykh, Clemens Eibner, Borislav Jovanović, Richard Pittioni, and Beno Rothenberg, amongst others. These include my research in Ireland and similar studies elsewhere, work that continues today with such initiatives as the HiMAT programme based in Innsbruck. This fieldwork is also linked to developments in science-based archaeology, such as the important application of lead isotope analysis to identify early mine sources pioneered by Noel Gale, Zofia Stos-Gale, Ernst Pernicka, and others.

It is often remarked that the study of prehistoric copper mines has for the most part been conducted in a contextual vacuum. This view, however accurate, partly reflects the specialist nature of a subject that requires expertise across several disciplines, including archaeology and geology, environmental sciences, and history. The prominence of geologists and engineers in the early years of research naturally led to an emphasis on technological process and the organizational aspects of mining. Sometimes dismissed as mechanistic in nature, such studies have nonetheless made a valuable contribution to the history of technology, and provide a basis for more anthropologically informed consideration of ancient mining. Researchers in this field have always been concerned with the cultural identity of early copper miners and the significance of their labour for a wider society. The most challenging, and yet interesting, part of this research is possibly the distinctive nature of the mining communities concerned, with their own traditions and physical separation from an outside world. The key to contextualizing their work is to follow the trail of metal as it leaves the mines and enters a wider sphere of practical use and symbolic value in society.

One of the purposes of the book is to make this specialist work accessible to a wider academic audience and so further place it within the mainstream of prehistory research in Europe. This is a proper time to acknowledge the work and contributions made by many researchers in this field. The Austrian research has been to the fore, but there is also the work of other pioneers across Europe, from Ireland to Russia. Their work laid the foundation for the research of my generation and those that will follow. In writing this book, I received much help from the following colleagues: Paul Ambert, Hélène Barge, Miguel Angel de Blas Cortina, Evgenij Chernykh, Brenda Craddock, Gert Goldenberg, Sian James, David Jenkins, Borislav Jovanović, Lina Kassianidou, Rüdiger Krause, Andy Lewis, Tim Mighall, Ignacio Montero Ruiz, Jose Miguel Nieto, Mark Pearce, Miljana Radivojević, Thomas Stöllner, and Simon Timberlake. I am particularly grateful to Nick Hogan and James O'Driscoll for help in preparing the illustrations and to many of the aforementioned who allowed me to use these images. As always, my wife Madeline improved my poor English and was a constant support, as indeed was my new best friend, Google Translate. I am grateful to University College Cork for providing me with the opportunity to carry out this research and to our Department of Archaeology who supported the project. I also acknowledge the support of Oxford University Press in publishing the book, in particular Mary Hobbins, Annie Rose, and Kizzy Taylor-Richelieu, as well as the anonymous referees for their helpful comments.

I have no doubt that a truly great book will one day be written on this most interesting of subjects. Hopefully, this publication is a step in that direction. For me personally, it represents a research interest that began three decades ago on the slopes on Mount Gabriel, followed by fieldwork at Ross Island, Derrycarhoon, and other early copper mines in south-west Ireland. This journey has also taken me to visit similar mines in some of the most beautiful landscapes of Europe, in the company of dedicated researchers and intrepid fieldworkers whose endeavours are presented in this book.

William O'Brien

Cork
January 2014

Contents

List of illustrations

List of tables

1

Europe: the birthplace of mining?

The origins of mining extend back several million years to when the first hominines in Africa, and subsequently in Europe and Asia, used their bare hands to collect loose rock to make stone tools. This gradually evolved into the systematic collection of rocks with desired properties, as well as an interest in earth minerals such as iron ochre. A desire for these substances led in some cases to more organized collection, involving surface quarrying and eventually underground mining. This activity increased significantly in Europe with the adoption of farming after 7000 BC in the Neolithic period. There is evidence for highly organized mining of flint and other hard rocks to make stone axes and other tools at centres such as Grimes Graves in England, Spiennes in Belgium, and Casa Montero near Madrid, among many others.

Mining is also the process by which metal was obtained from the Earth's surface. Metal objects were first made in the Middle East approximately 10,000 years ago; however, the oldest known copper mines are recorded in Europe. The discovery there of thousands of copper, bronze, and gold objects is a strong indicator that the mining of these metals was a widespread activity during the late prehistoric period. This began with the first use of copper and gold in what is known as the Chalcolithic ('Copper Age'), which occurred in different regions between the sixth and third millennia BC. Technical advances and growing demand for metal led to the widespread adoption of bronze by 2000 BC, or soon afterwards. This was made possible by the discovery of copper and tin sources in many parts of Europe. Some of these were mined intensively over long periods during the Bronze Age that followed, while others were worked on a smaller scale. The copper produced was supplied to areas without their own resources, creating trade networks that provided economic opportunities and played an important role in cultural exchanges across the Continent.

This book is concerned with the prehistory of copper mining and its important place in shaping European societies during the period 5500–500 BC. The archaeology of these early mine sites and landscapes is considered with reference to the technological processes of mineral extraction and metal production, and how that activity was organized in relation to the human participants and their distinctive communities. The wider implications of copper supply for prehistoric societies are examined, as is the impact that mining had on the natural environment. Before considering the archaeological evidence, it is necessary to discuss the natural occurrence of copper, to understand how its geological formation influenced the approach to mining at different locations over time.

THE GEOLOGICAL ENVIRONMENT

Though occasionally found as pure ('native') metal, copper most commonly occurs in the form of a rock mineral, where it is chemically combined with other elements. This may take the form of primary copper minerals, which can contain significant amounts of sulphur and other elements such as iron. Where these occur close to the Earth's surface, they can be oxidized in different ways to form secondary minerals, often with a higher copper content (Table 1). Where these primary and/or secondary metallic minerals occur in a concentrated form, this constitutes what is known in economic terms as 'ore'. This means the mineral deposit (orebody) is viable for mining at a particular time. The success of such ventures depended on the concentration of metal in the ore, known as its 'grade', the size of the ore deposit, and many other factors. The barren portion of an orebody from which the metallic minerals are separated is known as 'gangue'. The significance of these terms in the context of prehistoric mining will be discussed later in this chapter.

Most orebodies are heterogenous in their chemical and mineralogical composition. This results from complex geological processes involved in the initial formation and subsequent alteration of the mineral deposit. When

Table 1. The more common copper minerals found in prehistoric copper mines (with general estimates of copper content). There are in excess of 150 known copper minerals, most of which have rare occurrence and, unlike many of those listed, do not occur in sufficient concentration to qualify as ores.

Oxidized copper minerals	Cu %	Sulphidic copper minerals*	Cu %
Cuprite Cu_2O	89	Chalcopyrite $CuFeS_2$	35
Tenorite (melaconite) CuO	80	Bornite Cu_5FeS_4	63
Paramelaconite Cu_4O_3	80	Idaite Cu_3FeS_4	62
Malachite $Cu_2CO_3(OH)_2$	57	Chalcocite Cu_2S	80
Azurite $Cu_3(CO_3)_2(OH)_2$	55	Covellite CuS	66
Chrysocolla $CuSiO_32H_2O$	36	Tetrahedrite $Cu_{12}Sb_4S_{13}$	51
Atacamite $Cu_2Cl(OH)_3$	59	Tennantite $Cu_{12}As_4S_{13}$	51
Antlerite $Cu_3SO_4(OH)_4$	54	Enargite Cu_3AsS_4	48
Chalcanthite $Cu_4SO_4(OH)_6$	67	Famatinite Cu_3SbS_4	43

exposed to surface weathering, the upper part of a metalliferous orebody may develop distinct zones of oxidation, mostly through the action of percolating water. This can lead to the formation of what is known as secondary mineralization. Some of these copper minerals were available to the prehistoric miner, while other parts of an orebody could not be exploited for different reasons. In addition to metalliferous ores, these deposits also contain a significant amount of non-metallic mineral gangue, which is generally discarded as waste rock in mining operations.

Ore geology

Some discussion of the different types of orebody is necessary to understand the geological environment in which early miners worked. The classification of these deposits is complex due to their varied forms and different geological settings. Several criteria can be applied, including the physical shape of the deposit, its mineral associations, and the age and nature of the host geology, sometimes referred to as the 'country rock'. Most descriptive classifications are linked to some understanding of ore genesis, even though geologists do not always agree as to the processes involved. The following is an example of a broad classification of ore deposits based on geological associations and genesis (Table 2).

The formation of metalliferous orebodies involved complex geological processes ranging from magmatic concentration, contact metasomatism, and hydrothermal deposits to sedimentary, bacterial, submarine exhalative, and other processes (Fig. 1.1). This means that no two copper deposits are exactly alike, with considerable variation in mineralogy, texture, shape, size, and other features. Most were formed by hydrothermal fluids carrying metals and other minerals in solution from magma sources deep in the Earth. In some instances these fluids filled open cavities and crystallized to form an ore deposit, while this could also involve a replacement process within a host rock. A common distinction in the classification of metallic orebodies is between syngenetic and epigenetic deposits. The former are mineral deposits that were created at the

Table 2. Classification of copper deposits

1. Plutonic: includes ultra-mafic and mafic complexes, carbonatite and porphyry complexes, pyrometasomatic skarns
2. Hydrothermal: includes hydrothermal veins, replacement and breccia pipe ores
3. Volcanogenic: includes stratabound massive base metal sulphides and disseminated sulphides in aquagene tuffs and agglomerates
4. Sedimentary: includes deposits in continental red-beds and calc-arenites.

Source: Jacobsen 1975. More detailed examples are available in standard textbooks on ore geology (e.g. Jensen and Bateman 1981 and Robb 2005).

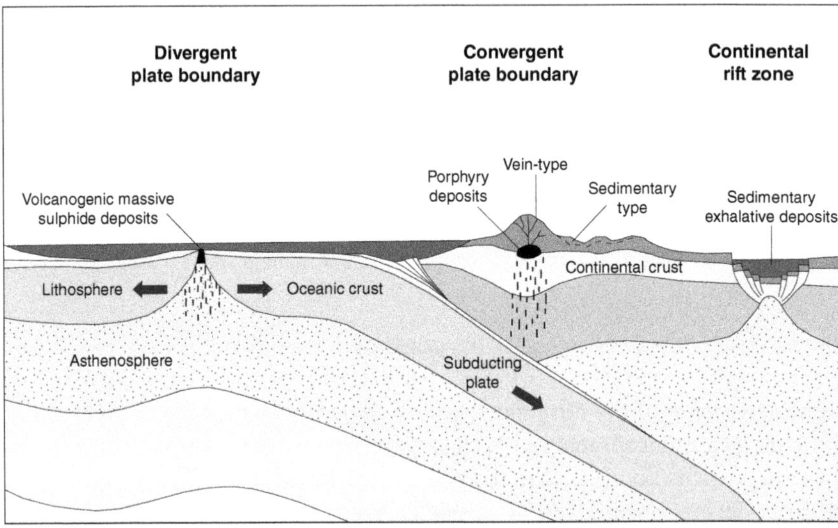

Fig. 1.1. Geological background to the formation of copper ore deposits

(*Source*: author).

same time as the country rock, whereas epigenetic orebodies formed much later than the host geology.

Genetic models and classifications are important in terms of characterization of orebodies and the metal they yielded, but would not have been part of any understanding prehistoric miners had of these mineral deposits. They were more concerned with the exposure and physical occurrence of the mineralization, how rich it was in terms of metal concentration and how easily this ore could be extracted. For such reasons, many orebodies were not accessible during the prehistoric period. In some instances only a portion of an orebody could be mined because of zonation within the mineral occurrence. There are also situations where there is more than one type of orebody present in a mine area.

While there is great variation in ore deposits, different types can be identified based on their physical features. Evans (1980) classified orebodies according to whether they are concordant or discordant with the lithological banding of the enclosing rocks. The former include what are known as stratiform orebodies, such as those in sedimentary geology where the mineralization conforms to the bedding structure of the rock. These can be epigenetic infillings of pore spaces, or replacement deposits, or else syngenetic ores formed by the exhalation of mineralizing solutions. They commonly occur in limestones as those rocks are highly soluble and reactive. Mineralization can form in particular beds or at certain stratigraphic interfaces, where the permeability of the host rock was increased by fracturing or processes such as dolomitization. Other fine-grained rocks, such as argillites, mudstone, shale, and slate, can be important hosts for concordant orebodies. Some of these are very extensive, the best example in Europe being the Kupferschiefer, a thin layer of metal-bearing slate of Upper Permian age, which occurs over a 600,000 square kilometre area extending from eastern England to Poland.

Concordant mineralization can also occur in coarser-grained sedimentary rocks. These include red-bed sequences where copper minerals are disseminated through permeable sandstone units. Other examples of concordant orebodies occur in igneous rocks, such as massive sulphide deposits associated with mafic volcanics, or vesicular filling deposits in basic lava flows. Many orebodies in volcaniclastic rocks formed where a stockwork of veins from a deep magma source provided channels by which mineralized fluids fed an overlying deposit of sulphide mineralization (Evans 1980).

Discordant orebodies cut across the geological banding of the host rock. They can be either regular or irregular in shape. Regular deposits may be tubular (pipes and mantos) in shape, but are commonly tabular (extensive in two dimensions). The latter includes veins, which are epigenetic orebodies formed by the infilling of a pre-existing open space by hydrothermal solutions. Veins were an important source of copper in prehistory, as they provided a localized deposit of highly concentrated mineralization. They were also relatively easy to find, as their outcrop can be highly visible against the country rock, marked by exposures of white quartz or calcite, and by colourful minerals produced by surface oxidation.

Hydrothermal veins typically form as a result of tectonic forces that fracture solid rock and create open spaces. Hot aqueous solutions rising up from magmatic sources fill these fissures, leading to the precipitation of metals and other minerals at a range of temperatures and pressures. Such veins are classified by their depth and temperature of formation into hypothermal, mesothermal, and epithermal deposits (after Lindgren 1933). The gangue minerals in these veins often reflect the country rock, with quartz being dominant where the host rocks are silicates and calcite where they are carbonates (Evans 1980: 145). Other gangue minerals include barite, chlorite, iron-bearing compounds, and a range of silicates, which miners had to separate from the metallic ore by a process known as beneficiation.

Veins are often inclined, with a defined hanging wall and footwall. There is great variation in vein thickness, from a few millimetres to as much as 100 metres. The vein thickness can pinch and swell depending on changes in the rock lithology that they pass through. They may occur singly or in groups, with varying degrees of surface exposure. Veins frequently form in fracture systems (fissure veins) caused by tectonic faulting, which is why they often branch and divide, and also can have a regular orientation. The infilling process involves the precipitation of metal-bearing sulphidic minerals at concentrations much higher than background geology. This may involve a single metalliferous mineral, but it is more common to find several occurring as part of a complex intergrowth of ore and gangue. The limits of the mineralization may be formed by the vein wall (contact with the country rock) or by assay boundaries within the vein.

There are several types of irregularly shaped discordant orebodies. These include deposits where the ore minerals are disseminated through the body of the host rock, as well as those formed in closed-spaced veinlets forming a stockwork that cuts through the host rock and across geological boundaries. In general, this mineralization fades gradually towards the margins and its limits are defined by metal content. Other irregular orebodies include those formed by the replacement of existing rocks. This is common in the case of carbonate-rich sediments such as limestone where they come into contact with igneous intrusives (contact metamorphism). These include so-called skarn deposits, which can be very irregular in shape. Somewhat different are so-called 'flats', which are horizontal ore deposits that branch out from vertical or inclined veins within carbonate host rocks occurring beneath an impervious rock type (Evans 1980).

While the physical shape of an orebody had a major bearing on how it was mined, this was also influenced by its physical exposure on the surface. The nature of the outcrop was determined in the first instance by geological structure; however, subsequent processes of weathering and erosion shaped the topographic expression of an orebody in a particular landform. This is not a major issue in modern operations, but the topographic expression of a mineralized outcrop had a significant bearing on how mining was conducted in prehistoric times.

Supergene enrichment

As stated previously, various processes of oxidation and weathering can affect the surface exposure of an orebody. The oxidizing action of percolating water is particularly significant, as it creates acidic solutions that can dissolve other minerals. Copper sulphides are soluble and, in extreme cases, the upper part of the orebody can be entirely leached of its metal content in what is known as the zone of oxidation (Fig. 1.2). This process intensifies in deposits with a high pyrite content where the mineral breaks down to form sulphuric acid, leaving a deposit of iron hydroxide at the surface known as a gossan, or what miners used to call an 'iron hat'.

As water percolates down through the zone of oxidation, it can precipitate secondary copper minerals, such as cuprite, malachite, or azurite, and, in some instances, native copper. Much of the dissolved copper will stay in solution until it reaches the water table, at which point conditions become gradually

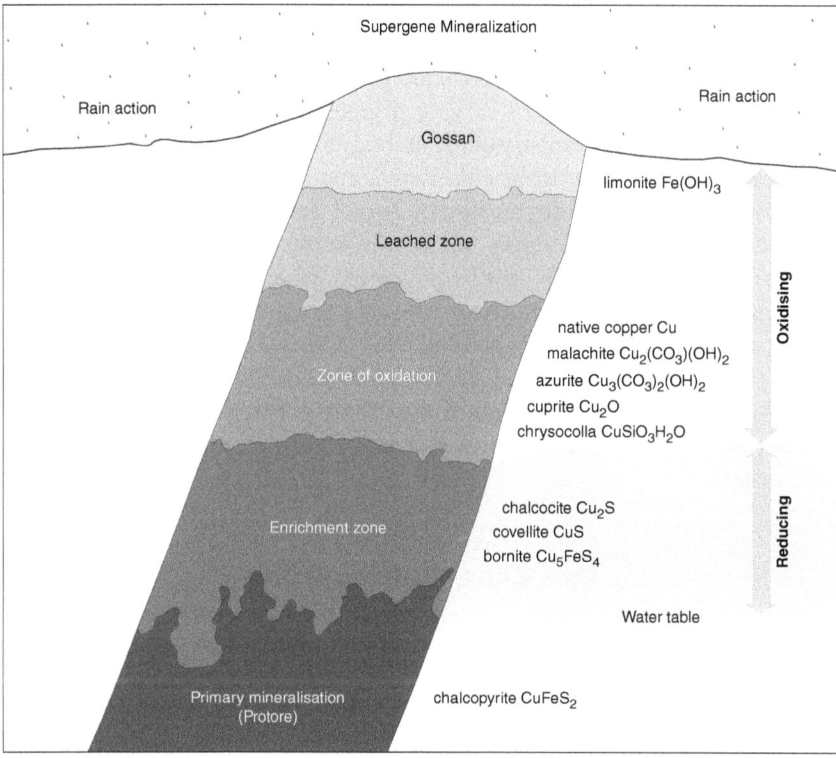

Fig. 1.2. Gossan formation and supergene processes in a metalliferous orebody (*Source*: author, based on Evans 1980).

reducing. This leads to various reactions that precipitate the dissolved metals and result in the replacement of primary by secondary sulphides (e.g. covellite, CuS, and chalcocite, Cu_2S), often with a significant increase in metal content. This zone of supergene enrichment overlies the primary mineralization, usually sulphides that may or may not be of ore grade.

It can take many thousands of years for significant deposits of secondary mineralization to form by such processes. It is important that the water table is deep and that the ground surface is not exposed to significant erosion. This explains why such deposits are more common in arid Mediterranean lands, such as southern Iberia and Cyprus, and are mostly absent in the highly glaciated landscapes of temperate Europe. Even where gossans do not develop, all orebodies are affected to varying degrees by surface oxidation. The most obvious indication is the presence of brightly coloured copper oxide and hydroxycarbonate minerals such as malachite and azurite. These green and blue-stained outcrops were an important indicator of copper for the prehistoric miner.

This model of depth zonation in an orebody is quite idealized as many examples do not have such clearly defined zones of mineral leaching and precipitation. It is not uncommon to find primary sulphide mineralization on or near the surface, mixed to varying degrees with secondary minerals as determined by the geochemistry of the orebody and local oxidizing conditions. While each ore deposit must be considered separately, this model has had an important influence on our understanding of early mining. It is generally accepted that prehistoric mining began with surface workings that removed secondary copper minerals, and even native copper, from the gossan or underlying zone of oxidation. The presence of the water table limited further mining, while the extraction of sulphidic mineralization in the reduction and primary zones required different smelting processes. The technology required for such deeper mining and complex smelting was first developed during the later Bronze Age, but was not in use in many parts of Europe until Roman times or later.

Case study: the Great Orme

This location near Llandudno on the north Wales coastline is one of the largest prehistoric copper mines in Europe, with an estimated 6 km of mine tunnels worked to a depth of 80 m. The formation of this orebody was controlled by geological structure, where the primary mineralization was directed along a series of north-south fractures and joints, and to a lesser extent along east-west fractures (Fig. 1.3). There are strong lithological controls, where the mineralization is restricted to dolomitized limestones, particularly those with a dominance of coarse-grained calcite crystals (calcarenites). This dolomitization is contained by confining horizons of non-calcareous lithologies, such as mudstone and, to a lesser extent, sandstone. Oxidation of the primary copper ore, chalcopyrite, produced large amounts of hydroxycarbonate minerals (malachite with some azurite), as well as hydrated iron oxides of goethite, partly altered chalcopyrite, and other minerals. This relates to a rotting process within the dolomite, especially in the immediate vicinity of the fractures, which is strongly developed nearer to the surface, but did extend in certain areas of the mine to depths of 130 m. Similar rotting occurs within some of the mudstones and towards the base of the sandstone where calcareous material occurs. The extent and shape of the majority of the early workings relate directly to the removal of secondary copper minerals in the zone of rotting of the dolomites, centred around the mineralized fractures and, to a lesser extent, in the adjacent mudstones (Lewis 1996).

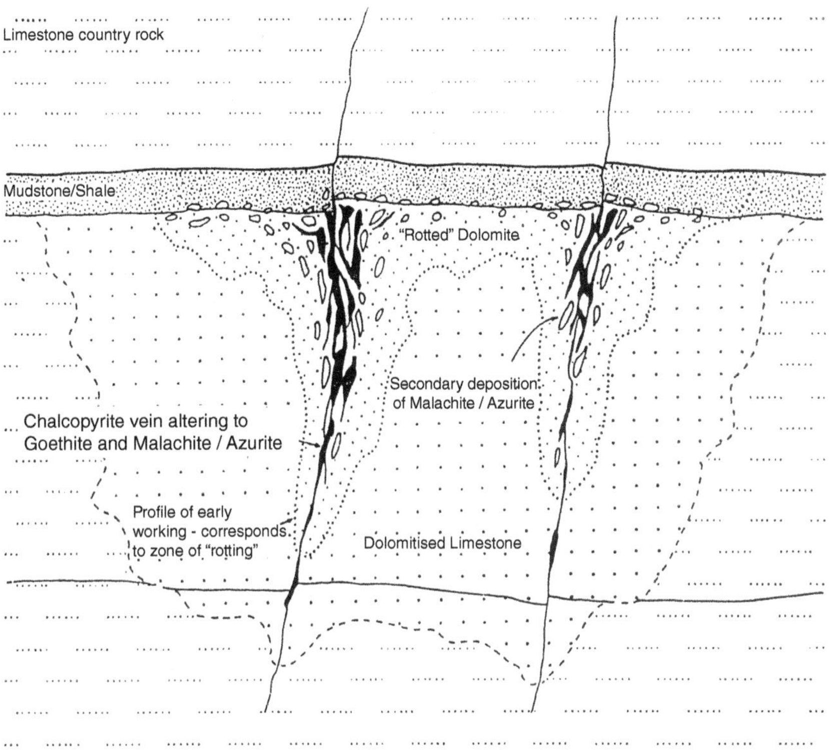

Fig. 1.3. The mineralization of the Great Orme mine, north Wales (*Source*: Andy Lewis).

The study of ore deposits

The scientific study of ore deposits is central to an understanding of the copper produced in prehistory. It is important in relation to how the chemical and isotopic profile of the mineralization compares to that of the finished metal. The characterization of orebodies is also important towards understanding the technology required to convert copper minerals to metal. In the past this has been approached quantitatively, with the aim of identifying characteristic trace element or stable isotope 'signatures'. This type of scientific finger-printing makes certain assumptions about the composition of orebodies that cannot always be proven. Problems arise due to the heterogeneity of ore deposits in terms of mineralogy and geochemistry, created by their complex geological formation and weathering environments.

Sampling strategies are important in terms of obtaining ores that are representative of those that were processed in prehistory. Too often sampling has been randomly applied to surface spoil, with no clear understanding of what constituted ore for a given period of extraction. By its residual nature, the material in spoil dumps is generally not representative of the ore type and grade extracted in prehistory. Ore deposits are highly variable in geochemical, and thus mineralogical, terms, which are expressed in spatial and temporal zoning (see Ixer 1999 for discussion). Careful sampling requires close collaboration between archaeologists and earth scientists, where an inter-disciplinary study uses determinative mineralogy in combination with chemical and lead isotope analyses to characterize an orebody.

It is also important to recognize the subsequent formation of secondary copper minerals in surface spoil, in what Ixer (1999) refers to as the supra-gossan zone (see Rehren 2009 for a rare example of chalcopyrite mineral formation in Neolithic crucibles from Switzerland). Groundwater action can significantly alter the geochemistry and mineralogy of the residual material, further complicating the process of sampling to determine the original ore. These problems can be addressed by paragenetic studies that will identify the mineralogy of the original ore assemblages, even if restoring the original geochemistry is more difficult.

Copper deposits in Europe

The continent of Europe has a complex and varied geology, ranging from rocks of Archaean age, some 1,900 million years old, to younger formations of the late Cretaceous and Tertiary periods (135–2.6 mya). The formation of copper and other base metal deposits was closely linked to tectonic processes at regional and global scale (Fig. 1.4). Several distinct orogenic belts can be identified where metalliferous orebodies formed (for reviews see Andrew 2003; Jankovic 1980; Large 2003). The first is the Fennoscandian shield of Archaean and Proterozoic age, comprising much of Finland, Sweden, and southern Norway. Though often presented in archaeological literature as poor in copper, this zone contains some of the largest copper deposits in Europe. These include enormous modern mines in districts such as Vihanti-Pyhäsalmi in Finland, and Skellefteå in Sweden where the mine of Aitik is one of the largest in Europe, with production in excess of 18 million tonnes of copper ore at 0.4 per cent grade, yielding 58,000 tonnes of metal.

The next significant metallogenic zone occurs in the Caledonides, a tectonic belt with Lower Palaeozoic sediments and granitic intrusives that stretches from northern parts of Ireland and Britain along the western side of Norway. There are many significant copper deposits in this zone, such as the vein systems of central Wales and the Pennines, as well as the volcanic-hosted

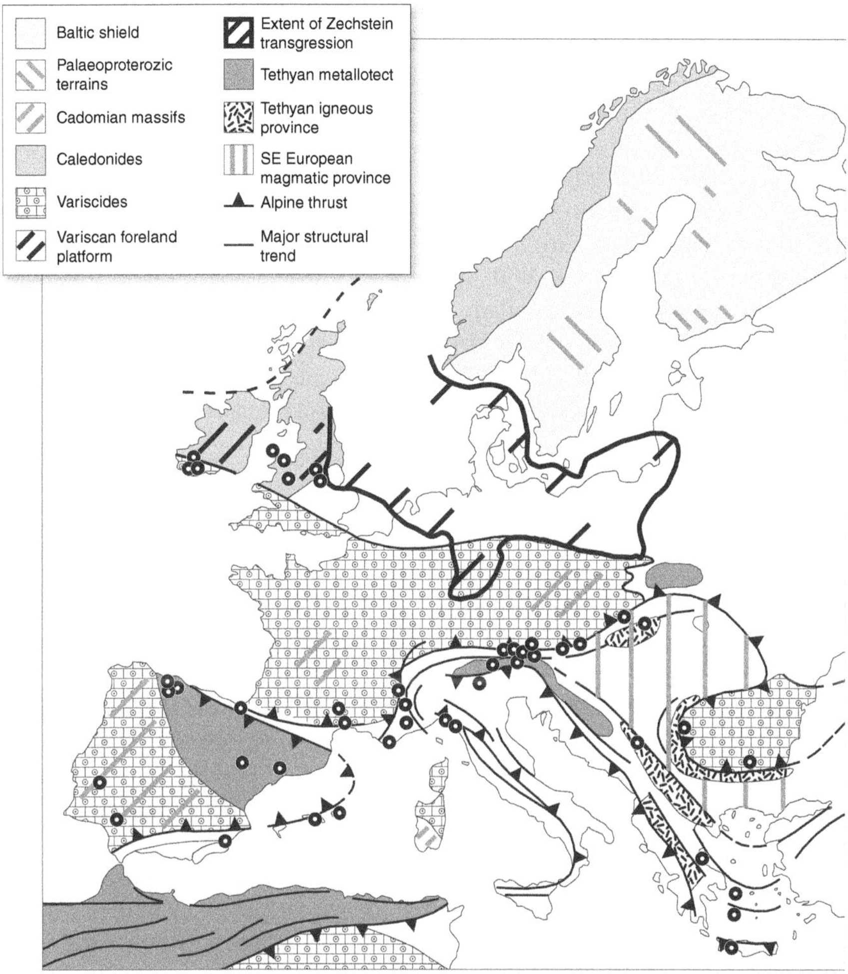

Fig. 1.4. Base metal metallogeny and the tectonic evolution of Europe, showing the known distribution of prehistoric copper mines (open circles)

(*Source*: author, adapted from Large 2003).

massive sulphide deposits such as those at Avoca in Ireland, Parys Mountain in Wales, and at Lokken and Stekkenjokk in Norway.

The Variscides is a third important metallogenic zone, which also has a long history of copper mining. This orogenic belt stretches from the massifs of Spain, northern France and south-west England, through the Schwarzvald and Ardennes of Germany, the Erzgebirge and Bohemian massifs of the Czech and Slovak republics, to the northern and eastern Carpathians. These various massifs comprise pre-Cambrian to early Palaeozoic volcano-sedimentary blocks, separated by basins infilled by Palaeozoic sediments. Copper resources in this zone include volcanic-hosted massive sulphide deposits, such as the Iberian Pyrite Belt of Spain and Portugal, as well as sediment-hosted exhalative orebodies of the Rammelsberg type in Germany.

The post-Variscan platform of north-central Europe is another important zone of copper mineralization. This includes the aforementioned Kupferschiefer stretching from eastern England to Poland. Finally, the Alpine-Tethyan (Carpatho-Balkan) Arc is a complex orogenic belt of rocks of Mesozoic to Cenozoic age that includes the Alpine and Carpatho-Balkan mountain ranges and parts of the Mediterranean. It contains some of the world's most important copper deposits in the form of porphyry deposits, such as those at Bor in Serbia and Resck in Hungary; polymineralic veins like those at Baia Mare in Romania; and volcanic-hosted massive sulphides of the Cyprus type.

Mining trends

Copper mining has an important place in the industrial heritage of Europe. This is partly because copper mineralization is so widespread across the Continent. Most of the known deposits have been mined at various times over the past 8,000 years, even though the earliest historical records only date from Roman and medieval periods. In recent decades Europe has produced only around 5 per cent of world copper; however, at the height of the Industrial Revolution it was one of the leading producers of the metal.

Many of the largest copper deposits in Europe were first mined in the modern era when the technology required to process low-grade mineralization was first developed. Prior to the twentieth century, vein deposits were the most important sources of metallic ores. This was because only ores with greater than 3 per cent copper content were economically viable and these mostly occur in vein systems. Advances in mining and ore treatment mean that rocks with as little as 0.4 per cent copper content can be processed today. Due to the high capital investment required in modern mining, the emphasis has shifted away from vein deposits to massive orebodies of lower grade. This shift in mining strategy means that much of the data produced by modern geological survey and mineral exploration is not directly relevant to an understanding of copper resources in prehistory. For this reason, it is generally difficult to obtain detailed mapping for any region of the type of small surface ore deposits that were mined during the prehistoric period.

THE HUMAN ENVIRONMENT

> Ore is a metalliferous mineral, or an aggregate of metalliferous minerals, more or less mixed with gangue, which from the standpoint of the miner can be won at a profit, or from the standpoint of the metallurgist can be treated at a profit.
>
> (Kemp 1909)

This continues to be a widely accepted definition of ore in the mining industry. The term has geological and technological aspects, in that metal content has to be physically present in the rock at sufficient concentration for extraction using prevailing methods. There is also an economic consideration that emphasizes the winning of metal at a profit. While this is expressed in mostly financial terms in modern mining, the concept of profit in pre-industrial societies is better understood in terms of a material that can be mined for the benefit of human groups. This is not to deny the important exchange value of metal in prehistoric societies, but rather to emphasize the broader societal context of this activity.

Concepts of ore resources

The availability of ore resources in prehistory was partly determined by the interaction between the physical occurrence of the mineralization and the technological capability to exploit it. While modern market forces control the production of base metals, the situation for prehistoric societies would have been entirely different. Where modern mining is more concerned with tonnage, surface access to even small amounts of copper mineral was significant for the prehistoric miner. All early mines would still have operated on the basis of a 'cut-off grade', i.e. where ores with a low copper content or a complex mineralogy could not be extracted. This was a technological constraint, determined by the prevailing methods of ore treatment and smelting.

One of the difficulties of looking at the resource potential of metalliferous ores is their heterogenous nature. This explains why the processes involved in their extraction and treatment are so different from the quarrying of solid rock. There are many problems in using information derived from modern mineral exploration and geological survey. This data is generally more concerned with ore tonnage at depth than with small surface occurrences of oxidized ore that were of interest to early miners. A modern interpretation of ore richness is often expressed in terms of bulk grade and tonnage, whereas in prehistory higher grades and surface accessibility were critical. For this reason the modern concept of ore is not appropriate when assessing the resource potential of different mineral occurrences for the prehistoric miner.

The copper minerals had to be present in a sufficient concentration to be extracted and smelted. Several mineral occurrences do not satisfy this requirement and should not be used in any resource analysis of early metallurgy. A good example is the supposed contribution that native copper (Scott and Francis 1981) or secondary copper arsenates (Pollard et al. 1990; Budd et al. 1992) made to the origins of Irish metallurgy. The natural occurrence of these copper sources in Ireland is extremely limited; in effect, they are mineralogical curiosities and could not have constituted a viable resource in prehistory.

To obtain a concentrated source of copper minerals in all periods it was necessary to mine a bedrock deposit. Several factors combined to make this feasible in prehistory, including the composition, grade, and physical exposure of the orebody. Some orebodies must be excluded from this consideration. An example is the Kupferschiefer, one of the largest orebodies in Europe, stretching from eastern England across Holland and Germany as far as Poland. This thin layer of black shale hosts a large quantity and variety of heavy metals, including copper, lead, zinc, and precious metals. This mineralization was not accessible to prehistoric copper miners due to its low grade and depth. The higher metal concentrations are restricted to specific regions of the depositional basin, such as southern Poland where copper grades up to 1.5 per cent are recorded at depths of 600–1500 m (Evans 1980: 130–2, fig. 13.1).

Also excluded are the many massive sulphide orebodies known in Scandinavia. There is no evidence that this fine low-grade mineralization could have been extracted in prehistoric times. This creates an unusual situation that there were apparently few copper resources available to early metallurgists in Sweden, a country with massive reserves of copper and major mine operations in historic and modern times. These include the Great Copper Mountain of Falun, worked more or less continuously from the eleventh century to the late twentieth century.

Prehistoric miners targeted accessible surface deposits of ore where the copper minerals were present in a sufficient concentration to be extracted. This was only possible in a bedrock environment, as copper does not generally concentrate in surface drift deposits. The metal has a considerable mobility in solution, and is not subject to the type of density concentration processes that create alluvial or eluvial deposits of gold and tin. The use of stream and soil geochemistry to demonstrate the availability of copper in prehistory (e.g. Briggs 1976: 275–6) is problematic for this reason. The very low concentrations of metal recorded by such surveys are useful in modern mineral prospection, but have no significance when applied to early mining. Similarly, while the expedient use of mineralized boulders in surface drift can never be ruled out, this did not constitute a viable resource for metal production in prehistory (*contra* Briggs 1976). Other secondary occurrences of copper, such as bog deposits, are so rare that they can be excluded from any analysis of early copper resources.

The prehistoric mine record reveals an ability to process different ore types from a range of geological environments. While many early mines were located on rich concentrated deposits of mineralization, others exploited mineral occurrences that, in terms of grade and tonnage, would not qualify as ore in modern times. An ability to adapt to different geological environments and mineral occurrences was an important factor in the spread of copper mining across Europe. In emphasizing large mines, it is easy to lose sight of the many minor occurrences of surface mineralization exploited during the prehistoric period. A modern perspective might view such minerals as a desperation resource, mined only when the richer sources of surface mineralization were exhausted, at a time when sulphide ore could not be processed. This would be a mistake, however, as prehistoric mining must be understood within a different technological and economic context than that of today. Geology and technological capability were not the only determinants of resource availability in that period.

An industry?

Various models that emphasize the industrial context of production have long influenced our understanding of copper supply in prehistory. For example, the view has often been expressed that different types of metal source became available as technology evolved to meet a growing demand for metal. This placed an emphasis on the optimization of copper resources, with different ore sources coming on line as the necessary technology developed to exploit them. These ideas have been used to explain the move from an early use of native copper to the extraction of surface oxidized mineralization, leading to deeper mining of supergene ore (oxides and secondary sulphides) before the use of primary sulphides became possible. There used to be a widely held view that the latter were first worked late in the Bronze Age, due to an 'industrial depression' caused by the depletion of oxide zone deposits (e.g. Charles 1985).

Prehistorians now reject such broad techno-historical explanations, considering technology to be secondary to social and economic influences on the organization of early metal supply. A view of metal sourcing as technologically constrained has been replaced by an emphasis on social control of resources and metallurgical knowledge. This includes the perceived ownership of the resource, the motivation to invest time and energy to exploit it, and the broader significance of the resulting metal. Not all regions with copper deposits responded in the same way to the opportunities presented by the new technology. Some areas exploited metal resources at a low level, while for others copper mining was a key economic activity that shaped the development of their societies over many centuries. There were also regions that were less enthusiastic about mining due perhaps to their cultural outlook, economic priorities, or an inability to mobilize resources and make the necessary time commitments. Even with direct access to copper deposits, these groups chose not to engage in copper mining or primary metal production, opting instead to obtain supplies through trade, as did those areas with no ore resources.

Any decision to engage in a labour-intensive and technologically complex activity such as copper production had important repercussions for the human groups in question. Much depended on the scale of mining and the overall time commitment in relation to other labour-intensive activities, such as farming, herding, fishing, salt production etc. There is much evidence that large-scale copper mining in particular regions of prehistoric Europe was sustained over many generations, leading to the formation of distinctive communities with their own traditions and economic base. Such operations are indicative of production for exchange purposes, trading copper to other groups for economic gain. The extent to which these mining communities were entirely autonomous is unclear, particularly as they are often situated on the mountainous periphery of major settlement zones. Other prehistoric mines were on a smaller scale; often seasonal ventures involving local groups of miner-farmers who produced enough metal to satisfy the needs of a community with some external trade. Each prehistoric mine must, therefore, be considered in its own socio-economic context, as different organizational models can apply depending on the circumstances in question.

A HISTORY OF RESEARCH

> If we may judge from the number of ancient mine excavations, which are still visible in almost every part of Ireland, it would appear, that an ardent spirit for mining adventure must have pervaded this country at some very remote period. In many cases no tradition that can be depended upon, now remains of the time, or people, by whom the greater part of these works were originally commenced: they are generally attributed to the Danes . . . with very little reason. It is worthy of remark, that many of our mining operations exhibit appearances similar to the surface workings of the most ancient mines in Cornwall, which are generally attributed to the Phoenicians . . .
>
> (Sir Richard Griffith 1828)

The study of prehistoric copper mines began in many parts of Europe with discoveries made by the mining industry during the eighteenth and nineteenth centuries. A surge in mineral exploration at the height of the Industrial Revolution uncovered older workings at many locations, where fire, stone, and wooden tools were used in shallow surface workings. The 'primitive' character of these mines, together with an absence of historical records and local traditions, suggested that they were of considerable antiquity. Depending on the local antiquarian tradition, these 'old man workings' were often ascribed to historical groups such as the Celts, Danes, or Romans.

As the above quotation illustrates, such associations were already being questioned in the early nineteenth century, influenced by ideas concerning the antiquity of humankind in the emerging discipline of archaeology. This is apparent in a letter written in 1850 commenting on the age of old copper mines, again in southern Ireland:

> By whom, or in what age this and similar excavations were made appears to be a matter of conjecture, but I believe the general received opinions are that they are the works of the Danes, however it is evident to me that this excavation was wrought previous to the era of powder and iron, as no traces of either can be detected in the workings . . . The theory of the learned Dane Worsaae is that implements of stone belong to one people, of bronze to a second, and of iron to a third. The first to a race which colonized our shores existed three thousand years ago, were the authors of cromlechs and giant's chambers, and were subdued by the second race (which in Denmark was rather of Gothic than of Celtic origin, but in England were Celts) who occupied the Country, used weapons and instruments of bronze, and were in their turn subdued by a third race, which brought Roman fashions and employed iron. I am therefore led to attribute the working of these ancient mines, not to the Danes, but to the Celtic ancestors of the tribes occupying this country at the dawn of modern history . . .
>
> (Thomas 1850, cited in O'Brien 2003)

The discovery of ancient copper mines was first recorded in mining and antiquarian literature of the eighteenth and nineteenth centuries. Many of these accounts were written by miners, curious about the history of their profession and the origin of primitive workings and tools found in the course of their work. By the late nineteenth century this curiosity had extended to more extensive investigations, the first of which were conducted in Austria and Britain. The latter include the discovery in the 1870s of stone hammers and other primitive tools in 'superficial pits' at Alderley Edge copper mine near Manchester (Fig. 1.5). The investigations there of Professor Boyd Dawkins (1875), Dr Sainter (1878), and Roeder and Graves (1905), concluded that the earliest phase of mining was Bronze Age, and involved combined fire-setting and stone hammer technology that was distinctive from Roman and later mining. Their site records are particularly valuable because, by the time mining finally ended at Alderley Edge in 1919, many of the ancient workings and spoil deposits had been destroyed.

Developments in Austria are of particular importance in the history of early mine studies. These began with investigations in the 1870s in the Mitterberg mine district, near Salzburg, by the archaeologist Matthäus Much (1878, 1879), and subsequent work there by the researchers Georg Kyrle (1916)

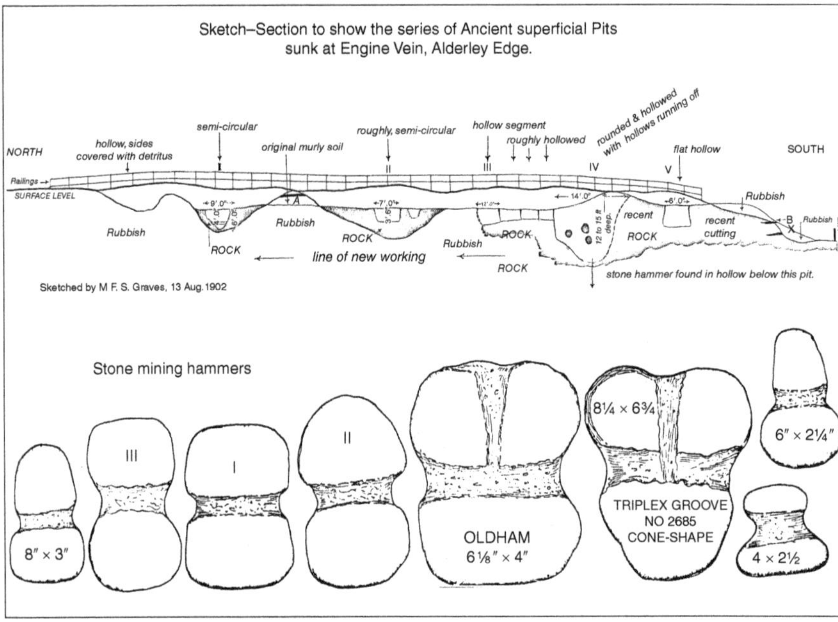

Fig. 1.5. Stone hammers and antiquarian survey of Alderley Edge mine, England (*Source*: Roeder 1901).

and Oliver Klose (1918). The scientific investigation of the Mitterberg mines began in the 1920s with the work of two mine engineers, Karl Zschocke and Ernst Preuschen. The publication of their work (Zschocke and Preuschen 1932) marks a new stage in this field of research, where detailed recording of underground workings, surface features, and equipment finds was combined with analysis of the geological setting, technology, and mining environment. Preuschen enjoyed a fruitful collaboration over the next three decades with the archaeologist Richard Pittioni, carrying out studies of the Kelchalm mines of North Tyrol, and other early mines (e.g. Preuschen and Pittioni 1937, 1954; see also Preuschen 1967). Pittioni wrote extensively on early copper mining in Austria, also bringing these mines to wider attention with a number of English-language publications (e.g. Pittioni 1948, 1951). This was a period when the study of early copper mines gradually entered the mainstream of prehistory studies in Europe.

The discoveries made by these early researchers in Austria created an interest in early copper mining in Europe that continues to the present day. Their influence may be seen as far away as Ireland, where in 1929 a geologist named Tom Duffy discovered an intact Bronze Age mining landscape preserved in blanket bog on a hill called Mount Gabriel (Duffy 1932). Archaeologists were quick to make a connection with the wealth of early metalwork in Ireland, however uncertainty about the date of these and other 'Danes Mines' would continue for many years (see Mahr 1937: 366).

In 1935 the British Association for the Advancement of Science established a committee to investigate early metal mining in Wales. The secretary was Oliver Davies, an authority on Roman mining in Europe (Davies 1935). Davies carried out a systematic survey of early copper mines in mid and north Wales (Davies 1937). He also trenched spoil tips containing stone hammers at Nant yr Eira (Davies 1938b), Parys Mountain (Davies 1939), Cwmystwyth (Davies 1947), and the Great Orme (Davies 1948), mines that are now central to an understanding of Bronze Age mining in Britain. With no datable artefacts, and without the advantage of radiocarbon dating, he concluded that the stone hammer phase at these mines was of the Roman or 'Old Celtic' period.

Oliver Davies had an interest in early copper mines in other parts of Europe. He visited the Rio Tinto mines in south-west Spain, describing cave-like openings with radiating tunnels he regarded as pre-Roman in date (Davies 1935). He also reviewed the evidence for early mines in the Balkans, and recorded examples at Karabajir and Rosenbajir in south-east Bulgaria and at Jarmovac in south-west Serbia (Davies 1938a). He also wrote about early copper mines in Cyprus and the Aegean (Davies 1928–30, 1932).

Recent investigations

The modern period of research in north-west Europe commenced in 1966, when a geologist in Ireland, John Jackson, with the assistance of Richard Pittioni, submitted a charcoal sample from the Mount Gabriel mines to the Vienna radiocarbon laboratory (Jackson 1968). The resulting date (VRI-66; 3450±120 BP) was the first obtained by applying this method to the dating of a prehistoric copper mine. Since then, radiocarbon dating has had a major impact, making it possible to date mine workings with 'primitive' technology where datable artefacts are absent. The application of this method was greatly helped by the large amount of sample material available in these mines, be it wood and bone tools, or the residues of wood fuel used in fire-setting. Radiocarbon results are now available for most early copper mines in Europe. Research is now at a stage where radiocarbon dating of individual operations can be refined by statistical (Bayesian) methods, with recent tree-ring research on Austrian mines going even further in terms of absolute chronology (Pichler et al. 2009).

This was an era in which scientific methods were widely used to provenance the source of prehistoric metalwork (see The Characterization of Ore Resources section later in this chapter). These projects revealed significant gaps in our understanding of early ore sources, which limited any scientific correlation with finished metalwork. Such problems began to be addressed from the late 1960s, when the first of many research projects on prehistoric copper mines began in Europe. Some of these studies centred on the older discoveries, however many new sites were also found. They include the oldest known copper mines in Europe, located in the Balkans where evidence was obtained for an autonomous development of metallurgy during the sixth and fifth millennia BC (Renfrew 1969). Of particular importance was the investigation carried out in 1968–86 by Borislav Jovanović at the mine of Rudna Glava in Serbia. Some 40 workings in this site were mined *c.*5400–4600 BC by groups of the Vinča culture (Jovanović 1979). Another important discovery was made at Ai Bunar in southern Bulgaria where survey and excavation work were carried out from 1971–4 by a Bulgarian/Soviet archaeological team (Chernykh 1978a). These copper mines, along with the remarkable discovery in 1972 of sophisticated gold objects dating to the mid fifth millennium BC at the Varna cemetery in Bulgaria (Ivanov and Avramova 2000; Slavchev 2010), demonstrate the importance of this part of Europe in terms of early metal sources.

Research also continued on the Austrian mines, notably the work of Clemens Eibner who investigated several mining and smelting sites on the Mitterberg (Eibner 1972, 1974, 1979, 1982, 1992). With Hubert Presslinger, he also conducted numerous surveys and excavations of smelting sites in eastern Austria (e.g. Presslinger and Eibner 1989, 1993, 1996). During the 1960s and 1970s Ernst Preuschen recorded several early smelting sites and potential mine sources in the Trentino-South Tyrol region of northern Italy (Preuschen 1962, 1973).

Elsewhere in Europe, the pace of research has gradually increased over the past four decades (see Weisgerber and Pernicka 1995; Groer 2008). In Britain, this began in 1976 with explorations at the Great Orme mine in north Wales, where a radiocarbon result (HAR-4845; 2940+80 BP) was obtained for charcoal associated with stone hammers and bone tools (James 1988). This led to a major investigation by the Great Orme Exploration Society during the late 1980s, with surface excavations carried out by Gwynedd Archaeological Trust in 1989 (Dutton and Fasham 1994; Lewis 1996). This work culminated in the opening of a visitor centre at the mine, the first public presentation of a prehistoric copper mine in Europe. Also in Wales, the Early Mines Research Group, led by Simon Timberlake, was formed in the late 1980s. These researchers conducted trial excavations at Parys Mountain, Cwmystwyth, and Nant yr Eira, obtaining radiocarbon evidence of Early/Middle Bronze Age copper mining. The focus of their research was a detailed investigation of Cwmystwyth mine in mid Wales (Timberlake 2003b), with subsequent studies at Alderley Edge mine near Manchester (Timberlake and Prag 2005) and Ecton mine in Staffordshire (Timberlake 2013).

Research in Ireland centred on the copper mines of the south-west region. This began with survey and excavation of the Mount Gabriel mines, which was subsequently extended to similar sites in the same area (O'Brien 1987, 1994, 2003). This was followed by an investigation of a copper mine of the Beaker culture at Ross Island in Kerry (O'Brien 2004), and more recently of later Bronze Age workings at Derrycarhoon in Cork (O'Brien and Hogan 2012).

Many recent projects in Europe began with the focus on a single mine, however, most also took a wider landscape perspective to place these operations in context. One of the earliest examples is the Huelva Archaeo-Metallurgical Project, undertaken from 1973–6 in south-west Spain (Rothenberg and Blanco-Freijeiro 1981). The project was modelled on an earlier investigation of prehistoric copper mines in the Timna valley of southern Israel (Rothenberg 1972).

Forty sites were recorded in the general vicinity of Rio Tinto in Huelva province, including a number of ancient mine workings and surface finds of smelting slag. Recent surveys have increased the recorded number of prehistoric copper mines and smelting sites in this region (Hunt Ortiz 2003).

In the 1980s important discoveries were made in the Cantabrian mountains of northern Spain with research led by Miguel de Blas Cortina (Fig. 1.6). These include the discovery of Chalcolithic and Bronze Age workings at the mine

Fig. 1.6. Modern research: Miguel Ángel de Blas Cortina in the El Aramo mine, Asturias, Spain

(*Source*: author).

sites of El Aramo and El Milagro in Asturias, and La Profunda in León (De Blas Cortina 1992, 1998, 1999, 2003, 2005).

There were also important discoveries in south-east France during the same period. A major programme of research directed by Paul Ambert commenced in the Cabrières area of the Languedoc. This led to the discovery of several copper mines and metal production sites dating to the third millennium BC (Ambert 1984, 1990, 1995, 1996a, 1996b, 1998). Another early copper mine in the same region was found at Bouco-Payrol. During the same period a French team working in the French Alps discovered an important Bronze Age copper mine at Saint-Véran (Barge et al. 1998a).

This is also the era when a major complex of Bronze Age copper mines was discovered at Kargaly in the south Russian Urals. In 1989 a programme of research led by Evgenij Chernykh of the Russian Academy of Sciences, with the support of Spanish colleagues, examined what is probably the largest prehistoric copper mine in the world. The investigation of this unique mining landscape included the settlement site of Gorny, located directly within the mining complex (Chernykh 1992, 1998b, 2004a).

In recent decades there has also been considerable research on early metal sources in regions bordering the east and central Mediterranean. Prehistoric copper mines are recorded in Italy, including in Calabria in the south and Lombardy and the Trentino-South Tyrol regions of the north. The oldest known copper mines in western Europe are recorded in the mountainous region of Liguria in north-west Italy, where the mines of Monte Loreto and Libiola are dated to the fourth millennium BC (Maggi and Pearce 2005).

The potential of Cyprus and the Aegean islands as sources of copper has been confirmed through the investigation of ancient mines and smelting sites, and the analysis of ingots and metalwork using the technique of lead isotope analysis (see The Characterization of Ore Resources section later in this chapter). With very rich deposits of copper, Cyprus was a major producer and distributer of copper in the later prehistoric period. During the later second millennium BC the island produced oxhide ingots of copper that were exchanged across the east and central Mediterranean, into Egypt and the Middle East. The very name of copper derives from this island, as the Roman natural historian, Pliny the Elder, writing in the first century AD used the phrase *aes cyprium* ('metal from Cyprus') to denote pure copper. The term was later abbreviated to *cyprium* and by the fourth century AD had become *cuprum* and the variant *coprum*, from which derives the European words for the metal.

Apart from the earlier work in Austria, it is fair to say that research on prehistoric copper mining was largely neglected in Europe until the 1970s. This is partly explained by the historical focus of early mine studies, in a field dominated by geologists and mining historians, who were often slow to recognize the antiquity of copper mining. Prehistorians, for their part, were reluctant to research a field that requires geological expertise and specialist knowledge to investigate the complex history of many mines. Also, there was a general pessimism about the potential of early metal mines to survive the large-scale industrial mining of recent centuries. This has largely been dispelled by the aforementioned projects conducted in recent decades. These site investigations confirm the survival of early mine workings and associated features in situations even where there has been considerable disturbance by later mining. It is also the case that some prehistoric copper mines were completely destroyed by later mining, particularly by the type of opencast mining recorded in areas such as south-west Spain (Figs 1.7 and 1.8).

The long tradition of mining research in the Eastern Alps continues to the present day. In 2007 a major programme of research, the 'History of Mining Activities in the Tyrol and Adjacent Areas' (HiMAT) was established in the University of Innsbrück (Anreiter et al. 2010; Goldenberg et al. 2011). There is also a new programme of research at the Mitterberg mines (Stöllner 2009; Stöllner et al. 2011), with other projects elsewhere in Austria (e.g. Klemm 2003; Krause et al. 2011).

Discoveries continue to be made in the well-established early mining centres in Europe, such as the recent investigation of Ecton copper mine in England (Timberlake 2013) and Derrycarhoon mine in Ireland (O'Brien and Hogan 2012). New projects are also underway in areas where mines were not known, such as Scandinavia (e.g. Melheim 2012; Ling et al. 2013) and

Fig. 1.7. The largest historic copper open-pit mine in Europe: the Corta Atalaya, Rio Tinto, Huelva, Spain

(*Source*: author).

Fig. 1.8. Historic mining landscapes and destruction of field evidence: Peña de Hierro, Rio Tinto, Huelva, Spain

(*Source*: author).

Pyrenean France (Beyrie and Kammenthaler 2008), details of which are presented in later chapters.

Going back to the seminal Austrian research from the 1920s, the study of prehistoric copper mines has generally involved a close collaboration of archaeologists, geologists, and mining specialists (see Weisgerber 1989, 1990). This inter-disciplinary approach continues today, with greater involvement by archaeologists in a specialist field that has entered the mainstream of prehistory research. Field survey has developed from traditional methods to the use of total station and remote sensing methods, including lidar, geophysics, and satellite imagery, as well as 3-dimensional laser scanning of underground workings. Excavation techniques have also progressed, with fine resolution digging combined with detailed stratigraphic recording and sampling strategies. Experimental studies, looking at fire-setting, ore beneficiation, and smelting, are routinely incorporated into many research projects. There is also much interest in the environmental impact of early mining, using pollen records from peat bogs (e.g. Mighall et al. 2000, 2012), Alpine varves (e.g. Guyard et al. 2007), and alluvial sedimentation (e.g. Le Blanc 2005; Delgado et al. 2012).

Finally, it should be recognized that the study of prehistoric copper mining in Europe has been influenced by similar research carried out in the Middle East in recent decades. These include Rothenberg's investigations at Timna, and subsequent research at Feinan in southern Jordan (Hauptmann 2003). Studies were undertaken in northern Turkey (Giles and Kuijpers 1974), Anatolia (De Jesus 1980), northern Oman (Weisgerber 1988), and west-central Iran (Holzer and Momenzadeh 1971), to mention but a few. Some of these mines exchanged copper into trade networks across the east Mediterranean juncture of Europe, Asia, and Africa. The complexity of such metal exchanges is evident in the composition of the Bronze Age cargoes found in the Uluburun and Cape Gelidonya shipwrecks off the southern coastline of Turkey (Bass 1967; Pulak 1998).

The characterization of ore resources

Researchers have long been concerned with matching the copper produced in prehistoric Europe to the mine sources from which it was derived. Such connections, if they can be made, offer insights into trade networks and other aspects of prehistoric economy. Broadly speaking, there have been two approaches to this issue:

1) Metal to mine: identifying the geological (and possible mine) sources by matching metal 'types', established by scientific analysis of finished objects, to copper deposits of similar chemistry and/or isotopic signature.
2) Mine to metal: following the flow of copper from a known mine source that has been subject to scientific investigation, to locations where primary production (smelting sites) and metalworking (workshops) have been identified, and the subsequent circulation and history of metal products in society.

The first involves an examination of objects produced by a particular metal-working tradition over a defined geographic region. The chemical composition of these metal objects is analyzed by laboratory methods, to identify the concentration of major, minor, and trace elements contained in the metal. This can lead to the identification of a distinctive metal composition. The distribution of objects made with this metal defines a 'metal circulation zone', within which particular mine sources may be located. A geological survey must be undertaken to identify copper deposits that could produce this type of metal. Fieldwork at these mineralized locations may reveal old copper mines, which can be archaeologically investigated to recover evidence of their age, technology, and ore types. Radiocarbon dating can reveal whether these mines are contemporary with the metal objects in question, while geological analysis can reveal if the copper ores could be smelted to produce metal of similar composition. This requires a careful comparison between the chemical composition of the ores and that of the metal objects. Such studies can be highly problematic, owing to the heterogeneous nature of the orebody, the complex partitioning of chemical elements in the transformation of ore to metal by smelting, and the effects of processes such as casting and other thermal treatments (see Tylecote et al. 1977). Recycling is also a major issue, as the supply of copper in prehistory not only included primary copper smelted direct from a mine source, but also secondary metal produced from scrap, as well as mixtures of both types from the same or different mines.

The idea that chemical analysis can be a guide to the provenance of metals dates from the nineteenth century (Göbel 1842 and Wibel 1865, in Pernicka 1999). This approach was given a considerable impetus by the development of spectrographic methods that allowed minor and trace element concentrations to be accurately determined. The early use of this technique sought to connect mine sources to the circulation of metal in central Europe (Witter 1938; Otto and Witter 1952; Neuninger et al. 1957). Some of this research emphasized the importance of the Eastern Alps as an early source of copper; however, attention was also drawn to potential supply from other ore deposits, such as those in central Germany, the Slovakian mountains, and the Balkans.

By the 1960s, chemical analysis was being used to identify early metal types in other parts of Europe. The largest of these studies was the Stuttgart programme, which analyzed some 20,000 prehistoric metal objects from across Europe (Junghans et al. 1960, 1968, 1974). These chemical assays were used to identify the circulation of distinct metal groups, which could potentially be traced back to their geological source. With some exceptions (e.g. Coghlan and Case 1957; Coghlan et al. 1963), this type of blind source provenancing was not a success. This was mainly because of complex changes that occurred during the conversion of ore to metal by smelting and by subsequent thermal processes such as casting. The research also exposed a limited understanding of relevant sources of copper ores and their geochemistry, which limited any direct comparison with the analysis of finished metalwork (see Pernicka 1999 for discussion).

The perceived failure of the Stuttgart programme in terms of source provenancing of prehistoric metalwork cast a shadow over this area of research for many years. There were important successes in later research, such as the analysis of Copper Age metalwork in the Balkans and its relationship to ore sources (Pernicka et al. 1993, 1997), but mostly it proved difficult to establish direct links with particular ore sources, let alone identify specific mines. The development in the 1960s of a new laboratory method, lead isotope analysis, seemed to offer a way forward in the scientific identification of ore sources (Brill and Wampler 1965, 1967). This is based on the observation that lead isotope ratios in a metal object, including copper/bronze objects with trace amounts of lead, relate to the geological age of the ore source. Unlike chemical composition, these lead isotope ratios are not affected by the production process (smelting), which offers the possibility of making a connection between the primary metal and its ore source.

The first large-scale application of the technique in Europe was an analysis of the trade in metals in the Mediterranean Bronze Age, conducted by a research team from Oxford University (Gale and Stos-Gale 1982, 1986). The initial results were the subject of considerable controversy (see Pollard 2009 and Gale 2009 for review), which debated the validity of the technique and its relevance to the study of copper sources in the Mediterranean (Budd et al. 1993, 1995, 1996; Tite 1996). The results of the Oxford team have been affirmed by recent research (Hauptmann 2009; Gale and Stos-Gale 2012). It is now accepted that in certain cases lead isotope analysis can provenance metal to a specific ore source. Examples include the Apliki deposit in Cyprus, believed to be the source for all post-1400 BC oxhide ingots in the Mediterranean (Gale and Stos-Gale 2012), or the confirmation of Ross Island mine as the earliest source of copper in Ireland (Northover et al. 2001). The limitations of the method are also understood, including that metal objects may carry a lead isotope signature that relates to a broad geological area with many ore deposits, as is the case with most Bronze Age metalwork found in Britain (Rohl and Needham 1998).

Lead isotope analysis is now an important tool in source provenancing of prehistoric metalwork, with major studies conducted or underway in Bulgaria (Gale et al. 2003), Portugal (Müller and Cardoso 2008), Scandinavia (Ling et al. 2013), to name but a few. The limitations of the Stuttgart programme and the 'lead isotope debate' did question the validity of these scientific approaches to the identification of metal sources in antiquity (reviewed by Pernicka 1998). The entire field has been re-invigorated by new research on the mining sites themselves. With new discoveries and detailed mine studies, it is increasingly possible to use source-specific information to examine the ore-to-metal relationship in more detail, and connect this with the objects produced in a metal circulation zone (see Ixer and Budd 1998 in relation to the British Isles). This provides a greater insight into the types of metal produced by various mine sources in different periods.

THE ARCHAEOLOGY OF PREHISTORIC COPPER MINES

The study of prehistoric copper mining may be approached in a number of ways. As already discussed, scientific analysis has the potential to match metal objects to primary copper production, and hence to particular ore sources and possibly mines. This approach can be combined with an investigation of contemporary settlements that have on-site evidence of copper metallurgy. The discovery of smelting sites is of particular importance, as these can be located close to mine sources. Information from these production centres, and from consumption contexts in settlements, burials, and ritual deposits, can throw light on the organization of copper mining and smelting, and the relationship between supply and demand within a metal circulation zone.

Any analysis based on geographical proximity, linking ore deposits—and thus mines—to metal finds, settlements, or ritual sites in the landscape, can be highly problematic. These studies provide a basis for further research, but lack credibility until prima facie evidence of mining is discovered. This involves the search for early mines and their direct investigation by geoarchaeological methods, culminating in excavation and detailed sampling. In some cases these mines can be connected directly with smelting sites and settlement contexts, allowing a complete sequence of production and supply to be established (Fig. 1.9). More often, important elements of this chain are missing, making it difficult to understand the wider context of the mine.

Survival and recognition

For a long time research in this field was hindered by a belief that physical evidence of early mining could not survive the intensive re-working of ore deposits in later periods. This pessimism has now been dispelled by the many successful projects undertaken across Europe in recent decades, which demonstrate the significant potential for the survival of early mine archaeology in those circumstances. This is not to deny that many important sites were destroyed, damaged, or concealed by large-scale mining during the Roman, medieval, or modern industrial era. Historical records of early mine discoveries in Europe mostly date from the early nineteenth century and the reporting of such finds was inconsistent. Much depends on the particular history and scale of mining at a given location. It is only necessary to consider the impact of open-pit mining in the Cyprus orefields, or at Rio Tinto in Spain, to appreciate that many important mines have been lost (Fig 1.7). However, where opencast extraction was not practiced there is considerable potential for the survival of physical evidence of earlier mine operations. This is borne out by evidence from recorded prehistoric copper mines in Europe, most of which were re-worked in historic times.

The discovery of early copper mines is affected by other factors. Sea-level change and marine erosion might explain why there are so few prehistoric copper mines on exposures of copper mineralization along the coastline of Europe. The recognition of mines in mountainous terrain, often in areas with extensive forest cover or bog growth, or where landslides have occurred, can be difficult. This is particularly true of early mines in temperate Europe,

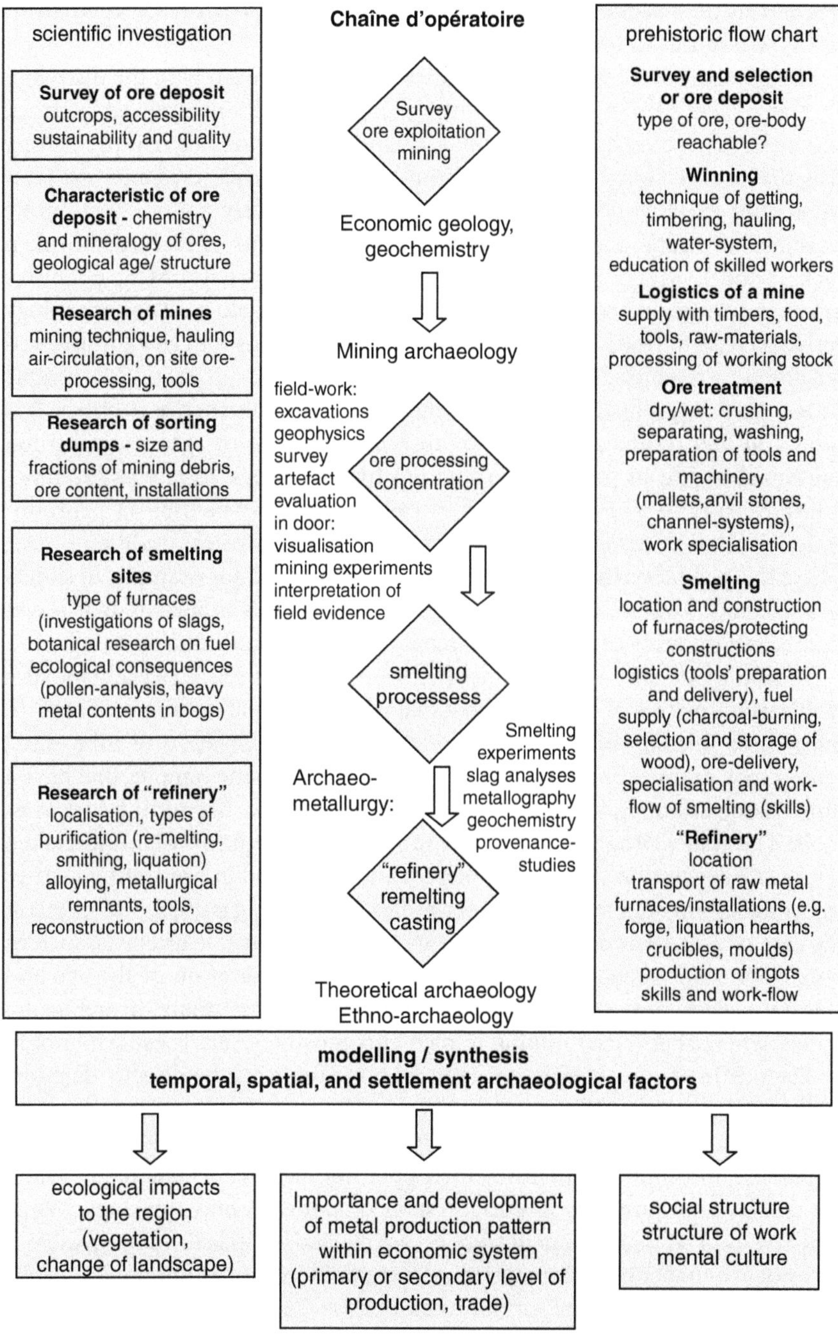

Fig. 1.9. A *chaîne opératoire* of prehistoric ore mining

(*Source*: Thomas Stöllner 2003a).

whereas in the Mediterranean zone they may occur in erosive arid environments where surface evidence may be more visible.

The recognition of early copper mines in the field began with the discovery of man-made openings of a primitive character on mineralized bedrock exposures. These are associated with early mining methods such as fire-setting and the use of stone, wooden, and/or bone tools, with evidence for later techniques employing iron tools or explosives obviously absent. Such 'mine workings' are generally associated with surface deposits of discarded broken rock ('spoil'), which may contains artefacts of early date. Most primitive mines, however, cannot be dated on surface features alone. The technology employed can indicate their antiquity; however, this must be critically assessed for different regions. For example, fire-setting is a rock extraction technique widely used in prehistoric mining, generally leaving distinctive wallrock features and fuel residues. It is not an absolute indicator of age as the method continued in use in parts of Europe into historic times, as late as the nineteenth century in Norway (see Willies 1994). It can still be used as a chronological marker in certain regions where there is no historical tradition of use, where this can be verified by archaeological assessment; for example, in southwest Ireland. The use of fire in rock extraction is also important as it leaves significant amounts of wood or charcoal that can be radiocarbon dated.

The dating of early copper mines has been controversial, especially for sites with several phases of mining where diagnostic material culture is absent. In most cases this can be resolved through scientific excavation of mine sites, followed by radiocarbon dating of wood, charcoal, or bone samples that have a direct association with the process of mining, either as fire-setting fuel or as tools. The many programmes of scientific dating that have been undertaken refute any suggestion that the use of peat and bog wood in fire-setting negates the radiocarbon data for early copper mines (*contra* Briggs 1983, 1984, 1991in the case of Britain and Ireland.) The presence of distinctive artefacts, such as pottery or stone/metal axeheads, can be a further indication of the age and cultural associations of a mine. Such finds are not common in prehistoric mines, where specialized mining implements mostly occur. These tend not to be diagnostic of age, though wooden and bone tools can be directly dated by the radiocarbon method.

The discovery of stone hammers in mine contexts is probably the most important indicator of prehistoric copper mining. They are often found broken in large numbers in surface spoil or underground workings. While other types of stone implement were occasionally used in later mining, there is no evidence that cobble hammerstones were used in copper mining in Europe into historic times. It is important to emphasize that the discovery of similar implements outside of mine sites does not constitute a priori evidence of early mining. Put simply, a stone hammer is only a mining tool when found in a mine context.

It should be noted that recent research on early copper mines has emphasized those examples where fire-setting and stone hammers were employed. This type of technology leaves a distinctive archaeological record and it will always be difficult to identify early mines where such techniques were not employed. This may explain why in some regions it has proved easier to identify Chalcolithic and earlier Bronze Age mines with so-called primitive technology, than mines of later prehistoric date that used different methods. The absence of stone hammers, in particular, removes an important surface indicator of early copper mining, while the application of radiocarbon dating will depend on whether fire-setting, bone and/or wood tools and equipment were in use.

The physical evidence

The archaeology of prehistoric mining is distinctive to the activity itself and its physical environment. Much depends on the time and energy commitment at a given location. Some measure of rock extraction was involved, and so most mines comprise one or more surface openings that may lead into underground workings. These can vary in size and shape depending on the geological setting, the approach to mining, and the scale of operation. They include shallow and deep surface pitting and trenches, vertical or inclined shafts, and tunnels that are generally accompanied by surface dumps of broken rock spoil.

While the term 'mining' can refer to the physical act of rock extraction, the process involves a sequence of four closely related activities:

1. *Prospection*: the search for surface indicators of copper mineralization suitable for mining. This activity may be inferred from the discovery of trial pits and the subsequent mining strategy employed.
2. *Rock extraction*: the removal of mineralized rock to varying depths beneath the ground surface, leaving mine workings of different size and types.
3. *Ore beneficiation*: the concentration of mineralized rock for smelting, leaving deposits of broken rock spoil in the general area of the mine workings.
4. *Smelting*: the production of primary copper metal through heat-treatment of ore concentrate in furnaces.

The first three are carried out within the mining site, a discrete location on the landscape that is both geologically and topographically defined. Depending on local tradition, and factors such as the availability of fuel, the smelting of copper ore may have been undertaken in the mine site or some distance away. Subsequent metallurgical process, such as the refining and alloying of copper, the casting of objects in moulds, and other methods of fabrication, could be undertaken in permanent settlements or specialized workshop settings in the general vicinity of the mine or some considerable distance away.

Mine workings and spoil deposits are the most visible elements of a Bronze Age mine. These sites also had a number of activity areas that are not well exposed today. They include hut shelters, cooking hearths, equipment and fuel stores, as well as places where copper ore was smelted to metal in furnaces. Many of these facilities were located within what is often termed a 'mine camp', generally located close to the mine workings and used on a temporary or permanent basis. The identification of these habitations and work places continues to be a major challenge to the archaeological investigation of early mine sites.

The preservation of this archaeological evidence can vary considerably, depending on the geological environment, surface conditions, and underground drainage. Many prehistoric copper mines in temperate Europe flooded soon after they were abandoned, resulting in waterlogged preservation of wooden and other organic materials. There can also be excellent preservation of organic materials, particularly bone artefacts, where mining was carried out in calcareous geology. The circumstances under which mine camps were abandoned were also important in archaeological site formation (Stevenson 1982). The final abandonment of most prehistoric copper mines seems to have been well-organized, as serviceable equipment or metal implements are rarely found. There may have been some caching of stone hammers and other work tools between different mining seasons.

PREHISTORIC COPPER MINES IN EUROPE

The study of early copper sources has been advanced in many parts of Europe, with major research projects completed in Russia, the Balkans, Cyprus, Austria, France, Spain, Britain, and Ireland. This reflects the widespread geological occurrence of copper deposits across the Continent. These, however, are not evenly distributed, which partly explains why it took several millennia for the knowledge and practice of copper mining to spread across Europe, *c.*5000–2000 BC. This chapter has already examined why it was not possible to exploit all copper deposits during the prehistoric period, as many occur at great depth, have a low metal content, or have a geochemistry/mineralogy that requires complex treatment. Conversely, many locations where copper was extracted in prehistory were not economically viable in historic times, which can contribute to the survival of such mines. The distribution of mines also reflects the efforts of some source regions to control metal supply by limiting the spread of mining and metallurgical expertise.

The known distribution of early copper mines reflects the survival and recognition of site evidence, combined with the results of research projects in the modern era (Fig. 1.10). While the latter have contributed significantly to

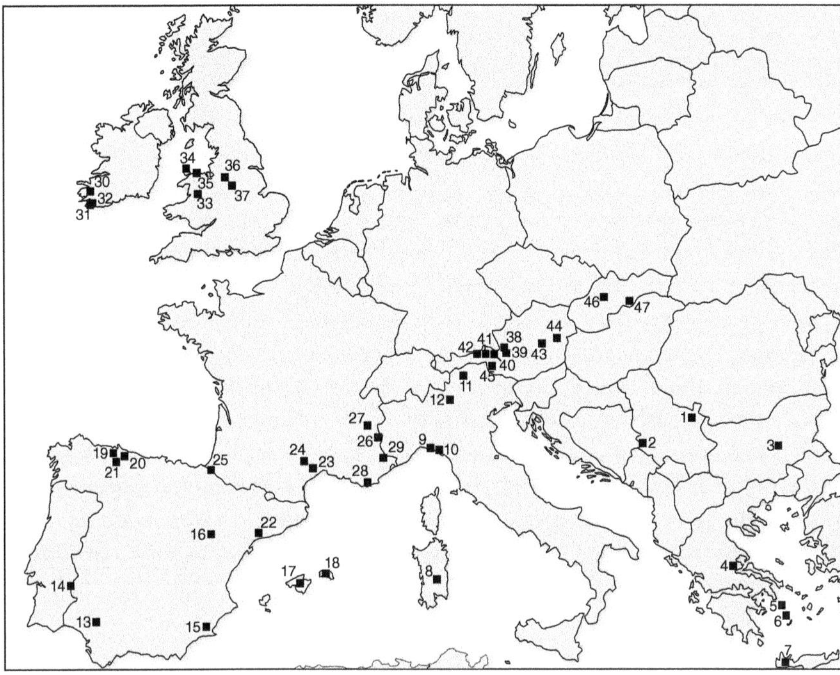

Fig. 1.10. Distribution of known sites or districts where prehistoric copper mines have been identified and, in most cases, radiocarbon dated

* indicates some areas with evidence of primary ore smelting where mines are likely to exist. (*Source:* author).

Balkans (Chapter 2)

1. **Rudna Glava**, Serbia; 2. **Jarmovac**, Serbia; 3. **Ai Bunar**, Bulgaria.

East Mediterranean (Chapter 3)

4. **Othrys Mountains**, Greece*; 5. **Kythnos***; 6. **Seriphos*** (see Fig. 3.1 for location of these Aegean islands); 7. **Troodos Mountains**, Cyprus (see Fig. 3.2 for site details).

Central Mediterranean (Chapter 3)

8. **Funtana Raminosa**, Sardinia*; 9. **Libiola**, eastern Liguria; 10. **Monte Loreto**, eastern Liguria; 11. **Trentino***; 12. **Campolungo**, Lombardy.

Iberia (Chapter 4)

13. **Chinflon**, Huelva (see Fig. 4.1 for detail on south-west Spain); 14. **Mocissos**, Portugal; 15. **Sierra de Orihuela**, south-east Spain; 16. **Loma de la Tejeria**, Teruel*; 17. **Mallorca***; 18. **Mitja Lluna**, Minorca; 19. **El Aramo**, Asturias; 20. **El Milagro**, Asturias; 21. **La Profunda**, León; 22. **Montsant**, Tarragona.

Southern France (Chapter 5)

23. **Cabrières**, Languedoc; 24. **Bouco-Payrol**, Aveyron; 25. **Causiat**, western Pyrenees; 26. **Saint-Véran**, French Alps; 27. **Les Rousses**, Grandes Rousses Massif; 28. **Maraval**, Toulon; 29. **Clue de Roua**, Alpes-Maritimes (see Fig. 5.1 for detail).

Ireland and Britain (Chapter 6)

30. **Ross Island**, Co. Kerry; 31. **Mount Gabriel**, Co. Cork; 32. **Derrycarhoon**, Co. Cork; 33. **Cwmystwyth**, mid Wales; 34. **Parys Mountain**, Anglesey; 35. **Great Orme**, north Wales; 36. **Alderley Edge**, Manchester; 37. **Ecton**, Staffordshire (see Fig. 6.1 for detail).

Austria (Chapter 7)

38. **Mitterberg**, Salzburg; 39. **St. Veit**, Salzburg; 40. **Saalfelden-Becken**, Salzburg; 41. **Kitzbühel-Kelchalm**, North Tyrol; 42. **Schwaz-Brixlegg**, North Tyrol; 43. **Eisenerz**, Upper Styria*; 44. **Prein**, Lower Austria*; 45. **Virgental**, East Tyrol* (see Fig. 7.1 for detail).

Slovakia

46. **Spania Dolina-Piesky**, Lower Tatras Mountains; 47. **Spanie Pole**, Rimavská Sobota region.

Not shown: Kargaly mines of Russian Urals

knowledge, the identification and dating of early copper mines continues to be difficult in several parts of Europe. The absence of evidence in south-east Spain, north-west France, south-west England and Scotland, Germany, Sweden, and Norway is notable, as these are metalliferous areas with strong metalworking traditions in the Bronze Age. With further fieldwork it is likely that early copper mines will be discovered in those regions, but, for reasons outlined in this chapter, probably not in all areas.

There were important variations of scale between prehistoric copper mines in Europe. Some sources were able at different times to establish dominant positions in metal supply. This is true of the Mitterberg mine district in Austria, the Kargaly mines in Russia, or the Cypriot orebodies, all of which established supra-regional monopolies in copper supply over long periods. The scale of production extends from what can be considered a prehistoric mining 'district', to the operation of large individual mines, such as Ross Island, El Aramo, or the Great Orme, which were dominant sources of copper in their respective regions, down to small operations that were of local significance.

Copper supply in prehistoric Europe

It is possible to distinguish three major types of copper mining in prehistoric Europe, based on the type of ore resources exploited and the prevailing technology. The earliest mining of copper ore involved the extraction of secondary minerals, such as malachite and azurite, from surface zones of oxidation. These minerals are rich in copper and easy to smelt using a primitive hearth or crucible technology. They were first used in the Chalcolithic, and where locally available continued to be exploited throughout the Bronze Age. There are a large number of such mines known across Europe, including most of the British and Irish examples. The use of these ores declined significantly by the later second millennium BC, suggesting widespread exhaustion of accessible deposits.

The second mining tradition involved the use of fahlore, namely copper sulpharsenides of the tennantite-tetrahedrite series. These were rich in copper, and like the oxidized ores could be smelted using a low temperature, non-slagging technology. Fahlore mining probably commenced in central Europe in the fourth millennium BC, and is particularly associated with the spread of Beaker culture metallurgy in western Europe in the third millennium BC. Relevant mining sites include Cabrières in southern France and Ross Island in Ireland. The use of fahlore declined significantly with the adoption of tin bronze around 2000 BC, but was developed further in central Europe during the Late Bronze Age, at mines such as Brixlegg and Mauk in North Tyrol, Austria.

The final copper mining tradition in Europe centred on the extraction of primary sulphide ore, particularly copper-iron sulphides such as bornite and chalcopyrite. Their earliest exploitation is not well dated, possibly beginning in the late Chalcolithic/Early Bronze Age at mines such as St Véran in France and in the Mitterberg complex in Austria. Many operations once thought to have processed sulphidic ore, for example the mines of central Wales, are now believed to have extracted oxidized minerals. While copper-iron sulphides, such as chalcopyrite, can be reduced with primitive processes, the efficient recovery of metal required an advanced furnace technology that was only developed at a later stage in the Bronze Age.

These mining traditions often overlapped in different parts of Europe. This would have occurred where separate mines with different ore types were worked at the same time using different technologies, and even in the same mine operation where different ore types were encountered. There is a general chronological trend from the mining of native copper and oxidized ores, to the extraction of fahlores, to the extraction of copper-iron sulphides. This does not reflect depth zonation within orebodies, as was once thought, but rather the development of new technologies and mining strategies that allowed different types of ore deposit to be mined. This occurred in the case of the fahlores due to the ease of extraction and perceived qualities of the resulting metal. In contrast, the development of a technology to process copper-iron sulphides was probably driven by a diminishing supply of oxidized ore and fahlore.

There is no simple linear progression of mining strategy, but a more complicated picture of local variability set against some broad trends. Different regions had their own mining traditions that were both historically and environmentally contingent, as well as being determined by cultural choice and economic circumstances. The importance of copper mining was driven by the fact that this metal maintained its position as a valuable commodity across Europe through the Bronze Age. That said, there was a notable decline in copper mining activity across the Continent during the first millennium BC, which must be connected to the changed status of this metal in the Age of Iron.

This book

The following chapters review the evidence for prehistoric copper mining in Europe on a regional basis. The survey commences with the earliest known copper mines, which date to the sixth millennium BC in the Balkans (chapter 2). This is followed by a consideration of early copper mining in the east and central Mediterranean, including the importance of Cyprus from where the name of this metal derives (chapter 3). The focus then turns to the west Mediterranean, and the mineral resources of the Iberian peninsula (chapter 4), followed by an examination of early copper mines in southern France (chapter 5). The next area to be considered is northern Europe, including my own work in Ireland and those of colleagues in Britain, with reference also to the potential for early copper mines in Scandinavia (chapter 6). The survey continues with some of the largest prehistoric copper mines known in Europe, including those discovered in the home of early copper mining studies, Austria, and the remarkable Kargaly operations in the Russian Urals (chapter 7). The book concludes with a general consideration of the technology and organization of these mines (chapter 8) and their broader economic, environmental, and societal context (chapters 9 and 10).

2

South-east Europe

Copper was the first metal used by humans, a practice that began at different times in various parts of the world. The earliest evidence comes from the Near East around 10,000 years ago, when some early farming communities started to experiment with surface finds of native copper. Initially collected for their golden colour, it was soon discovered that these small pieces of pure copper could be cold-hammered into desired shapes, making them different from rock minerals. This first occurred in areas such as northern Iraq and eastern Anatolia where native copper occurs naturally. By 7000 BC there is evidence from sites such as Cayönü in Anatolia for the heating of native copper (annealing) to improve the production of beads, awls, and other small objects (Muhly 1988, 1989). In time, this led to another important discovery, namely that native copper could be melted and poured into moulds at temperatures around 1083° C. It is not certain when this first occurred, but most probably in the sixth millennium BC (see Pernicka and Anthony 2010 for overview).

One of the reasons for the slow development of metallurgy in the Near East was the scarcity of native copper. The growing interest in metal eventually led to experimentation with copper minerals, such as malachite or azurite (Wertime 1973). These were initially used for non-metallurgical purposes, with malachite beads dating to the eleventh millennium BC known from a number of sites, including Shanidar Cave in northern Iraq (Solecki 1969). They were first recognized during the search for native copper, when rock outcrops were discovered bearing the distinctive green or blue staining produced by oxidation of copper minerals. The extraction of these surface minerals must have led in some instances to underground mining.

It is not certain when copper ore was first smelted in the Near East. The dating of copper smelting slag at Catal Höyük in south-central Anatolia to the seventh millennium BC remains contentious. The earliest secure evidence comes from the later fifth millennium BC, at sites such as Norsuntepe in south-east Anatolia and Abu Matar in the northern Negev, Israel (Pernicka 1990; Golden et al. 2001). By the late fifth millennium BC, copper metallurgy was well established at a number of centres in Anatolia, Iran, and the Levant. These early metalworkers were familiar with the mining and smelting of copper ores.

There were also important developments in casting, with the early use of simple open moulds followed in the later fifth millennium BC by the sophisticated lost-wax method, as famously seen in the Nahal Mishmar hoard and other finds in the Levant west of the Jordan valley (Bar-Adon 1980). This also indicates the first steps towards alloying, through the conscious selection of arsenic-rich copper ore to produce metal with superior technical qualities (Pernicka and Anthony 2010).

THE FIRST METALLURGY IN EUROPE

For many years it was assumed that metallurgy was introduced to Europe through colonization, trade, and/or the movement of specialists from more advanced cultures in the Middle East and Anatolia, reaching first the Aegean and from there into south-east Europe. This *ex oriente lux* view was challenged in the 1960s on the basis of major inconsistencies arising from the earliest radiocarbon results (Renfrew 1969). The new dates revealed that the earliest Balkan metallurgy significantly pre-dates that in the Aegean, while the smelting of copper ores in south-east Europe may even have occurred earlier than in the Middle East. Renfrew proposed an independent origin for copper mining and metallurgy in south-east Europe, following a similar line of development to the Near East, from the cold hammering and annealing of native copper to the mining and smelting of copper ores, and eventually tin alloying. He concluded that metallurgy was not a single invention, but rather a number of distinct and separate discoveries. These ideas received significant support with the discovery around that time of copper mines in Serbia and Bulgaria dated to the fifth millennium BC.

The south-eastern part of Europe, specifically the Balkans and Carpathian Basin, has the oldest evidence of copper metallurgy in Europe with objects of copper metal dating as early as 6000 BC. The earliest copper objects were made of native copper and include a number of awls, fishhooks, and rolled wire beads found in Early Neolithic settlements. From 5000 BC this metalworking expanded to larger implements, including axes, chisels, shaft-hole hammer-axes, and axe-adzes. The variety of metalwork in circulation by 4000 BC reflects the different roles that copper played in these early farming societies, both as functional objects and as items for social display. The prestige value of metal, both copper and gold, is considered especially significant in the emergence of social hierarchies at that time.

The origins of copper metallurgy in south-east Europe remain contentious. The early use of copper trinkets may have been associated with other elements of the 'Neolithic package' that originated in the Near East; however, no direct links can be established between the two areas. Renfrew (1969) observed that the pryotechnological knowledge necessary for smelting copper was already present in the ceramic traditions of the fifth millennium BC. This is evident in the production of graphite-decorated pottery, and also in the dark-burnished ceramic wares that are distinctive of the Vinča culture. The widespread occurrence of copper ores in the mountain ranges of the central Balkans, as well as experience gained in the underground extraction of flint and other hard rocks in this region during the Neolithic, were also significant in the local evolution of copper mining (Jovanović 1979, 1980a, 1980b, 1995). This mining and pyrotechnological expertise, the availability of copper resources, and the emergence of craft specialization in proto-urban and socially differentiated tell settlements, all contributed to metallurgical innovations in this part of Europe. The picture remains complicated, particularly as the possibility of trade and other contacts with early metal users in Anatolia means that an early transfer of metallurgical knowledge into south-east Europe from that direction cannot be excluded.

EARLY COPPER MINING IN SERBIA

The south-eastern part of Europe is a particularly metalliferous region, with numerous occurrences of copper and other metals in the west Carpathian and Transylvanian mountain belts. The most important copper deposits in the central Balkans are in Serbia, with the mines at Bor and Majdanpek particularly significant in the modern era (Krajnović and Janković 1995). Many of these ore deposits are exposed in hill and mountain terrain in a variety of geological settings. Some have well-developed gossans and zones of surface oxidation, which explains why the earliest use of copper was based on secondary copper minerals and also native copper.

A distinction may be made between the use of copper minerals for non-metalliferous purposes during the Middle Neolithic and the first production of copper metal from the same minerals in the Late Neolithic. The latter was a precursor to the beginning of the Copper Age, or what is generally termed the Eneolithic in that part of Europe. There are several finds of copper minerals from settlements and burials of Middle Neolithic (Starčevo-Körös-Cris cultures) contexts across the Balkans (see Borić 2009: fn. 1; Kienlin 2012: 129). These include beads of malachite and azurite associated with the phase 3 occupation at Lepenski Vir in the Danube Gorge (Srejović 1972). Beads, fishhooks, and awls of metallic copper are recorded from Early and Middle Neolithic contexts, including the settlements of Iernunt and Balomir in Transylvania, Obre 1 in Bosnia, Divostin and Selevac in Serbia, and Ducharova 1 in Bulgaria. These finds indicate that farming cultures of the sixth millennium BC in south-east Europe were acquainted with both native copper and oxidized copper minerals, and collected these to make beads and other personal ornaments (Glumac and Tringham 1990; Glumac and Todd 1991).

The number of copper metal finds in this part of Europe greatly increases during the early fifth millennium BC, with the emergence of the Vinča culture of the central and northern Balkans, and neighbouring groups such as Lengyel, Tisza, and others in the Carpathian Basin. The Vinča culture is well known for its distinctive pottery, anthropomorphic figurines, and tell architecture. Small objects of copper metal and caches of malachite are known from many Vinča settlements and burials, including the type-site of Vinča-Belo Brdo. The first use of copper minerals in that culture may have been more concerned with the aesthetic, using these minerals as ornaments and pigments; however, there is a clear emphasis on the production of metal by 5000 BC. The earliest evidence of copper smelting slag dates to the later sixth millennium BC, coming from the Vinča settlement of Belovode in eastern Serbia (Radivojević et al. 2010). A number of copper mines connected to this culture have also been identified, the best known of which lies in the Saska valley, *c.*20 km south-east of the town of Majdanpek in north-east Serbia.

Rudna Glava

This is one of the oldest copper mines known in Europe, dating to the late sixth/early fifth millennia BC. It was discovered on the slopes of the Okna hill during opencast extraction of iron ore in the modern era (Jovanović 1971). A series of small mines worked with stone hammers and antler picks are exposed along the steep north face of the opencast (Fig. 2.1). Archaeological investigations were undertaken in 1968–86, directed by Borislav Jovanović. Some 40 mine workings were investigated, which were connected by pottery finds to the Vinča culture. Evidence of Roman copper mining in the fourth century AD was also uncovered (Jovanović 1971, 1979,1982).

Rudna Glava (literally 'ore head') is an iron ore deposit of skarn type, occurring in a geology of metamorphic and Hercynian magmatic rocks. The latter occurs within a lens-shaped deposit formed at the contact between the granodiorite and marble rocks in narrow hydrothermal veins. The iron ore consists of magnetite and pyrrhotite, associated with small amounts of copper mineralization, principally chalcopyrite. The ore deposit is capped by a 20 m thick gossan, which overlies a zone of secondary mineralization. This contained oxidized copper ore (malachite, azurite, and cuprite), with malachite being the main ore extracted in prehistory (Jovanović and Ottaway 1976).

Some 32 individual workings of Chalcolithic date are recorded, though others probably were destroyed by the recent mining of iron ore. They occur in an east–west line over a distance of 50 m along the steep northern slopes of the limestone massif (Fig. 2.2). The workings are not mine shafts in the

Fig. 2.1. Location of Chalcolithic copper mines at Rudna Glava, Serbia (*Source*: B. Jovanović).

modern sense, but instead represent the emptied channels of the ore veins. This determined the overall shape and size of these workings, which are visible on the surface as small pits and irregular trenches ranging 0.5–2.5 m in width. They were worked to depths of 15–25 m, mostly determined by the position of the water-table. The size and shape of individual workings also depended on the concentration of oxidized ore present in each vein (Fig. 2.3). Some workings are close together on the surface following the same fissures and ore veins. They tend to narrow significantly with depth, often branching into short inclined tunnels and narrow crevices. None of the surface workings

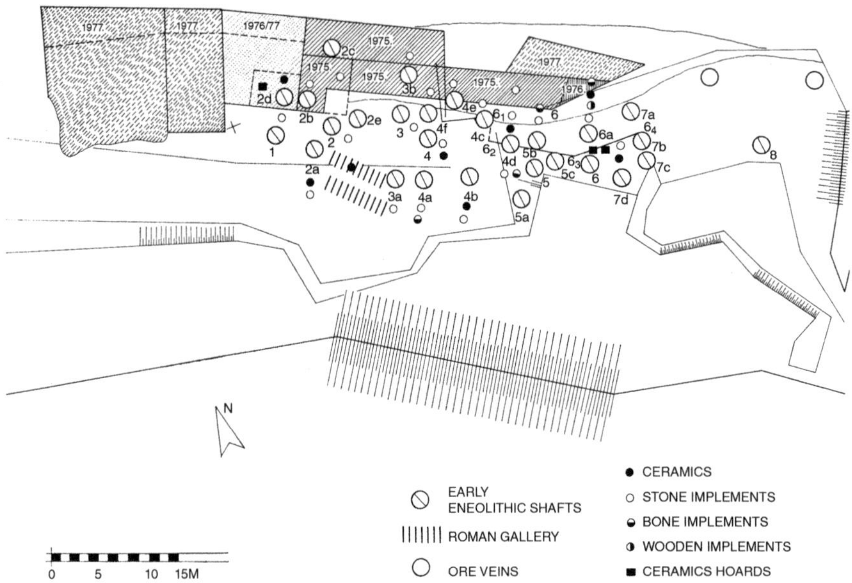

Fig. 2.2. Distribution of early mine workings at Rudna Glava
(*Source*: Jovanović 1979).

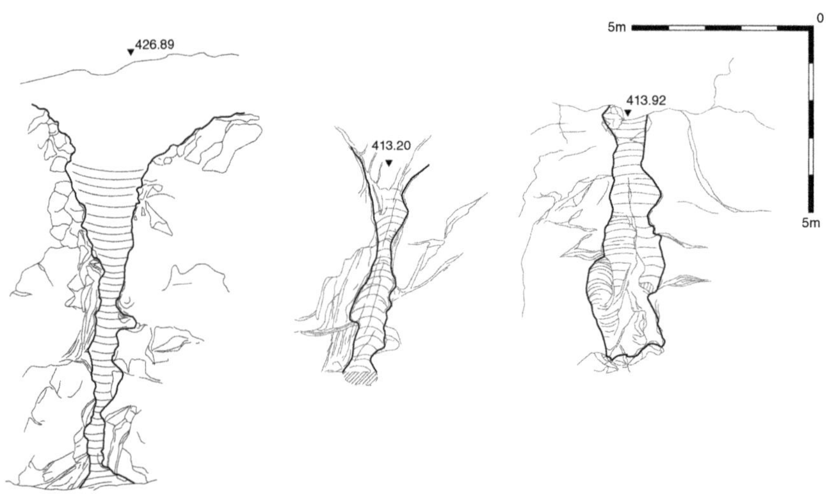

Fig. 2.3. Profiles of prehistoric mine workings at Rudna Glava
(*Source*: Jovanović 1982).

inter-connect with each other, and seemingly there was no attempt to explore beyond the natural configuration of the veins.

The mining strategy at Rudna Glava was closely adapted to the geological setting and exposure of the ore veins. Some of these veins were already exposed at outcrop, with the bright colours of the copper carbonate minerals clearly visible against the white limestone rock face. The removal of soil was also necessary to fully expose some of these mineral veins. Their occurrence along a steep rock face also meant the miners had to dig into this slope to create small access platforms to conduct mining.

The ore veins were progressively worked across the steep rock face. The miners took care to extract as much copper ore as possible from each vein, while removing as little of the country rock as possible. This was done using stone hammers and antler picks. Some 200 heavily worn stone hammers were found on the access platforms, in the mine infill, and in rock crevices underground. These are naturally rounded cobbles of various shapes, ranging 1–4 kg in weight and 10–20 cm in length. They are mostly of gabbro and other volcanic rocks, and may have been sourced in the bed of the river Saska below the Rudna Glava massif (Jovanović 1979). Most were modified for a haft, with medial rilling or side abrasions.

The stone hammers were used with antler picks and possibly bone tools to extract the softer vein mineralization. A number of heavily worn antlers were found in the mine. These were used as picks to prise out vein material in the highly fractured limestone. There are no finds of copper mining tools, nor any metal tool-marks on the mine walls. Fire-setting may have been employed on occasion to extract the harder limestone rock. This is uncertain, however, as the use of fire is not particularly suited to rock extraction in narrow vertical workings.

The extraction of these veins did not require artificial supports inside the workings. Stone walls were built to retain rock spoil at the surface, where space was limited on the narrow access platforms. The miners probably used ropes to haul sacks of ore out of the veins, possibly using horizontal timbers placed across the mine openings. The rock extract was finely crushed on the surface platforms, with hand-sorting of copper minerals to produce a concentrate for smelting. Each working was backfilled with rock debris once the vein was exhausted.

The dating of these mines was initially based mainly on the discovery of pottery of the Vinča culture. Three separate collections of pottery (hoards 1–3), as well as a number of individual vessels, were found inside the mine workings and on adjacent access platforms (Fig. 2.4). The pottery consists of large conical vessels with or without handles, used as amphorae in the transport of water. A number of large, wide-mouthed vases were also discovered, made of the same black burnished ware. Some of this pottery was found intact, but most vessels were broken. A number of these ceramic finds were discovered with stone hammers and antler tools.

Two of these pottery hoards were discovered on the access platform to Mine 6a, where they had been carefully placed in narrow rock crevices (Fig. 2.5). Hoard 1 consisted of a flat-rimmed conical pot and a large amphora jug with handles and channelled spiral decoration on the shoulder. Hoard 2 contained

Fig. 2.4. Cache of prehistoric pottery in Rudna Glava mine
(*Source*: B. Jovanović).

Fig. 2.5. Cache of prehistoric pottery in Rudna Glava mine
(*Source*: B. Jovanović).

three tall amphorae, each with three small handles and narrow necks. Hoard 3 consisted of three large amphorae, including one with channelled decoration, found together with a stone hammer, antler pick, and a zoomorphic altar. These finds can be dated closely as similar pottery is known from radiocarbon-dated contexts on excavated settlements of the Vinča culture in Serbia. The earliest pottery at Rudna Glava dates to the transitional Gradac phase between the Early and Late Vinča culture, while other pottery is characteristic of the Late Vinča culture.

A recent programme of radiocarbon dating has provided dates for antler and bone tools found in eight different contexts in this mine (Borić 2009). The earliest result is for an antler tool from Shaft 4f dated 6080–6010 BC, albeit with certain reservations as to the analysis. The result is of interest as it is contemporary with some of the finds of copper minerals from Neolithic settlements (Starčevo-Körös-Cris groups) in the Balkans (listed at the beginning of this main section, Early Copper Mining in Serbia, above), including those from the famous Lepenski Vir settlement located 30 km from the mine. If the date is accepted, it points to an early phase of malachite extraction at Rudna Glava for non-metallurgical purposes (body ornament etc.) during the Middle Neolithic period.

The main series of dates from Rudna Glava indicate that mining to produce copper metal was underway by 5400–5300 BC, and continued at a low level in the following centuries. The mining intensified in the early fifth millennium BC in the so-called Gradac phase of the Vinča culture, finally ending around 4700–4600 BC. A date of 2910–2880 BC for an antler mattock from Shaft 10 may indicate some resumption of copper mining at Rudna Glava in the Late Copper Age (Jovanović 1979).

Rudna Glava is one of the few prehistoric copper mines known in Europe where pottery was used to any extent. The ceramic finds are impressive because of the size and quality of the pottery, which finds close parallels in settlement contexts of the Vinča culture. The significance of the mine pottery is uncertain. The excavator suggested that these large vessels were used to hold water for sudden quenching of heated rock surfaces (Jovanović 1979). The miners may also have required water for washing and drinking. Another possibility is that these vessels carried water for use in the gravity separation of the crushed ore to extract the heavier magnetite content from the secondary copper minerals. This would explain the presence of the large ceramic bowls; however, there are no recorded sediment residues to support this theory.

Another possibility is that the deposition of pottery at the entrance to these mines, as part of the backfilling process, was a ritual act connected with some fundamental beliefs of the miners. This is supported by the discovery of three broken ceramic altars of zoomorphic design at the base of separate mine workings. These were initially interpreted as representing deer or ibex, but may also represent the stylized head of a ram with chevron and curvilinear line decoration (Gimbutas 1983). They may have been used as lamps, but are best interpreted as small altars. The excavator suggested they were used in propitiatory offerings for the protection of the miners and in thanksgiving to an earth goddess for the ore. This may also explain why each working was backfilled on the completion of mining. A symbolic connection between the use of red deer antlers as mining tools and these zoomorphic altars has also been proposed (Borić 2009).

No evidence of settlement directly connected to the Rudna Glava mine has been discovered, nor any indication that ore smelting or metalworking was carried out at the site. This may reflect problems of archaeological visibility and survival in the modern mining landscape. There are a significant number of contemporary settlements with finds of copper metal in Serbia. These include large numbers of copper implements and ornaments from Pločnik near Prokuplje, as well as the discovery of finely ground malachite ore at the Late Vinča settlement of Fafos near Kosovska Mitrovica (Jovanović 1979).

The closest known settlement of the Early Vinča culture is at Belovode near the town of Petrovac, some 50 km from Rudna Glava (Radivojević et al. 2010). There are similar stone hammers from both locations, with a large amount of copper ore (malachite and azurite) found at Belovode, much of it heat-altered and found mixed with charcoal and ash. The discovery of copper slag is a convincing indication of metallurgical processes. Radiocarbon dating confirms that the occupation of this site, *c.*5350–4650 BC, is contemporary with the mining at Rudna Glava. The copper ore at Belovode has no direct lead isotope connection to that mine, but has some similarity to a nearby ore source at Velika Brestovica (Radivojević et al. 2010). The malachite ore found at Belovode also has a significant lead isotope correlation with mineralization at Zdrelo, 10 km away. A number of trench workings, as yet undated, are recorded on copper mineralized outcrops in that area (Sljivar 2003, 2006).

The scale of production at Rudna Glava may have been quite limited, with one estimate of annual production in the order of 250–300 kg of copper (Jovanović 2003). It has proved difficult to trace the circulation of copper from this mine in the wider Balkan region. An early programme of lead isotope analysis revealed that 95 per cent of analyzed artefacts from the Copper Age Bulgaria and Serbia fell into nine distinct geological sources (Pernicka et al. 1993, 1997). Surprisingly, the initial study could not match these artefacts to copper mined at Rudna Glava. The same analyses did indicate the existence of potential sources at the large copper deposits of Majdanpek and Bor in eastern Serbia. The difficulty is that modern opencast mining at these locations seems to have removed any trace of prehistoric workings.

The decline of the Rudna Glava mine around 4600 BC did not mark the end of copper production in the central Balkans. Other copper mines may have been exploited at that time. These include the site of Mali Sturac located in the

Rudnik mountain range of central Serbia (Jovanović 1983). This is a lead-zinc-copper orebody in volcanic geology, with a thick gossan containing secondary copper minerals. This ore was mined for different metals in the Roman, medieval, and modern eras. Traces of prehistoric workings are recorded across a 250 m by 50 m area on the western slopes of Mali Sturac (Jovanović 1989: plate 2.1). Several mine workings are exposed by erosion on the top of the hill slope, with others covered by deposits of rock spoil. A large number of grooved stone hammers were found near the exposed workings (Jovanović 1989: plate 2.2; also Bogosavljević 1995), as well as pottery of Late Eneolithic and Early Bronze Age date.

Another copper mine of the Vinča culture has recently been identified at Jarmovac near Priboj-on-Lim in south-west Serbia (Derikonjić et al. 2011). The site was first recorded by Oliver Davies, in an early survey of ancient mines in the central Balkans (Davies 1938a). A number of early workings are recorded on outcropping veins of copper ore along the Jarmovac river valley. These include workings at Majdan where a vertical shaft and horizontal tunnel followed a malachite-rich vein. Davies discovered a grooved hammerstone of the Rudna Glava/Mali Sturac type at this location. Mine workings are also recorded at Curak, 50 km to the east of Majdan, where some 42 grooved hammerstones and smaller grinding stones are recorded. There is no independent date for either of the mines at Jarmovac, though a connection with a Vinča culture settlement at Kaludersko polje, 300 m to the west, has been suggested (Derikonjić et al. 2011).

EARLY COPPER MINING IN BULGARIA

There are numerous copper deposits in Bulgaria, many of which are of considerable size. The potential for prehistoric mining was explored in 1971–4 by a Bulgarian/Russian team who recorded ancient copper mining at several locations (Chernykh 1978a). In most cases there is no direct dating evidence (see Gale et al. 2003: 155 for discussion). The most important evidence comes from the southern foothills of the Sredna Gora Mountains in Thrace, southern Bulgaria.

Ai Bunar

This is a hilly, forested area located 8 km north-west of Stara Zagora. The site was discovered in 1934 when a local engineer, Ivan Azmanov, explored some of the early workings and collected artefacts and mining tools. The results of this survey were never published, and so the full significance of the site was not

Fig. 2.6. Excavation of Chalcolithic copper mine at Ai Bunar, Bulgaria (*Source*: E. Chernykh).

appreciated until the 1970s when further excavations were carried out (Fig. 2.6; Chernykh 1978a, 1978b).

The mineralization at Ai Bunar consists of a series of near-vertical veins of hydrothermal origin within a limestone, marl, and dolomite geology. They extend in a north-east/south-west direction across the small hills of Chasanov Trap, Malka Tumba, and Cairlyskata Usojna, where early workings have been discovered. The mineralization is polymetallic and includes small amounts of primary copper sulphide. The latter is of low grade and instead the early miners extracted secondary copper minerals, principally malachite and azurite, from a well-developed oxidation zone up to 25 m in depth (Pernicka et al. 1997: 57).

Mine workings of Chalcolithic date have been identified at 11 separate locations over a distance of 1.5 km, with the probability that others remain to be found (Chernykh 1978a, 1978b). With no mining in the modern era, most of these workings have survived intact since the prehistoric period. This is also because the ancient workings were backfilled with rock spoil as soon as the mining ended, leaving only shallow depressions today. The backfilling of the mines involved considerable effort and is difficult to explain on practical grounds connected to safety concerns or a desire to conceal the mines. One possibility suggested by the excavator, Evgenij Chernykh, is that it was a ritual performed to expiate the act of removing of minerals from the earth (Chernykh 1978b).

Described in the literature as wedge-shaped open-pits and shafts, the early workings at Ai Bunar are better characterized as trench mines that followed the strike exposure of the mineralized veins (Fig. 2.7). They average 15–30 m in length, with some examples 80–110 m (Chernykh 1978b: table 1). The form

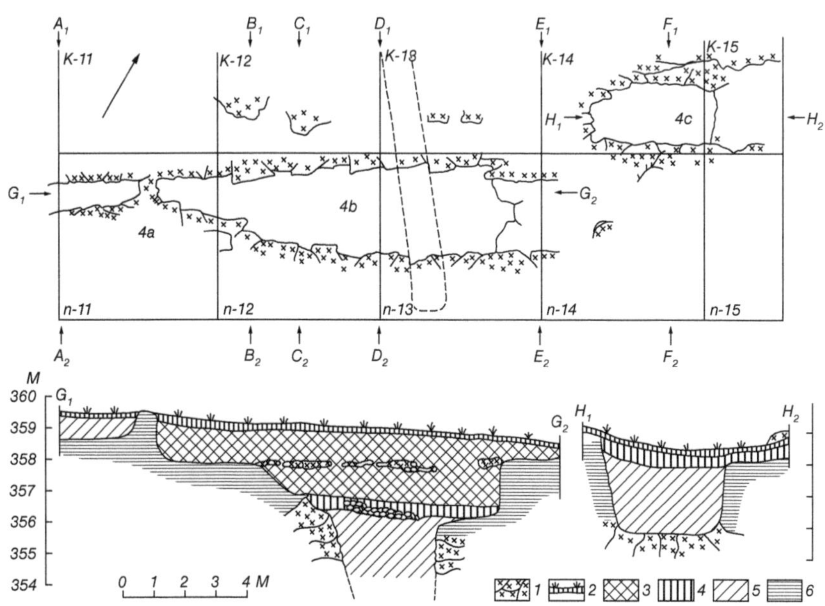

Plan of the excavation of mining site No. 4a, b, c, and stratigraphy of the layers.
1 — limestone and dolomite
2 — top layer (humus)
3 — occupation layer of the house
4 — dark grey/brown layer with rubble
5 — spoil
6 — iron-enriched veins

Fig. 2.7. Plan and stratification of mine 4, Ai Bunar

(*Source*: Chernykh 1978b).

of the workings was determined by the width and shape of the vein. Where these were massive, the workings could be up to 10 m in width, but could be as narrow as 0.5–1 m where thin veins were extracted. Some of the workings extend laterally to form true underground workings; however, these are of limited extent. The depth of many of the workings has not been established. Some are only 2–4 m deep, while others can reach 15–30 m to the water-table. One example (Mine 2) was investigated to a depth of 18–20 m, before flooding and safety considerations led to its abandonment without reaching the bottom (Chernykh 1978b). This mine on the southern slope of Chasanov Trap was re-excavated in 1972, when quantities of Chalcolithic pottery were found within an 80 m long by 5–10 m wide opencast (Chernykh 1978b).

Two other workings were excavated at that time, with some surprising results. Investigation of Mine 4 close to the summit of Malka Tumba hill exposed a group of three short trenches. One of these (Mine 4b) measured 12 m in length, 2.5 m wide, and was excavated to a depth of 6 m, but not bottomed. This working had been backfilled with rock spoil as soon as mining had finished. The partly filled trench was then used for human burial, with the discovery of a double grave containing an adult male and adult female (see Chapter 9). There were no grave-goods apart from some sherds of pottery that date the burial to the Chalcolithic. Excavation uncovered the remains of a hut structure overlying this burial. It is associated with pottery of Late Bronze/Early Iron Age date and includes the find of a grooved 'macehead' that could well be a mining hammer (Chernykh 1978b: plate 16). This raises the possibility of a later phase of prehistoric mining at Ai Bunar. There are no radiocarbon dates available for this mine at the present time.

The excavation of Mine 3, to the north-east of Mine 4, exposed a large double trench where two ore veins joined together. This working was 50–80 m in length and 3–10 m wide, and was investigated to a depth of 15 m (Chernykh 1978b: figs 4 and 5). As with Mine 4, there is evidence of deliberate backfilling with waste rock once the mining ended. There is also evidence of Chalcolithic occupation in the partly filled mine trench, with hearths and pottery but no built structures. The unburnt burial of an adult male was recorded at this level (Chernykh 1978b: fig. 6). There were no grave-goods; however, the grave contained a large amount of malachite thought to be a ritual deposit with this burial.

A number of artefacts were recorded during the investigation of this mine in the 1930s, and again in the 1970s. Some 20 fragments of antler were found, including some with remnants of sleeves used to fasten handles. These were used as picks to prise out mineralized rock from the vein and loosen the limestone wallrock. This seems to have been the main method of rock extraction as fire-setting was not used to any extent in these mines. Another feature is the limited use of stone hammers, with none recorded from the workings and only a single grooved example from the entire mine area

(Chernykh 1978b: plate 19). It was discovered in 1974 at a location 30 m away from Mine 3, where it may have been used in the concentration of copper ore in the mine area. The latter is not well understood as the waste rock from these processes was backfilled into the abandoned workings.

In the absence of stone hammers the miners may have used copper tools for the extraction and beneficiation of mineralized rock. Significantly, there are two copper implements recorded from the mine, both found by Azmanov in the 1930s (Fig. 2.8). The first is a shaft-hole hammer-axe measuring 128 mm in length, of a type described as a gad in recent mining literature. The second copper implement is a shaft-hole axe-adze, 168 mm in length, which, like the hammer-axe, is heavily worn. Both objects are of types made by Chalcolithic copper producers in the Balkan/Carpathian region (e.g. Renfrew 1969: plate 5).

Also of significance is the discovery of Chalcolithic pottery in a number of excavated mine workings. Two basic types were used; with large coarse ware vessels and a smaller number of graphite-decorated fine ware vessels. The latter is well known from excavated settlements of the Gumelnita/Karanovo VI culture in Bulgaria (Chernykh 1978b: plate 18). In the absence of radiocarbon dates, this pottery association dates the mine to the period 4800–4300 BC. There is also analytical evidence to connect the mine with copper used in Bulgaria during the Karanovo V period as early as 5000 BC (Gale et al. 1991, 2003).

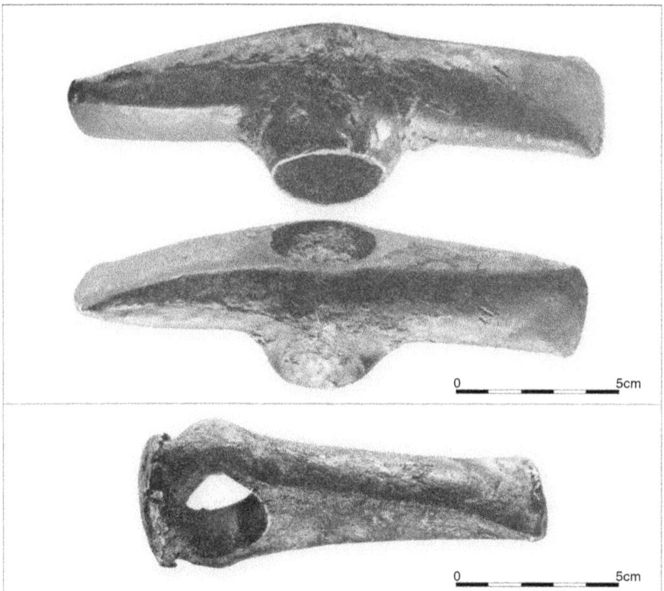

Fig. 2.8. Copper axe-adze (1) and copper hammer-axe (2) from mine 1, Ai Bunar (*Source*: Chernykh 1978b).

The scale of copper mining at Ai Bunar is considerable compared to Rudna Glava. It is estimated that some 20,000–30,000 tonnes of rock were extracted from the eleven known workings, containing about 2000–3000 tonnes of malachite ore that may have yielded up to 500 tonnes of copper metal (Chernykh 1982; 1998a).

The organization of this production is uncertain as no smelting sites or metal workshops have been identified in the vicinity of the mine. While these may yet be found, there is evidence for the transport of copper ore to settlements farther away, suggesting some separation of these activities within the chain of production.

The research team at Ai Bunar discovered a number of tell settlements within a 15 km radius of the mine, where oxidized copper ore that can be chemically linked to the mine was found in layers containing Gumelnita culture (Karanovo VI) pottery. Two of these settlements also have ore found in association with Maritza culture (Karanovo V) occupation, suggesting an earlier start to the Ai Bunar mine. While these settlements produced copper minerals and finished metal objects, surprisingly there is no evidence of on-site smelting or metalworking. In a number of cases the ground copper carbonate ore was found with ceramic mortars used in the grinding of these minerals. Chernykh (1978b) suggests that this may have been connected with the preparation of mineral pigment rather than metal production.

This is supported by the fact that the chemical and isotopic signature of metalwork from the settlement zone around Ai Bunar does not match the ore signature from that mine (Chernykh 1978b: 216; Gale et al. 1991). There is, however, analytical evidence to indicate that copper produced with Ai Bunar ore achieved a wider circulation in the Balkan/Carpathian zone (Pernicka et al. 1997). These same lead isotope results indicate that Ai Bunar was not the only Chalcolithic copper mine in Bulgaria. Other sources include a mine at Prochorovo in Sliven district, 70 km east of Ai Bunar where pottery of Karanovo V–VI period was found (Chernykh 1978b: 216). Finally, there is also the possibility of mining in that area during the Bronze Age, possibly including Ai Bunar itself. In 1974 two burials accompanied by Bronze Age pottery were discovered in the fill of a mine shaft at Tymnjanka copper mine, 4 km west of Ai Bunar (Chernykh 1978b).

Lead isotope analyses confirm that Ai Bunar was not the only early copper mine in the region. The results for Late Copper Age metalwork from Varna and other sites suggest possible ore sources in south-east Bulgaria, around Burgas, Medni Rud, and Rosen (Pernicka et al. 1997; Gale et al. 2003). Reference can be made to the discovery of ancient copper workings in those areas, including Karabajir, near Burgas on the Black Sea coast, where stone hammers were used to extract malachite ore from three ore veins up to 2 m wide (Davies 1938a). The most southerly of these veins was worked for a distance of a kilometre by narrow trenches, 50 m long and up to 40 m deep, which were connected underground by rock arches. The other two veins were mined using rows of shafts placed 8–15 m apart and probably connected underground. These early workings are not dated, however Davies suggested the Karabajir mine may date to the Early Iron Age.

Davies also drew attention to another early copper mine in this part of Bulgaria, located at Rosenbajir, 24 km from Burgas. Again, there is evidence of trench mining and vertical shafts connected to underground galleries, the date of which is uncertain as there was Roman mining at the site (in Shepherd 1980: 190). Finally, ancient workings have been identified along the Medni Rud ridge of the northern Strandza Mountains in Bulgaria (Leshtakov 2010). These mines are not dated; however, a nearby settlement at Cernomorec has evidence for an entire chain of production from ore to metal (Barthelheim and Krause: personal communication).

DISCUSSION

South-east Europe was one of the few places in the world making cast tools of smelted copper during the fifth millennium BC. Also notable is the amount of copper in circulation, with almost five tonnes of copper recorded archaeologically for that period (Pernicka et al. 1997; Pernicka and Anthony 2010). The original amount in circulation must have been many times greater, indicating mining and smelting of copper ore on a scale much greater than the earlier use of native copper.

Rudna Glava and Ai Bunar provide evidence for relatively advanced mining techniques in the Balkans from around 5000 BC. Neither mine belongs to the very beginning of copper production in the Balkan/Carpathian region, which may be linked to the collection of native copper and secondary copper minerals (malachite and azurite) in the late seventh and sixth millennia BC. Early farming groups collected these brightly coloured materials to make beads and other personal ornaments. This led to the discovery of copper ore deposits at several locations, such as Rudna Glava, which were taken advantage of once the basic techniques of smelting were developed in the late sixth millennium BC.

The earliest copper mining in the Balkans may have developed from an older tradition of hard rock extraction (Jovanović 1980a). Surface extraction of flint and other rock types is well known from the Palaeolithic and Mesolithic in both central and eastern Europe. Numerous flint mines of Neolithic date are also recorded; for example, at Mauer near Vienna or Krivo Polje in central Serbia (Jovanović 1980a). Jovanović argued that this older tradition of hard rock quarrying and underground mining was an important technological influence on the emergence of early copper mines such as Rudna Glava.

Following the initial use of native copper and malachite pigment, there is evidence of large-scale mining and smelting of oxidized copper ore in the Balkans during the fifth millennium BC. Chernykh (1978a) referred to this as a 'metal-boom' period. The scale and diversity of metal production increased progressively during the fifth millennium BC, underpinned by significant output from mines in Serbia and Bulgaria. This continued to around 3300 BC when there was a widespread decline in metallurgy. Some have attributed this to the exhaustion of oxidized copper ores (see Taylor 1999 for an alternative explanation).

Some insight into these changing patterns of copper supply in prehistoric Bulgaria is provided by lead isotope analysis (Pernicka et al. 1997; Gale et al. 2003). The results emphasize the importance of Bulgarian and Serbian ore sources for early Balkan metalworking, with no evidence of any significant supply of copper from external sources. This affirms the autonomous development of early copper metallurgy in the Balkans, which is also indicated by the typology of the metal products (see Todorova 1978).

To summarize, the origins of copper production in Europe can be traced from developments in south-east Europe during the sixth and fifth millennia BC, involving culture groups such as the Vinča and Gumelnita who were engaged in copper mining. While contacts with Anatolia cannot be excluded in the early stages, the evidence points to an independent development of copper mining and metallurgy. This occurred at a time of increasing craft specialization in tell-centred societies that were undergoing a major transformation of culture and belief.

By the later fifth millennium BC the centre of innovation gradually moved into the central Balkans and Carpathian Basin associated with late Vinča and Bodrogkeresztúr culture groups (Kienlin 2012). From there, small numbers of first copper objects were traded into the north Alpine region of central Europe. It eventually led to a wider use of copper in that region by 3800 BC, associated with culture groups such as the Mondsee in Austria, Altheim in southern Germany, and the Cortaillod and Pfyn in Switzerland. This involved the continued supply of metal from the Carpathian Basin; however, an eventual transfer of metallurgical knowledge through these contacts led to groups such as the Mondsee becoming involved in copper mining. The details of this transfer of metallurgical knowledge will be examined in Chapter 7, when the emergence of the Eastern Alps as the most important copper mining centre in Bronze Age Europe is considered.

3

Eastern and central Mediterranean

GREECE AND THE AEGEAN ISLANDS

Copper objects first circulated on the Greek mainland during the fifth millennium BC and shortly after in the islands of the southern Aegean (Zachos 2007). The earliest metalwork of Late Neolithic date comprised small objects such as awls, beads, and bracelets. Metal use gradually expanded during the Chalcolithic stage that followed, with production of larger items such as axeheads. There are parallels with the development of early metallurgy in the Balkans, however there was much less copper in circulation. This may be explained by the absence of early copper mines comparable to Rudna Glava or Ai Bunar in either Greece or the Aegean islands.

The use of metal in the Aegean expanded significantly during the third millennium BC, with the emergence of a flourishing culture that had extensive seafaring contacts (Renfrew 1972). The importance of maritime trade in this region dates from the Neolithic when the island of Melos was a major source of obsidian across the east Mediterranean. Lead isotope analysis confirms that the copper, lead, and silver used by the Cycladic culture of the Early Bronze Age came from ore sources on many of those islands (Stos-Gale 1989). These metals were traded widely across the Aegean, with supply also into mainland Greece. While no copper mines have been identified, lead/silver workings of this period are recorded at Lavrion and at Ayios Sostis on Siphnos (Wagner et al. 1980).

There are numerous deposits of copper ore and other metals in mainland Greece. No prehistoric copper mines have been identified; however, the potential has been examined by lead isotope analysis (Gale and Stos-Gale 2002: fig. 1). An examination of various ore deposits in northern Greece, including examples in Thrace and eastern Macedonia, Thasos, the Pangeon Mountains, and Chalkidki did not reveal any likely sources of copper in prehistory. Samples were also taken in east-central Greece, from mineralization in the Othrys Mountains where there are several indications of ancient mining. Radiocarbon dates indicate copper mining at various locations there during the first millennium BC (Gale and Stos-Gale 2002: table 3). There is no direct evidence of prehistoric mining, however, lead isotope results do suggest that copper ores in the Othrys Mountains were exploited to a small extent during the Early Bronze Age, and again by the Mycenaeans during the Late Bronze Age (Gale and Stos-Gale 2002: fig. 3).

Another potential source is the Lavrion mine area in Attica, recorded in historical sources as a major source of silver and lead during the first millennium BC. There are many deposits of copper ore in that area; however, no definitive evidence of prehistoric mining has been identified. This is possibly due to the destruction caused by lead and silver extraction during the Archaic/Classical, Roman, and early modern era. The potential is indicated by the discovery of a copper-smelting furnace of Early Bronze Age date at Raphina near Lavrion (Theocharis 1952, 1954). More significant are the results of lead isotope analyses, which reveal that 70 per cent of metalwork analyzed from Late Bronze Age sites in the Aegean matches with copper ores from Lavrion (Stos-Gale 2000: fig. 3.5). While the field evidence is still lacking, these ore deposits seem to have been a major source of copper during the Late Bronze Age, with some use in the Early Bronze Age (Gale and Stos-Gale 2002).

In the absence of dated workings, the most important indicator of prehistoric copper mining is the discovery of smelting evidence in the vicinity of mineralized outcrops. Deposits of copper slag of Early Bronze Age date are recorded at 19 locations extending from the southern Greek mainland to the eastern Cyclades, as far south as Crete (Fig. 3.1). The largest concentration occurs on the islands of Kythnos and Seriphos, with early slag deposits also recorded on Aigina, Keos, Keros, and Siphnos (Catapotis 2007: fig. 12.1, table 12.1).

The most significant production of copper may have come from Kythnos. Lead isotope results suggest that this island supplied copper across the Aegean during the Early Bronze Age (Stos-Gale 1989; Bassiakos and Philaniotou 2007). While there are no massive orebodies on the island, there are numerous small occurrences of oxidized copper mineralization associated with deposits of iron ore. Mining in the modern era led to the discovery of ancient workings at several locations, many of which are now destroyed. The copper ores are found mostly in the north-west, north-east, and eastern parts of the island, often on exposed coastal headlands in proximity to deposits of early copper-smelting slag (Bassiakos and Philaniotou 2007: fig 2.1).

The best known of these smelting sites is Skouries on the north-east side of the island, where there is a deposit of some 3,000 tonnes of smelting slag. Radiocarbon dating and pottery finds indicate an Early Bronze Age date for this site (Stos-Gale 1989). The copper ore may have come from malachite-stained outcrops some 400–500 m to the west of this smelting location. Another possible source is located 2 km to the south at Cape Tzoulis in the Ayios Ioannis area, where surface finds near an opencast mine indicate activity during the Early Bronze Age (Hadjianastasiou and MacGillivray 1988).

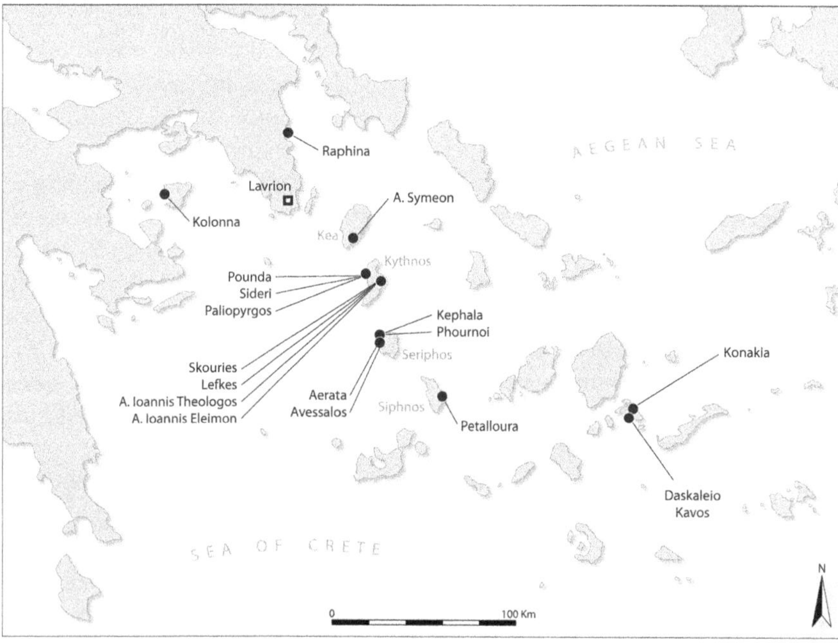

Fig. 3.1. Copper smelting sites of probable Early Bronze Age date in the Aegean islands. The location of the Lavrion ore deposit is also shown

(*Source*: author, based on Catapotis 2007 with additions).

Copper mineralization has been identified close to other smelting sites on Kythnos. These include the slag heaps at Pounta, Siden, and Paliopyrgos on the north-west side of the island, which are close to opencast mines at Aspra Kellia and Petra (Catapotis 2007; Bassiakos and Tselios 2012: 155).

Deposits of Bronze Age copper-smelting slag are recorded on other Aegean islands, notably Seriphos, Siphnos, Kea, Thassos, Andros, and Paros. Siphnos was the main source of silver and lead during the Early Bronze Age, where the site of Ayios Sostis was a significant producer of these metals. Copper smelting is recorded at the sites of Kephala and Phournoi on the northern coastline of Seriphos (Georgakopoulou et al. 2011). This island has important deposits of secondary iron minerals that were mined in the modern era, with indications also of accompanying copper mineralization. The most significant of these lie in the southern part of the island, whereas most of the early slag heaps occur along the northern coastline. This may indicate the destruction of early smelting sites in southern areas. Field survey has located copper minerals near the Kephala slag heap, however, no early mines have been identified. As with Kythnos, the emphasis seems to have been on small-scale surface extraction of secondary copper minerals at different locations on the island.

The overall picture is one of small-scale, dispersed copper mining on a number of Aegean islands during the third millennium BC. This was based on the use of small surface deposits of oxidized copper minerals, principally malachite, azurite, and chrysocolla, with no evidence for the smelting of sulphidic ore. In most cases copper smelting was carried out close to the mineralized outcrops, with the scale of mining indicated by the small size of the slag heaps. There are few details of the mines themselves; however, opencast workings may have been common. One possible example is that identified at Aspra Kellia on Kythnos (Bassiakos and Philaniotou 2007: fig 2.5). The method of mining is uncertain, as these workings have not been investigated to any extent. It is unlikely that fire-setting was used given the shortage of wood on these islands. This may indicate a reliance on stone hammers, however these are notably absent from both the mineralized outcrops and the smelting sites.

In addition to this mode of dispersed production, there is evidence of centralized smelting where copper ore from several mine sources was transported from a distance of a few kilometres. This is likely in the case of the Skouries smelting centre on Kythnos, and for the slag deposits of Kephala and Avessalos on Seriphos (Catapotis 2007). The site of Chrysokamino in eastern Crete is another example of a smelting centre located some distance from potential mine sources (Betancourt 2006). There was a brief phase of extractive metallurgy in that site during the Early Bronze Age using copper ore imported from the Cyclades (Gale and Stos-Gale 2002). There are no prehistoric copper mines recorded on Crete, where there are few deposits of copper ore. Lead isotope results suggest that mineralization at Miamou and Chrysostomos, on the southern side of the island, might have been exploited at the beginning of the Bronze Age (Gale and Stos-Gale 1986: fig. 7; Gale N. 1990). The same analyses also reveal a complex pattern of copper supply to Minoan Crete, based on sources in mainland Greece, Anatolia, Cyprus, and the Near East.

The pattern of small-scale copper production in the Aegean islands ended around 2000 BC. This was probably due to the exhaustion of the available copper deposits (Bassiakos and Tselios 2012). Metalworking activity did continue across the Aegean during the second millennium BC, when the focus in terms of copper supply shifted to another island in the east Mediterranean.

CYPRUS: ISLAND OF COPPER

This island is recognized as one of the most copper-rich regions in the world. The Latin word for this metal (*cuprum*) is derived from *Aes Cyprium* (Cypriot Copper), a term used by Pliny in his *Natural History* to describe the pure metal rather than its alloys (HN XXXIV: 2–4).

Pliny's observation that copper and the art of mining were first discovered on the island, though incorrect, demonstrates the renown of the island's ore deposits in Late Antiquity. This mineral wealth, combined with its strategic location in the east Mediterranean, help to explain why the island came to have a major role in copper supply during the Bronze Age.

There are abundant traces of copper extraction on Cyprus dating from the Early Bronze Age to the Roman era. Ancient mine workings and deposits of smelting slag are recorded at many locations across the western end of the island, where rich deposits of copper ore occur on the lower slopes of the Troodos Mountains (Fig. 3.2). Many of these locations were re-worked by large-scale opencast mining in the modern era, with a resulting loss of information about the earlier operations. For this reason, there is little direct evidence of prehistoric copper mining; however, extensive operations may be inferred from the discovery of smelting and metalworking evidence in the vicinity of many ore deposits. Scientific studies, in particular lead isotope analysis, also help to connect the mineralization on the island to finished metal products in circulation during the Bronze Age.

The earliest copper production on Cyprus dates from the late fourth to mid third millennia BC. This was small in scale and probably based on surface occurrences of native copper and oxidized ores (Muhly 1989: fig. 1.1). Knapp (2012) has outlined how this first copper production developed within a broad sphere of metal exchange and other interactions during the Late Chalcolithic, when Cyprus was in contact with coastal Anatolia, the Aegean islands, and

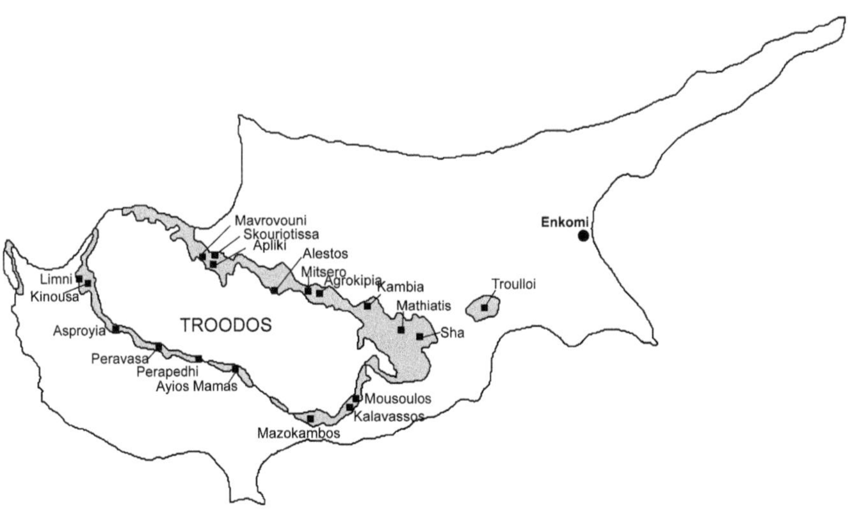

Fig. 3.2. Map of Cyprus showing location of ancient copper mines in the Troodos Mountains, as well as the location of the Bronze Age trading centre at Enkomi

(*Source*: Lina Kassianidou 2012a).

the southern Levant where copper metallurgy was already established. The exchange of metal led to a spread of metallurgical knowledge, and to the commencement of copper mining in the Troodos foothills by 2500 BC. The growing importance of Cypriot copper is evident in the centuries that followed, in its supply to pre-palatial Crete, and in the many references to 'Alashiyan' (Cyprus) copper contained in cuneiform texts from Syria dated to around 1900 BC (Knapp 1985).

Between 1600–1200 BC, the island became a major supplier of copper in the east Mediterranean, with trade extending into the Near East and Egypt, and as far west as Sardinia and Sicily. This is confirmed by lead isotope analysis of copper traded during that period, including some major cargoes lost off the southern coast of Turkey in the fourteenth and thirteenth centuries BC (Bass 1967; Pulak 1998, 2008). The scale of copper production on Cyprus during the Late Bronze Age was connected to mining operations at several locations in the Troodos Mountains.

This mountain range is underlain by rocks of the Troodos ophiolite complex. These represent a slice of ancient oceanic crust that formed by seafloor spreading processes in a deep ocean some 90 million years ago. Tectonic movements uplifted these marine sediments around 25 million years ago to form the Troodos Mountains (Constantinou 1982, 2012). The same processes led to the formation of numerous copper orebodies in the foothills of these mountains. These lie within the Lower Pillow Lava geological formation, which occurs on the margin of the Troodos ophiolite. Massive deposits of copper sulphides, principally chalcopyrite, were formed on the ancient sea floor by hydrothermal solution processes connected to Pillow Lava volcanicity. Some 30 orebodies have been identified, with major clusters in the mining districts of Skouriotissa, Mitsero, Kalavasos, and Limni. These deposits range from around 50,000 tonnes to 17 million tonnes of copper ore. The primary sulphide mineralization in these orebodies consist of pyrite (FeS_2) and chalcopyrite ($CuFeS_2$), with copper content typically in the 0.5–4.5 per cent range.

The uplifting of these orebodies through tectonic movements, and their subsequent weathering, exposed this mineralization to processes of surface oxidation. The action of percolating rainwater leached the mineralized outcrop, leading in time to the formation of limonitic gossans (see Fig. 1.2). The copper content was leached down in solution, where, on reaching the water table, it was precipitated to create a zone of supergene enrichment. This contained secondary sulphide mineralization made up of chalcopyrite, covellite, chalcocite, bornite, digenite, idaite, and tennantite, as well as copper oxides and sulphate minerals. Discoveries of native copper are also known, with only rare occurrences of copper hydroxycarbonate minerals such as malachite and azurite.

The early miners on Cyprus targeted the deposits of secondary enriched ores beneath the bright surface gossans. These ores have a conglomeratic structure and high porosity, which meant they could be extracted in large amounts using the most primitive of tools (Constantinou 1982, 2012).

Ancient mines

Ancient copper mines are recorded at numerous locations in the Pillow Lava formations of the Troodos Mountains. These include mines at Maurovouni, Apliki, Skouriotissa, Memi, Alestros, Mitsero, Agrokipia, Kambia, Kapedhes, Mathiatis, and Sha. On the southern side of the Troodos are the mines of Kalavasos, Parekklisha, Peravasa, and Vreccia, with the mines of Limni on the western side (Kassianidou 2012a: fig. 10.2).

The re-opening of these mines in the early twentieth century revealed evidence of earlier operations at many ore deposits. Ancient galleries, shafts, and adits, including many with timber supports, were found, along with ancient mining tools including ladders, ropes, hand windlasses, wedges, baskets, shovels, and oil lamps. Many such finds probably date to the Roman period, when extensive copper mining was conducted on the island.

Opencast mining in the modern era destroyed much of this evidence, though some records were made (Davies 1928–30; Bruce et al. 1937). No systematic study of this early mine archaeology has been undertaken nor have prehistoric workings been dated, with the exception of the mine at Ambelikou. One of the difficulties lies in distinguishing prehistoric mines of the Bronze Age from mines of the Hellenistic and Roman periods.

With many of the early mines now destroyed, the most important evidence for early copper production in Cyprus comes from the discovery of slag heaps in the general vicinity of the orebodies. Fieldwork has identified significant traces of smelting activity in the vicinity of many mine sites. Some 40 deposits of copper-smelting slag are recorded in the Pillow Lava formations along the margins of the Troodos Mountains (Constantinou 2012: fig. 2.4). These may contain a total of four million tonnes of slag, deposited over 3,000 years from the Bronze Age to Roman period when as much as 200,000 tonnes of copper was produced on the island (Constantinou 1982). These slag deposits can be located some distance away from the mine workings.

Cyprus was one of the first parts of Europe where copper was produced from the smelting of pyritic ores using an advanced slagging technology. This developed from an early technology of one-step smelting of roasted ore, which produced crude furnace conglomerate that had to be re-smelted in small furnaces, or else crushed to extract prills of copper from the viscous slag. This primitive technology was used during the Early Bronze Age when only small amounts of copper were produced on Cyprus. The scale of production changed with the adoption of new smelting methods in the sixteenth century BC. The introduction of a matte smelting process, employing bellows and tuyères at higher temperatures with improved fluxing, allowed for a more effective separation of metal and tapped slag. It is not known whether this was a local development or was the result of contacts with Anatolia where copper was extracted from sulphide ores at a relatively early stage.

The earliest evidence of copper mining, and probably smelting, on Cyprus comes from the Middle Bronze Age site of Ambelikou Aletri in the north-western Troodos. Ancient mine galleries with stone axes, and possibly hammerstones, discovered in 1942 are dated by pottery and radiocarbon analyses to 2044–1929 BC (Merrillees 1984; Webb and Frankel 2013). Finds from the site included copper ore, a blow-pipe, a crucible, and a mould, part of a complete chain of production in the vicinity of an ancient mine working (see also Knapp 1999: 100).

Politiko-*Phorades* is one of the earliest primary copper-smelting workshops known on Cyprus (Kassianidou 1999, 2012a; Knapp 2012). The site is located 500 m from a small deposit of copper ore at Kokkinorotsos, where there is a large gossan formation. Archaeological excavations uncovered an estimated three tonnes of primary copper-smelting slag and 50 complete tuyères, which are dated 1600–1400 BC. The remains of cylindrical furnace structures built of clay were also discovered at the site. The associated slag is different from earlier smelting remains on Cyprus, and indicate the early use of an advanced fayalitic technology connected to a matte smelting process. No evidence of ore processing was discovered at Politiko-*Phorades*. That activity was probably carried out in the mining site, with the sulphidic ore roasted at an intermediary location. It is likely that small-scale smelting operations such as Politiko-*Phorades* were common throughout the Troodos Pillow Lava zone during the Late Bronze Age (Knapp 1999).

Another settlement has been identified at Apliki-*Karamallos* in western Cyprus. This site lies on the high slopes of the Troodos foothills, close to the Apliki orebody (Fig. 3.3). It was excavated in 1938–9, when evidence of copper production was found in a settlement with house structures dated to the thirteenth century BC (Du Plat Taylor 1952). The site has been interpreted as a miner's village with storage facilities for food and with domestic activities such as spinning and weaving (Muhly 1989). The discovery of crucibles, smelting slag, and tuyères confirms that the residents of *Karamallos* were involved in copper smelting connected to the nearby Apliki mines, with ore beneficiation and roasting carried out in the vicinity. It has been suggested that this was an official residence from where the copper production was controlled (Knapp 2012: 22). There were probably many such centres producing copper in the Troodos Mountains during the Bronze Age. These inland production sites were connected to larger settlements along the coast, where secondary smelting and casting of copper was undertaken, centres such as the fortified town of Enkomi at the eastern end of the island dating *c.*1600–1200 BC (Kassianidou 2012a).

Fig. 3.3. View of Apliki mine from Skouriotissa
(*Source*: Lina Kassianidou).

Ironworking was adopted on Cyprus as early as the twelfth century BC, and began to be widely practised between 1050 and 750 BC. Copper production continued on the island during the first millennium BC into the Roman era. Bronze was still a valuable commodity with specialized uses for ornaments and weapons. Copper mining was controlled by various Iron Age kingdoms on the island, with some of the copper exported across the east Mediterranean (Kassianidou 2012b).

There is evidence of intensive copper mining on the island during this period, both from the mine workings and the deposits of smelting slag in their general vicinity (Weisgerber 1982; Kassianidou 2012b). Modern opencast mining has destroyed many mines of this period. Some evidence has survived; for example, wooden mine supports from Kokkinoyia are dated between the ninth and seventh centuries BC, while timber from the mines at Skouriotissa is also dated to the earlier first millennium BC (Weisgerber 1982; Zwicker 1986).

Important evidence of Iron Age copper production comes from Almyras on the north-eastern slopes of the Troodos Mountains, close to the ancient copper mines of Sha and Mathiatis (Fig. 3.4). The site was excavated in 1998–2000, with evidence for a complete chain of production, from mining to ore processing, copper smelting and copper working, in the period *c.*600–100 BC (Fasnacht and Kassianidou 1992; Fasnacht 1999). A small mining pit was discovered on a mineralized outcrop (Fig. 3.5), with finds of stone hammers and grinding stones

Fig. 3.4. Ancient copper mines at Mathiatis South
(*Source*: Lina Kassianidou).

Fig. 3.5. Ancient mine working close to copper production site at Almyras
(*Source*: Lina Kassianidou).

used in ore beneficiation. Evidence of copper smelting discovered 30 m from this mine includes a furnace dated to *c*.500 BC, the only near-complete example excavated on Cyprus.

In conclusion, the scale of copper production on Cyprus from the Late Bronze Age to Roman times had severe environmental consequences for the island. The production of some 200,000 tonnes of copper metal over those two millennia is estimated to have required approximately 60 million tonnes of charcoal for smelting fuel. One estimate suggests this required the felling of up to 1.2 million cubic metres of wood, equivalent to 150,000 square kilometres of forest on an island that is only 9,300 square kilometres in extent (Constantinou 1982). It is not possible to examine the environmental impact in more detail, as no pollen records are available for the island. The scale of production certainly points to careful woodland management, which was only possible because of the unique ability of the Troodos Mountains to regenerate forests due to higher rainfall and favourable growth conditions.

Cyprus and the Late Bronze Age trade in copper

Between 1600–1200 BC Cyprus became a major supplier of copper in the east Mediterranean, with trade networks extending into the Near East and Egypt, and as far west as Sardinia and Sicily. This is confirmed by lead isotope analysis of finished metalwork and raw copper ingots from this period. The latter include the 'oxhide' ingots, so-called because of their distinctive shape, each weighing up to 40 kg of pure copper. They have a wide trade distribution in the east and central Mediterranean with finds extending into the Balkans, Middle East, and Egypt. The largest number is recorded from Sardinia, followed by Crete and Cyprus (Lo Schiavo et al. 2009).

These oxhide ingots are famously recorded in two Late Bronze Age shipwrecks off the southern coast of modern Turkey. The Uluburun shipwreck was discovered in 1982 and is dated to around 1306 BC. It contained a cargo of exotic trade goods with connections across the central and east Mediterranean into north Africa, Egypt, and Mesopotamia. This included an estimated 10 tonnes of copper metal, in the form of 354 oxhide ingots and 120 bun and oval ingots, as well as one tonne of tin metal in similar ingot shapes (Bass 1986; Pulak 1998, 2008).

The Cape Gelidonya wreck discovered in 1960 dates to the later thirteenth century BC. It also contained a rich cargo of metal, with both copper and tin ingots (Bass 1967). The copper in both cargoes came from Cyprus; however, the exact source of the tin is uncertain (Stos 2009).

Lead isotope analysis indicates that almost all the oxhide ingots produced after 1400 BC were made with copper from Cyprus (Stos-Gale et al. 1997; Gale and Stos-Gale 2012). This includes the ingots on Sardinia and those found on the Uluburun and Cape Gelidonya wrecks. No ingot moulds or casting sites have been identified on the island, possibly because sand casting methods were used in their production. This trade in oxhide ingots continued to around the eleventh century BC.

Following some considerable debate around the scientific validity of lead isotope analysis as a source provenancing tool, as well as the selection of ore and metal samples used for these analyses, a Cypriot origin for these oxhide ingots is generally accepted (see Hauptmann 2009 for review). More controversial is a view that all this Cypriot ingot copper came from the Apliki ore deposit at the western end of the island. That mineralization has a distinctive lead isotope signature, which matches closely with 90 per cent or so of oxhide ingots analyzed within and outside of Cyprus (Gale and Stos-Gale 2012). While the scientific evidence is compelling, some researchers still question the likelihood of only one mine source for all of these ingots, at a time when copper ore was being produced from several deposits in the Troodos Mountains (see Knapp 2012).

The earliest evidence of mining at Apliki dates to around 400 BC, and comes from radiocarbon dating of shaft timberwork (Weisgerber 1982). There are similar dates from the nearby Skouriotissa mine (Zwicker 1986), however mining at both locations probably commenced at an earlier date. This is confirmed by evidence from the mining settlement of Apliki-Karamallos, which confirms that the nearby Apliki ores were being mined in the thirteenth century BC (Du Plat Taylor 1952). It is supported by lead isotope results that match smelting slag from the settlement with the Apliki ores. The prominence of this mine at that time may be due to the discovery of a large deposit of rich supergene mineralization. Lead isotope analysis confirms that some of the Apliki copper was used to make bronze artefacts on Cyprus; however, most of this metal was destined for export in a highly regulated way, using the standardized form of the oxhide ingot. The latter were probably produced in secondary smelting and casting centres in coastal settlements, with primary (matte) smelting undertaken at inland sites such as Politiko-*Phorades*, located closer to the mine sources (Knapp 2012: 18). Finally, these lead isotope analyses confirm that other ore deposits around the Troodos Mountains were mined during the Late Bronze Age, possibly on a smaller scale for local domestic consumption (Gale and Stos-Gale 2012).

In conclusion, the scale and sustained development of Bronze Age copper production on Cyprus is impressive. Stöllner (2003a: 437) suggests that can be explained by a number of factors. These included the presence of highly visible deposits of rich copper ore, the island setting that contributed to a natural concentration of production efforts and specialization, the central trading location in the east Mediterranean and, finally, its natural richness in fertile soil, climate, and regenerative forests.

SARDINIA AND CORSICA

The island of Sardinia is rich in metal sources, principally lead and silver, but also significant deposits of copper. There are many different styles of mineralization from Cambrian to Tertiary age (Marcello et al. 1978; Zuffardi 1989; Gale 1999: 112). The main sources of copper are in the south-west region, with smaller deposits in the north-west side of the island. The island is believed to have been an important source of copper in prehistory, when the island had a strategic location in the circulation of metal across the Mediterranean.

Sardinia has a long history of mineral extraction, extending back to the sixth millennium BC when obsidian was exploited (Tykot 1996). Trade in obsidian and other materials during the Neolithic established connections with the east Mediterranean, which led to the spread of copper metallurgy to Sardinia in the fourth millennium BC. The earliest use of copper in Sardinia can be linked to the native Ozieri culture dated 4250–3350 BC, where a case for an independent development of the new technology can be made (Skeates 1993).

The amount of copper used on the island during the Chalcolithic and Early Bronze Age was limited, however, this changed around 1500 BC within the emergence of a Bronze Age culture connected to Nuraghi settlements (Lo Schiavo 1988). This was a period when Sardinia had well-established connections with the Aegean and Cyprus, once thought to be connected to the activity of east Mediterranean prospectors in the central and west Mediterranean. It is now clear that copper was being traded to Sardinia from Cyprus as part of a complex trade network established in the fourteenth and thirteenth centuries BC (Lo Schiavo 2012).

A large number of oxhide ingots have been discovered on Sardinia, which lead isotope analyses confirm came from Cyprus (Stos-Gale 2000; Begemann et al. 2001). The same studies indicate the use of local copper ores in the production of plano-convex and flattened cone-type ingots, and in some of the bronze metalwork of the Nuraghi culture. Ancient copper mining is recorded in the Funtana Raminosa region, though no definite Bronze Age workings are recorded. Lead isotope analysis suggests that the Iglesiente-Sulcis region, also in south-west Sardinia, may have been another early source of copper.

Finally, copper metallurgy also appeared on the neighbouring island of Corsica in the fourth millennium BC (Skeates 1993). Copper deposits are known on the island; however, no mine workings of prehistoric age have been identified.

ITALY

While there is a possibility of earlier finds, the first significant use of copper in Italy occurred during the Late Neolithic (*c.*4500–4000 BC), when small tools and other items were in circulation in some northern regions (Skeates 1993). Many of these were made of native copper, which was either imported or obtained from unidentified sources in Italy. There has been considerable debate as to the origin of the new technology. Coastal connections with metal-producing cultures in Sardinia and Corsica, or alternatively, influences from the Swiss Alps and copper-using groups such as the Cortaillod culture, have been proposed. What is clear is that the earliest copper production in northern Italy was heavily influenced by contact with north Alpine groups, such as the Altheim and Pfyn cultures in southern Germany and Austria. A recent review suggests that metal technology from the Balkans and/or Carpathian Basin was transmitted to these groups in the east-central Alpine region during the third quarter of the fifth millennium BC (Dolfina 2013). Copper use spread rapidly though northern and central Italy into Sardinia by 4300 BC, reaching Corsica and southern Italy somewhat later.

Metalworking was fully established *c.*3600–3300 BC, when Early Copper Age groups in northern and central Italy, Sardinia, and Corsica produced larger weapons and tools of copper made from smelted ores (Dolfina 2013: fig. 12). Several distinct metalworking traditions using copper with arsenic and antimony impurity patterns emerged in that period, including the Rinaldone culture group of central Italy and the Remedello culture of the Po Valley and northern region. This continued down to the late third millennium BC when this type of copper production was gradually replaced by tin-bronze technology (Dolfina 2010: table 4).

While there may have been imports at different times from sources as far away as Cyprus and the Eastern Alps, or closer in the case of Sardinia and Corsica, much of the early copper used in Italy probably derived from indigenous sources. This is supported by evidence for primary ore smelting in many regions. Such production was based on the numerous occurrences of copper ores in different parts of Italy (Skeates 1993: fig. 1). These surface deposits are most numerous in west-central (Tuscany), north-west (Liguria, Valle d'Aosta, and Piedmont) and northern (Lombardy and Trentino/South Tyrol) regions, as well as the south-west (Calabria and northern Sicily). There is growing evidence that some of these copper ores were mined at different times from the Early Copper Age to the Iron Age.

North-west Italy

The oldest known copper mines in western Europe are recorded in the mountainous region of Liguria (Fig. 1.10). This region has provided early evidence of copper metal, including an awl, possibly of native copper, dated *c.*4000 BC found in the cave of Arene Candide in Savona province (Campana and Franceschi 1997: see also Campana et al. 1996). There is evidence of copper mining at Libiola dating to the later fourth millennium BC, while the mines at Monte Loreto were worked *c.*3500–2500 BC (Maggi and Pearce 2005; Pearce 2007).

These mines were first discovered during mineral exploration in the late nineteenth century, when they were recorded by the geologist, Arturo Issel. He identified ancient mine workings at several locations in the Ligurian Alps, including Val di Spine and Bargone in Casarza Ligure, Casali in Castiglione Chiavarese, and Monte Bardeneto in Statale (Issel 1879, 1892). The discovery of copper-smelting slag in a layer dated 3365–3045 BC at Valle Lagorara in Maissana, close to a copper mineralized outcrop, is also relevant (Maggi et al. 1995; Cortesogno and Gaggera 2002). The latter is of interest as there was contemporary extraction of jasper at this location.

Libiola

This mine is located on the eastern side of the Gromolo valley, 8 km north-east of Sestri Levante in eastern Liguria. It was one of the most important copper mines in Italy in the modern era. There are extensive deposits of low-grade sulphide ore at this location, hosted in pillow and brecciated basalts (Ferrario and Garuti 1980). This pyrite and chalcopyrite-rich mineralization mainly occurs as massive lenses (25–35 wt per cent sulphides) and stockwork-like epigenetic veins (Marescotti et al. 2010). The early miners probably extracted secondary copper minerals associated with a surface gossan.

The resumption of copper mining in the early modern era led to the discovery of ancient mine workings and implements. The latter included stone hammers, wooden wedges, a shovel, and an oak pick-handle found in an old mine tunnel (Issel 1879, 1892). Opencast mining and subsequent landfill into the modern era destroyed almost all of these early workings. A recent survey identified a number of old mine galleries on the face of the modern opencast, in the general area of the nineteenth-century discoveries (Fig. 3.6; McCullagh and Pearce 2004). Possible indications of fire-setting and primitive tools marks were identified. The antiquity of this mine is indicated by a radiocarbon date of 3490–3120 BC for the oak pick-handle found in the mine, now preserved at the Genoa-Pegli Museum (Maggi and Del Lucchese 1988).

Monte Loreto

This mine lies in the hinterland of Sestri Levante a few kilometres inland from Libiola mine, in the south of Genoa province. It is situated on the south-east slopes of the mountain (367 m ordnance datum (OD)), where a series of copper veins are exposed in a basalt geology with a Pillow Lava structure (Pearce 2007). These veins occur in faults at a contact between the basalt and serpentinite

Fig. 3.6. Libiola mine, eastern Liguria, Italy
(*Source*: Mark Pearce and Accordia Research Institute).

breccia formations. The primary copper ore is chalcopyrite; however, secondary copper minerals, such as malachite, were probably the main target of the mining.

The ancient mines were first uncovered during copper mining in the late nineteenth century. Issel (1892) recorded at that time that the copper ore had been extracted from the upper part of the mineral veins, leaving vertical trenches 20–30 m in depth that were no wider that the vein itself. Archaeological investigations commenced in 1996, with a programme of excavation and radiocarbon dating. This revealed that copper mining commenced on this mountain around 3600–3500 BC and continued to the mid third millennium BC (Maggi and Pearce 2005: table 1).

Excavations conducted at two locations on the mountain slope identified prehistoric mine workings, as well as evidence for the treatment of copper ores. The lower area comprised a series of vertical trenches where the miners emptied narrow veins of copper ore. Some of these worked-out fissures are only 0.3–0.4 m wide, one of which was extracted to a depth of 4 m. These workings seem to have been deliberately backfilled once the mining ended, with a complete infill sequence. Radiocarbon dates for charcoal indicate that these trench mines were worked between the mid fourth to early third millennia BC (Maggi and Pearce 2005; Pearce 2007).

Excavation was also carried out 150 m farther up the mountain at a mine dated 3645–3355 BC. This is a steeply inclined pit, 2 m in depth, with rock-cut notches on the walls used as hand-holds (Fig. 3.7). Excavation close to the mines identified a series of graded deposits of broken rock produced by the hand crushing and sorting of copper ore during the early–mid third millennium BC. One of these deposits overlay a 6 m long by 3 m wide by 1.5 m deep hollow, which is likely to be a shallow mine working. This feature contained spoil dated to 3630–3036 BC, with fire-reddened clasts and charcoal indicative of fire-setting (Fig. 3.8). Evidence of built structures was also uncovered in the vicinity of this mine, with a number of hearths, post-holes, and low stone walls dated to the early–mid third millennium BC.

While fire-setting may have been used to some extent, the miners relied on stone hammers to extract the mineralized veins, removing as little of the country rock as possible. These implements were also used in the concentration of copper in areas close to the workings. They consisted of sub-spheroidal cobbles of basalt, dolerite, gabbro, sandstone, and diorite gathered from streams and other sources near the mine (Fig. 3.9). Most were modified for hafting, in the form of side notching or single/double grooving.

Fig. 3.7. Copper mine working dated to fourth millennium BC, Monte Loreto, eastern Liguria

(*Source*: Mark Pearce and Roberto Maggi; copyright permission from Antiquity Publications).

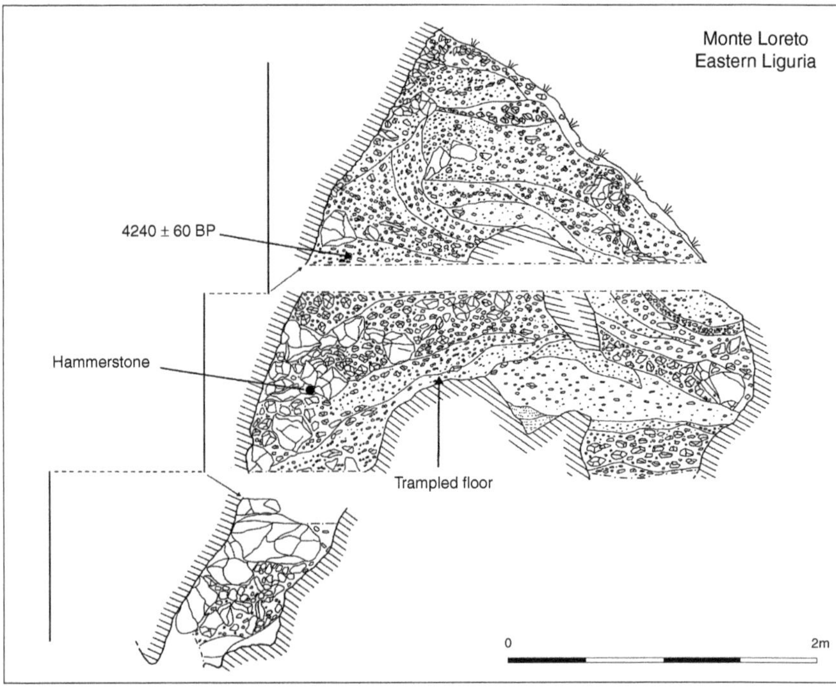

Fig. 3.8. Section across prehistoric copper mine (T5). Monte Loreto, eastern Liguria (*Source*: Maggi and Pearce 2005; copyright permission from Antiquity Publications).

North-east Italy

There is extensive evidence of early copper production in the provinces of Trentino-South Tyrol and Lombardy. There are many occurrences of copper ore in this southern Alpine region (see Skeates 1993 and Pearce 2007 for summary), several of which seem to have been exploited in the Copper and Bronze Age. Much of this centres on the valley of the river Adige and its tributaries, in the regions of Trentino and Alto Adige/south Tyrol. Veins of copper ore with polymetallic mineralization are recorded at numerous locations across this region, mostly at altitudes above 1,000 m. Many of these were mined in medieval and modern times, which has destroyed or concealed earlier workings. The most important evidence comes from Vetriolo on Monte Fronte, where opencasts, surface spoil, and copper-smelting slag of early date are recorded (Preuschen 1962, 1968b, 1973). Stone hammers and mortars were discovered at the site, as well as large grinding stones used in the processing of this ore (Preuschen 1962: plate 4). Pottery finds indicate this is a Bronze Age mine: however, its date range remains to be established.

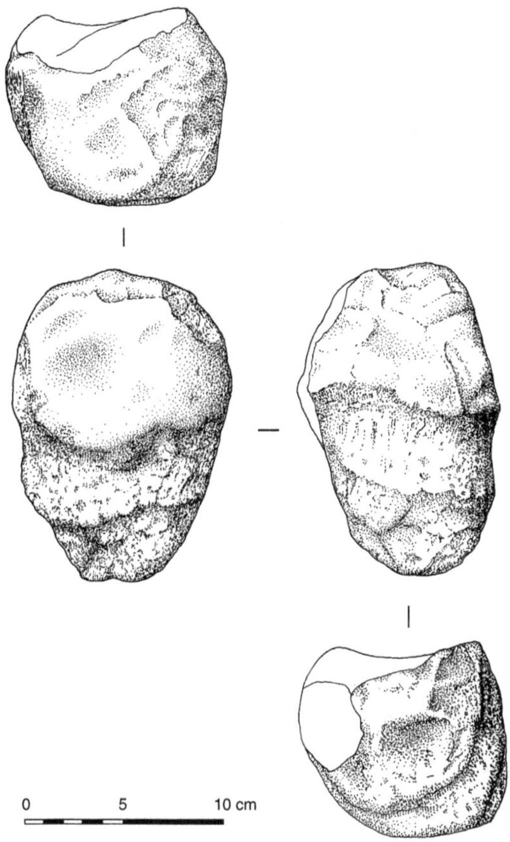

Fig. 3.9. Stone mining hammers from Monte Loreto, eastern Liguria
(*Source*: Maggi and Pearce 2005; copyright permission from Antiquity Publications).

Evidence of early mining is otherwise absent, mostly due to a lack of research. The existence of such mines must therefore be inferred from the discovery of copper-smelting sites in the vicinity of the ore deposits (see Preuschen 1973 for a review of the evidence). The earliest copper-smelting sites date to the Chalcolithic, *c*.3300–2900 BC, and mostly occur in the valley of the Adige (Cierny et al. 1998; Perini 2005). This was a time of widespread copper use among the Lagozza and Remedello cultures of north-east Italy, which includes the arsenicated copper axe of the famous Ice-man find in the Etsch valley.

This early activity was followed by an apparent lull in copper production in the Middle Bronze Age. This ended in the Late Bronze Age, with a large number of smelting sites recorded at higher altitudes in the mountain valleys on both sides of the Adige (Cierny et al. 1998: fig. 1; Pearce 2007). The scale of production from 1300–1000 BC was significant with an estimated 100 smelting sites alone recorded in the Val Sugana, east of the Adige. These include the well-known furnaces of Acqua Fredda in Bedollo, dated from the late fourteenth to twelfth centuries BC (Perini 1992; Stöllner 2009: fig. 12). An estimated 100 tonnes of smelting slag was recovered at the site, produced by a battery of stone-built furnaces that processed chalcopyrite ore. This centre is located close to the Redebus Pass, which connects the copper mining districts of Altipiano di Pinè, in the Valle dei Mòcheni (Pearce 2007).

North of the Trentino, the copper deposits of Bolzano province (Alto Adige/South Tyrol) have been mined at different times (see Baumgarten et al. 1998; Pearce 2007, 59–60). Again, the potential is indicated by finds of smelting slag of Copper Age and Bronze Age date in proximity to some of these ore deposits. This includes smelting dated to the third millennium BC at Tangasse and Milan in the Valle d'Isara, and production from the Bronze Age near the copper mine of Predoi in the Valle Aurina. In many cases the source of copper ore used for smelting is unclear, as in the case of the Fennhals site in the Adige Valley south of Bolzano (Pearce 2007). Also noteworthy is the large concentration of Late Bronze Age smelting sites on the Lavarone-Vezzena-Luserna plateau. These occur some distance from known copper sources, pointing to some separation of these activities.

Reference must also be made to the copper ores of Lombardy, in particular those in the provinces of Bergamo and Brescia to the west of the Trentino (Pearce 2007: 56–7). These include deposits in the Valcamonica in Brescia where evidence of Iron Age copper mining is recorded at Campolungo at an altitude of 1,550 m in the Val Grigna in Bienno. Investigations confirm large-scale mining of copper ore in the early Iron Age, *c.*800–400 BC (Ancel et al. 1998; Cucini Tizzoni et al. 2001; Tizzoni et al. 2003). A series of mine tunnels have been identified, along with surface deposits of rock spoil (Tizzoni et al. 2003: figs 6, 9, and 10). The copper occurs within quartz-sulphide veins that contain chalcopyrite ore, along with secondary sulphides such as digenite and covelline, as well as malachite, azurite, and other oxidation zone minerals. There is evidence for the use of stone hammers and mortars in the beneficiation of this ore, with the discovery of slag pointing to some smelting in the vicinity of the mine (Tizzoni et al. 2003: figs 2 and 8). The environmental impact of this mine will be considered in Chapter 9.

Stone hammers of similar type are recorded from a copper mine at Baita Cludona di Fundo, located 1 km south of Campolungo (Cucini Tizzoni et al. 2001; Mighall et al. 2003: fig. 1). There are further suggestions of Iron Age mining from the Sasso Moro district of the Val Malenco in Lanzada, Sondrio, where smelting sites dated to the mid first millennium BC are known (Pearce 2007: 57).

Central Italy

There is also a possibility of prehistoric copper mining in the volcanic geology of southern Etruria. A recent survey in the Monti della Tolfa (Allumiere, Latium) area led to the discovery of old mine workings and stone hammers at the site of Poggio Malinverno (Giardino and Steiniger 2011; Steiniger and Giardino 2012). There are finds of prehistoric pottery; however, no precise dating is yet available for these mines.

South-west Italy

There are numerous deposits of copper ore in the Calabria region of southern Italy (Skeates 1993: 47, fig. 1). An unusual copper mine has been investigated at Grotta della Monaca, a natural limestone cave located 600 m above sea level in the north-west part of this region (Larocca 2005, 2010). The cave is 355 m in length, and consists of a large entrance gallery, a huge central chamber, and a series of low passages at the rear of the cave. The cave was formed by karstic processes, which also led to the exposure of secondary iron and copper minerals on the interior surfaces.

The extraction of minerals in this cave began during the Upper Palaeolithic around 18,000 BC when the iron hydroxide, goethite, was extracted using bone and antler picks. The extraction of goethite resumed in the early fourth millennium BC, at a time when copper was discovered in the cave interior. These copper carbonates, principally malachite and some azurite, as well as other secondary copper minerals, were extracted at various times during the third millennium BC. The miners used grooved hammerstones and bone tools to scrape the copper minerals off the cave walls and break up blocks of mineralized rock. This continued until the Middle Bronze Age, when the interior of the cave was used for funerary purposes.

Grotta della Monaca remains the only known prehistoric copper mine in southern Italy. There is potential for new discoveries, particularly in Calabria where a recent survey recorded 48 finds of grooved hammerstones similar to those from Grotta della Monaca (Marino 2010: fig. 4). These include implements classified as hammer-axes, mallets, and picks. The majority are from the northern slopes of the Sila Massif in northern Calabria, with a second concentration in the Sierras of the south. There are metal-rich areas, with many of the hammerstones found in general proximity to deposits of copper ore (Marino 2010: fig. 8). While a connection with mining is not certain, the Grotta della Monaca context suggests that some of these implements may have been used for that purpose. The Grotta del Tesauro is another example of a karstic cave in the Upper Esaro river valley where secondary copper minerals may have been extracted in prehistory (Garavelli et al. 2012).

In conclusion, a recent review of the emergence of copper metallurgy in the central Mediterranean region stressed the importance of influences from the Balkans and/or Carpathian Basin (Dolfina 2013). These spread across the east-central Alpine region reaching northern Italy during the later fifth millennium BC, extending into the rest of the peninsula by around 4000 BC. Dolfina argues that the establishment of metalworking communities in west-central Italy, including the copper mining groups in Liguria, played a key role in the transmission of the new technology further west during the late fourth millennium BC. It explains the spread of metallurgy to Sardinia and Corsica, southern France, and possibly even Iberia. This view challenges suggestions that the new technology spread across the western Mediterranean, coming from southern Spain where it was independently developed in the fifth millennium BC (Pearce 2007). To consider this further, it is time to consider the evidence of early copper mining and metallurgy in Iberia.

4

Iberia and the western Mediterranean

The Iberian Peninsula is one the most mineralized parts of Europe, with a long history of metal mining from prehistoric and Roman to modern times. The earliest evidence for copper metallurgy dates to the fifth millennium BC; however, distinctive Chalcolithic metalworking traditions did not emerge in most regions until 3000 BC onwards. There are widespread occurrences of copper mineralization in Spain and Portugal, including many areas with deposits of lead, tin, silver, and gold. Copper deposits occur in the Galician and Cantabrian mountain ranges of northern Spain, extending east to the Pyrenees. They are also numerous in central Spain, in the provinces of Madrid, Avila, Salamanca, and Segovia in the Central Range, and also in the Toledo and Betic mountains of Cordoba. Farther south, there are major copper deposits in the so-called Pyrite Belt, extending from Seville to Huelva into southern Portugal, and also in the Penibetic range from Cartagena to Malaga crossing the sierras of Almeria (Rovira 2002: fig. 3c; see Delibes de Castro and Montero Ruiz 1999 for regional surveys of copper deposits and indications of early mining; also Gómez Ramos 1999; Hunt Ortiz 2003).

The widespread availability of ore deposits was a significant factor in the establishment of copper metallurgy in Iberia. How early is contentious, as is the means by which the new technology first developed in different parts of the peninsula. The older explanation of metal-seeking colonists from the east Mediterranean introducing this technology to southern Spain was replaced in the 1960s by a model that emphasized autonomous development (Renfrew 1967, 1973; Montero Ruiz 1994). This was based on the apparent antiquity of copper mining and metallurgy in Iberia and the distinctive technological processes that developed there relative to other parts of Europe.

The earliest indication of copper metallurgy in Iberia may come from the settlement of Cerro Virtud in Almeria, south-west Spain. A single sherd from a metallurgical crucible used to reduce oxidized copper ore was discovered in a layer dated to the early fifth millennium BC (Montero Ruiz and Ruíz Taboada 1996; Ruíz Taboada and Montero Ruiz 1999). The reliability of the find has been called into question, particularly as it predates other evidence of copper metallurgy in Iberia by a millennium or more (Roberts 2008, 2009; Roberts et al. 2009).

This is part of a wider debate surrounding the early development of copper metallurgy in Iberia. While the spectre of Aegean colonists is gone, many still believe in the spread of the new technology through interactions with early metal-using groups in the central and east Mediterranean (Roberts et al. 2009; also Dolfina 2013). It is true that copper mining and metallurgy were certainly present in north-west Italy by the fifth millennium; however, this was not the case for either southern France or the Balearic islands, which presumably would have been part of the same maritime contacts. A greater difficulty is the distinctive character of Chalcolithic metalworking in Spain and Portugal, such as the low-scale smelting of oxidized ores using ceramic oven-plates, and the limited use of copper ornaments in the absence of annealing expertise.

Murillo-Barroso and Montero (2012) argue that an independent invention of metallurgy in Iberia is the only satisfactory explanation from both the chronological and technological points of view (2012: 65). This presents a picture of a slow and essentially autonomous development of copper metallurgy in Iberia during the fourth and third millennia BC. Metal analyses indicate that was largely based on the smelting of oxidized copper ores, though the widespread use of arsenical copper may indicate a parallel development of fahlore metallurgy, as well as the use of other copper-arsenic minerals.

Despite the extensive distribution of copper ores and the widespread practice of metallurgy during the Chalcolithic and Bronze Age, there are relatively few prehistoric copper mines known in Iberia (see Domergue 1987). This cannot be explained by the absence of a mining tradition as there are many examples of hard rock extraction known from the Neolithic, such as the flint workings of Andalusia at sites such as Casa Montero near Madrid, and the famous variscite mines of Can Tintoner at Cava, Barcelona (Bosch 2005). It has been argued that the absence of early copper mines must be due to their small size and destruction caused by later mining (Rovira 2002). This is certainly likely in the case of the larger mine fields, particularly where opencast extraction was practiced in modern times. Smaller workings are likely to have survived where ore deposits were of less economic value in later times. Examples are known from many metalliferous regions; however, only a few early mines have been investigated in any detail. Some of these are surface workings, however, there is also evidence for intensive copper mining over long periods.

In some regions copper mining may be inferred from the evidence of on-site metallurgy at excavated settlements of the Chalcolithic and Bronze Age. This includes the well-known Millaran and Argaric cultures of south-east Spain, where the practice of crucible smelting can be linked to small-scale mining of local copper deposits.

SOUTH-WEST SPAIN

The largest concentration of prehistoric copper mines in Spain occurs in the south-west region, mostly in Huelva province and in the adjacent provinces of Badajoz, Cordoba, and Sevilla (Fig. 4.1). Most of these are within the Iberian

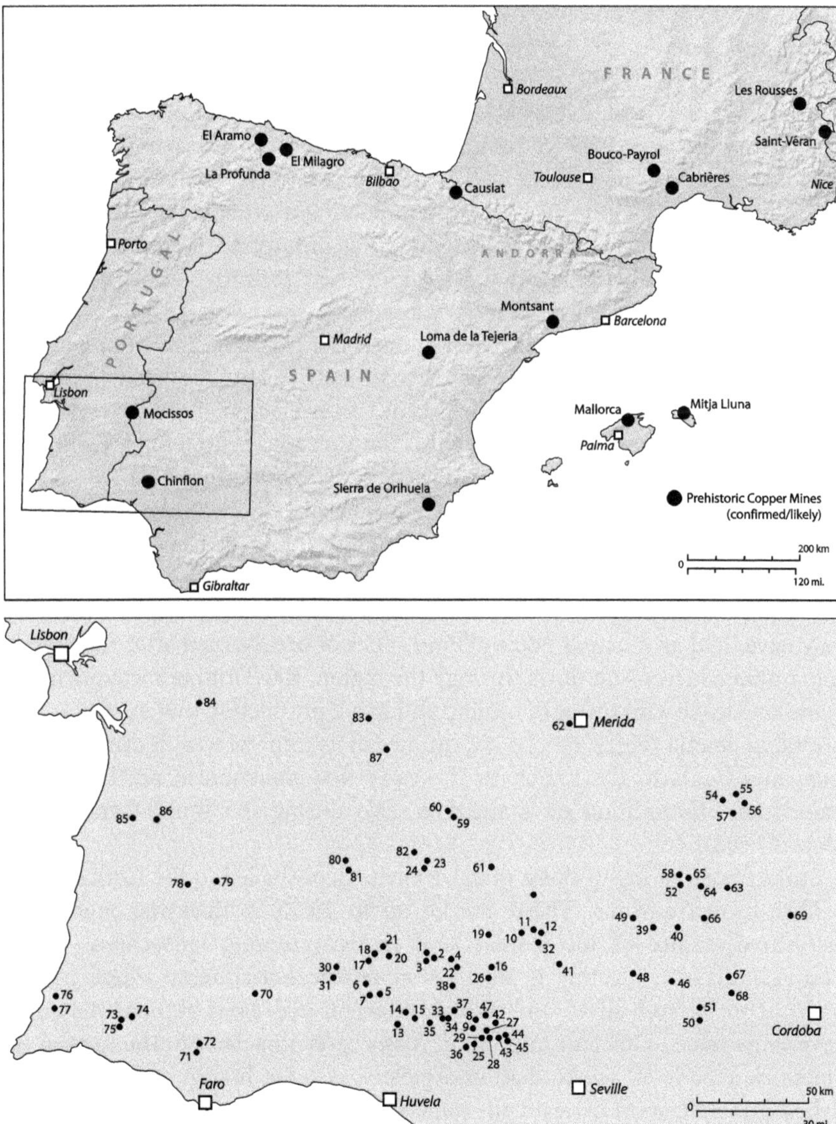

Fig. 4.1. Map of Iberia showing early copper mines

(*Source*: author; see Hunt Ortiz 2003 for names of numbered mines).

Pyrite Belt, one of the world's most important mining fields. An estimated 75 major orebodies have been discovered in an area that extends 250 km west from Sevilla through Huelva province into southern Portugal (Hunt Ortiz 2003). There is a long history of mining in that region, with copper, lead, silver, and gold extracted in great quantities in both the Roman period and modern era. These industrial scale operations also uncovered traces of older workings for copper and other metals. The earliest of these were connected to Chalcolithic and Bronze Age metallurgy in the region (reviewed by Hunt Ortiz and Hurtado Pérez 1999).

The first systematic investigation of early copper production in this region was undertaken in 1973–6 by the Huelva Archaeo-Metallurgical Project (Rothenberg and Blanco-Freijeiro 1981). Approximately 40 sites were recorded in the general vicinity of Rio Tinto, including a number of ancient mine workings and surface finds of smelting slag. Many of these mines were interpreted as Chalcolithic on the basis of their primitive appearance and the use of grooved hammerstones. Subsequent surveys have increased the number of possible prehistoric copper mines and smelting sites in this region to around 86 (Hunt Ortiz 2003), with more likely to be discovered.

The geology of the Pyrite Belt consists of Upper Devonian rocks overlain by a Lower Carboniferous geology known as the Volcanic-Sedimentary Complex. The latter contains massive, strata-bound, ore deposits that were formed by submarine volcanic activity in Lower Carboniferous times. There are different styles of mineralization, with the most significant in modern terms being massive pyritic orebodies of volcanogenetic and exhalative-sedimentary origin. The most important of these is at Rio Tinto where the original orebody may have held as much as 500 million tonnes of ore. Named after the metal-contaminated river that flows through the region, Rio Tinto is a unique man-made landscape with traces of mining and metal production that extend across several millennia (Fig. 1.8). The importance of its mineral wealth drew Phoenician merchants to the region in the early first millennium BC, laying the foundations for mining on a massive scale during the Roman era (Pérez Macías 1998).

Surface weathering of these massive pyrite deposits led to the formation of reddish-brown gossans. These can be up to 40 m in thickness and are a distinctive feature of the Roman and modern mining landscapes. These iron-rich deposits overlay a zone of supergene enrichment where metals such as copper, lead, silver, and gold are concentrated. Most of these orebodies have some trace of ancient mining. In many cases this dates to the Roman or Phoenician periods, when shaft-and-gallery systems honeycombed the enriched mineralization beneath the gossans (Rothenberg and Blanco-Freijeiro 1981: 168).

Evidence of prehistoric mining is scarce due to the intensive opencast mining of these pyritic deposits in the modern era (Fig 1.7). A good example is Aznalcollar mine in Sevilla province where opencast mining in the nineteenth century revealed prehistoric workings in the gossanized part of lode (Hunt Ortiz 2003: 104–7, fig. 76). Mining may have commenced there during the Chalcolithic with the exploitation of copper carbonates in the exposed gossan, however, this must be confirmed. Mining resumed in the Late Bronze Age/ Colonization period with extraction of silver-rich ore from beneath the gossan.

As for Rio Tinto, no prehistoric copper mines can be identified in this landscape, nor any evidence of copper metallurgy before the Roman era (Hunt Ortiz 2003: 125). The existence of early copper mines may be inferred from a number of stone hammer finds in the area, as well as antiquarian records of early mine workings. These include a visit by the mining archaeologist, Oliver Davies, in the 1920s, who noted large cave-like openings with radiating tunnels that may have been used in the pre-Roman era to extract enriched copper ore from beneath the surface gossans (Davies 1935). There is also a similar report from the Alto de la Mesa in Rio Tinto, where 'a ton of stone hammers' was supposedly found in a large cave beneath the gossan (Kennedy 1894 in Hunt Ortiz 2003). While those discoveries may relate to the extraction of silver-rich jarosite ore in the Early Iron Age, the possibility of earlier mining for copper remains. Unfortunately, the early workings at Rio Tinto were destroyed by modern opencast extraction. Indirect evidence is provided by a geochemical analysis of the Tinto estuary, which identified high levels of heavy metal contamination in sediments dated to the mid-third millennium BC (Le Blanc 2005). This will be discussed further in Chapter 9.

With much of the gossan-covered mineralization beyond the reach of Chalcolithic and Bronze Age miners, their efforts seem to have concentrated on the exploitation of small orebodies in the surrounding areas. Of particular importance was the extraction of oxidized copper ores from mineralized veins exposed in outcrops of tuffs and quartzitic rocks. The early miners followed these veins along their strike exposure either by continuous trenches or a line of close or widely spaced pits. These narrow workings were generally vertical and could connect with each other, but there were no true underground galleries. In most cases, this mining did not exceed 8 m in depth, with the mine trenches generally 0.4–1 m in width. Metal tools were not employed in these mines, apart from the probable use of copper/ bronze axes in the processing of wood. Rock was extracted by pounding the mine face with hafted or hand-held stone hammers. A range of pick-like tools made of antler, animal bone, or wood may also have been used, though these have rarely been recorded.

The treatment of copper ore on the surface involved simple crushing and hand-sorting using stone hammers and anvil stones, as well as grinders and stone mortars with hollows. The copper minerals mostly consisted of malachite and azurite, which could be easily hand-sorted by colour. This ore was generally smelted in the vicinity of the mine, using simple bowl furnaces at relatively low temperatures to produce prills of copper metal. The high iron content of this ore led to the production of primitive slags, as well as chemically impure metal (Hunt Ortiz 2003: 376–83).

Rothenberg and Blanco-Freijeiro (1981) presented this early copper mining and smelting in Huelva province as a 'Chalcolithic' horizon. While this was initially taken to imply a primitive stage of technology, the authors made the mistake of attributing a chronological primacy and cultural association to these sites. Influenced by earlier research at Timna (Rothenberg 1972), it was thought that the use of grooved stone hammers in copper mining in this region was also confined to the Chalcolithic. It is now known that these mining implements had a long chronology from the Early Chalcolithic to the Late Bronze Age. Similar hammers are recorded in metal production sites at Monte Romero, Castrejones, and Castillo de Dona Blanca dating to the eight century BC (Hunt Ortiz 2003: 283). These implements may have been used for mining in the Early Iron Age (Tartessian-orientalizing culture), however, there is no indication this continued into the Roman period, either there or anywhere else in Europe.

The dating of the earliest copper mines and smelting sites in Huelva and surrounding regions is uncertain. This is due to a lack of archaeological excavation and scientific dating, and the multi-period history of many mines. No copper mines of Chalcolithic date have been excavated; however, surface finds point to several possible examples. These include the mine at Potosi in Sevilla where pottery of Late Chalcolithic date was found with grooved hammerstones (Hunt Ortiz 2003: 58). Sherds of Beaker pottery are recorded from a small settlement adjacent to the mine at La Loba in Fuente Obejuna, Cordoba (Blazquez 1988). Finally, there is indirect evidence in the form of copper minerals discovered in a number of excavated Chalcolithic settlements, for example, at La Pijotilla in Badajoz and Amarguillo in Sevilla (Hunt Ortiz 2003: 277–8).

One of the early copper mines identified by the Huelva Survey is Cuchillares, located 6 km north of Rio Tinto (Fig 4.2; Rothenberg and Blanco-Freijeiro 1981: site 54). The surface mining there took the form of roughly hammered out cavities, narrow fissures, and irregular trenches. The mineralization consists of thin veins of secondary copper minerals (malachite and azurite). A large number of stone hammers were recovered on surface spoil below these ancient workings. These are mostly long narrow tools with no haft modification. They were regarded as more primitive than the grooved hammerstones from other mines in the region (Rothenberg and Blanco-Freijeiro 1981: 165, fig. 84). Unmodified hammers are recorded from a number of other mines, for example, at El Berrocal in Badajoz province where there may be surface finds of Chalcolithic pottery (Merideth 1998: 212–23). None of this indicates a date range for the Cuchillares mine as unmodified stone hammers may have been used for mining over a long period.

Fig. 4.2. Prehistoric trench workings, Site 54, Cuchillares, Huelva, Spain (*Source*: author).

Chinflon

This mine is located east of the village of El Pozuelo, 12 km south-west of Rio Tinto. It is on a prominent hill where east–west trending veins of quartz are exposed. These contain primary chalcopyrite ore and secondary copper minerals. The site was discovered in 1974 by the Huelva Project and excavated over three seasons in 1978–80 (Rothenberg and Blanco-Freijeiro 1980; Andrews 1994). It was identified as a type-site of Chalcolithic extractive metallurgy in south-west Iberia, involving the mining of copper ores from outcropping veins using simple trenches and stone hammer technology. This oxidized ore was then smelted in simple bowl furnaces at relatively low temperatures to produce prills of copper metal and primitive slags.

Rothenberg and Blanco-Freijeiro regarded the proximity of Chinflon to a group of early megalithic tombs at nearby El Pozuelo as culturally significant (Rothenberg and Blanco-Freijeiro 1981: fig. 19). They argued for a connection with Chalcolithic copper production, however, those tombs are not closely dated, nor do they have definite associations with local metalworking. A further difficulty is the dating evidence from the mine, which now seems to have been first worked in the Late Bronze Age.

The early mines at Chinflon consist of short irregular trenches following the exposed quartz veins (Fig. 4.3). These are recorded at three separate locations in the site, where there are also modern workings. One of the primitive trenches (Mine 3) has been investigated in detail (Fig. 4.4). This working has an overall length of 14.5 m and averages 0.6–0.9 m width. It is divided into three sections, one of which (Mine 3b) was completely excavated to a depth of 12.5 m. This particular working is divided into three separate cavities, which are interconnected at depth but separated by rock bridges at the surface. The latter were used as working platforms for the hoisting of rock out the mine. Some marks on the interior walls may be hand-holds or connected to the use of wooden stemples (Hunt Ortiz 2003: 68–76).

Excavation revealed that waste rock from this mine was stored underground, while the upper part of the infill was deposited from adjacent workings after it was abandoned (Fig. 4.5). The absence of charcoal indicates that fire-setting was not used. Instead, the mine was worked using stone hammers, marks from which are visible on the interior walls. A large number of grooved hammerstones were found in the infill sequence, with evidence also for the use of hollowed mortar slabs and grinding stones. These implements are also exposed on the surface spoil heaps adjacent to these trench mines.

The excavated mine is significant because waterlogged conditions in the lower interior resulted in the preservation of several items of wooden equipment. These include part of a notched ladder, an object that may be a platter or shovel blade, and three wooden branches with pointed ends, one of which may have been part of a notched tree-trunk ladder. A pottery bowl and a clay lamp

Fig. 4.3. Trench mine 1 at Chinflon Huelva, Spain

(*Source*: author).

Fig. 4.4. Trench mine 2 at Chinflon

(*Source*: author).

were found, as was a human femur possibly from a disturbed burial on the surface (Andrews 1994: 17).

The dating of Chinflon mine has been contentious. In their initial interpretation Rothenberg and Blanco-Freijeiro (1980, 1981) identified three phases of copper mining at the site, namely Chalcolithic trench mines, Late Bronze Age shafts of circular form with metal tool marks, and deeper shafts from the modern era. The subsequent excavation results and radiocarbon dates from Mine 3b indicate two phases of mining (Late Bronze Age and modern), with no evidence for Chalcolithic operations. Radiocarbon dating of wood from the lower fill indicates an early date of 1138–892 BC for this mine, while a date of 770–330 BC from the upper fill may relate to mining in adjacent workings at a slightly later date (in Hunt Ortiz 2003: 76).

The Huelva Project also excavated a settlement area located 20 m north of Mine 3. The site has been heavily eroded; however, some post-holes and possible hearths indicate a small settlement connected to the mine. Two phases were initially identified, beginning with a metallurgical area and storage pits of Chalcolithic date, which was re-occupied by Late Bronze Age miners who erected temporary shelters (Rothenberg and Blanco-Freijeiro 1980: fig. 7). Subsequent interpretations suggest that the main occupation dates to the Late Bronze Age, *c.*900–800 BC, with some activity in the early second millennium BC (Pellicer and Hurtado 1980; Hunt Ortiz 2003).

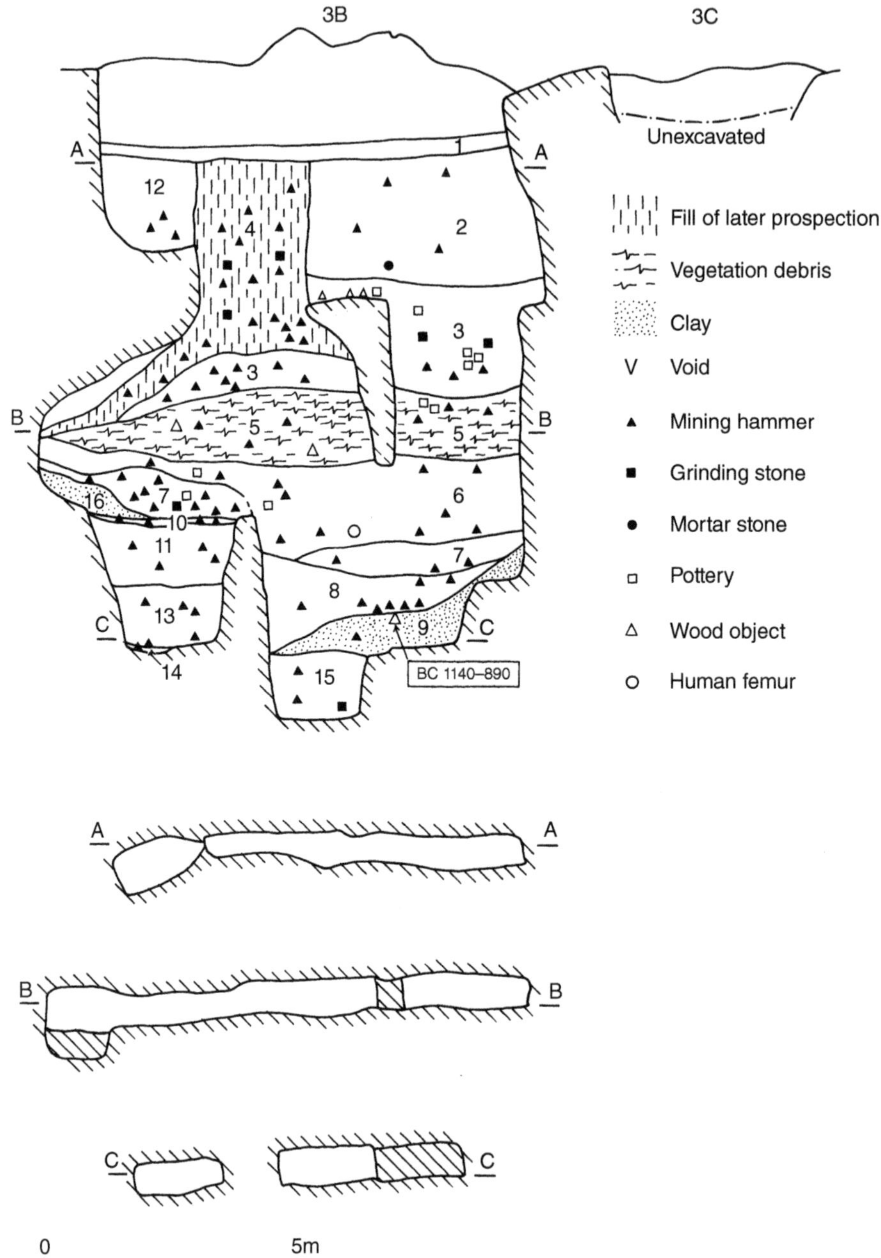

Fig. 4.5. Stratigraphic section, mine 3 at Chinflon

(*Source*: Andrews 1994).

In conclusion, available dating evidence indicates that Chinflon mine was first worked during the Late Bronze Age, *c.*1200–700 BC. The possibility of earlier workings cannot be excluded, but there is no evidence to support copper mining or metallurgy at Chinflon during the Chalcolithic, nor any connection with the El Pozuelo group of tombs. Other researchers continue to propose a link between these dolmen builders and the earliest copper mining and metallurgy in the region during the third millennium BC (e.g. Piñon Varela 1986; Nocete et al. 1993; but see Rodríguez Casal 1991).

While Chinflon is securely dated, the method of mining at that site is unlikely to have been exclusive to the Late Bronze Age. Many similar workings are known in the region, however, most lack independent dating. They include Masegoso (Rothenberg and Blanco-Freijeiro 1981: site 27) where grooved stone hammers are recorded. There is a similar mine at Las Navas in Berrocal, where pottery of Late Bronze Age date has been recorded (Hunt Ortiz 2003: 77). Trench workings of the Chinflon type are also recorded at Cueva del Monje mine in Paterna del Campo, where there is also evidence of Iron Age/ Phoenician mining (Rothenberg and Blanco-Freijeiro 1981: site 35). Hunt Ortiz lists a group of 15 primitive copper mines north of El Chorrito in the municipality of Paterna del Campo (Hunt Ortiz 2003: 79–96; Rothenberg and Blanco-Freijeiro 1981: site 52). These are mostly surface trenches associated with use of grooved hammerstones, with Late Bronze Age pottery from one mine (Site 16).

Other similar mines include El Piconcillo mine in Fuente Ovejuna, Cordoba, where there are trench workings 1–1.4 m wide and up to 11 m deep, again associated with the use of grooved stone hammers and stone mortars (Hunt Ortiz 2003: 13). At Redondilla mine in Alanis, Sevilla, there is a 20 m long by 6 m deep trench formed by two workings separated by a 2 m wide rock arch, again associated with grooved stone hammers (Hunt Ortiz 2003: 60). The aforementioned La Loba mine also has evidence of Bronze Age trench mining involving the use of stone hammers. Excavation of a small settlement next to the mine revealed huts dating to the Middle Bronze Age, as well as copper smelting slags and metal objects (Blazquez 1988).

Reference can also be made to the mining/metallurgical settlement discovered at Cabezo Juré, also located in the Iberian Pyrite Belt of south-west Spain (Nocete 2006). Investigations there uncovered a large settlement with numerous houses built on hill terraces, close to metallurgical areas and facilities for water storage. There is evidence for all stages of copper production, from the processing of ores to the smelting and casting of copper. Lead isotope analysis confirms that some of this copper ore came from the Tharsis deposit, less than a kilometre away, where primitive mines have been identified (Nocete et al. 2005: fig. 2.1). This was based on secondary copper ores (carbonates, oxides, and sulphides) from the supergene-enriched portion of that orebody.

Cabezo Juré was part of a specialized copper production in the Iberian Pyrite Belt from early in the third millennium BC. This can also be linked to a settlement at Valencina de la Concepción near Seville, where large-scale smelting of copper ore was undertaken. Nocete et al. (2008) suggest that intensification of copper production was connected with a major settlement expansion in the Guadaquivir valley, where major centres such as Valencina de la Concepción were supplied with ore and smelted metal from mines in the Pyrite Belt, a situation that continued to around 2000 BC (2008).

In conclusion, south-west Spain was undoubtedly an important source of copper (and other metals) during the Bronze Age and probably also during the Chalcolithic. Rothenberg and Blanco-Freijeiro (1981) proposed that the earliest copper metallurgy in this region was based on surface extraction of very pure sources of oxidized copper minerals. Once these were exhausted, there followed a hiatus in mining during the Late Chalcolithic/ Early Bronze Age before an ability to extract supergene mineralization and produce arsenical copper was developed. This model is no longer accepted, partly because arsenical copper was present from the earliest stages of metallurgy in southern Iberia. Furthermore, analyses confirm that much of the ore from these primitive trench mines had significant arsenic content.

It is obvious that small vein-style mineralization was more accessible for prehistoric miners than the massive pyritic orebodies with their hard gossans. Yet, some of the latter were occasionally exploited where copper mineralization was exposed at the surface. This possibly occurred at the Aznalcollar mine and in some of the Rio Tinto deposits where the early mines are now destroyed. The smaller vein workings on the margins of these massive orebodies are more likely to have survived—mines such as Cuchillares in the Rio Tinto area or Cabezo Hueca near Tharsis (Hunt Ortiz 2003: 373).

There is a significant change in metal mining in the south-west region with the Phoenician colonization of the early first millennium BC. The introduction of iron led to improvements in mining methods, with the first significant exploitation of supergene mineralization from the major gossan orebodies. This included the mining of secondary copper sulphides such as chalcopyrite and covellite, as well as the extraction of silver from argento-jarosite ores using cupellation technology. The ancient mining zone of the Aznalcollar opencast, the 'protohistoric gallery' of the Corta Lago in Rio Tinto and the Monte Romero metallurgical site are typical of this period (Hunt Ortiz 2003: 392). Metal mining intensified with the Roman conquest from the second century BC and continued at an industrial scale over the following five centuries, extracting many millions of tonnes of copper and lead-silver ore.

PORTUGAL

The mineral wealth of Huelva extends into southern Portugal where there are numerous copper deposits. While few early mines have been investigated in that region, there are indications of an early copper production based on surface oxidized mineralization, similar to that recorded in Huelva. Finds of grooved stone hammers and copper/ bronze implements are recorded from a number of copper mines. Eight such sites have been identified in the Faro district, with five in Beja, and two each in Evora and Setubal (Hunt Ortiz 2003: 137–43). There are reports of copper axes from the mines of Vendinha do Esteval and Santo Estevâo in Faro district. Grooved stone hammers were discovered at the Los Algares lode in the mine of Aljustrel, Beja, with a copper arrowhead recorded from that mine, as well as reports of Beaker pottery (Domergue 1987: 30).

A limited number of copper deposits are known in central and northern Portugal; however, there is little direct evidence of prehistoric mining. Copper metallurgy was first established in that region in the early third millennium BC, as recorded in the pre-Beaker and Beaker phases of the Vila Nova de São Pedro (VNSP) culture. Copper metallurgy is recorded at several fortified settlements of the VNSP culture in Portuguese Estremadura, including Zambujal, Leceia, and Vila Nova de São Pedro itself. The metal concerned is an arsenical copper with low trace element concentrations, of a type that is very typical for the Chalcolithic in the Iberian Peninsula. It is not certain whether native copper or mineral ores containing arsenic were used to produce this metal. Chemical analysis indicate the most likely source is a rich oxidized copper ore containing secondary arsenopyrite, rather than a fahlore, This ore was smelted under poorly reducing conditions, as indicated by slag found at Zambujal and Vila Nova de São Pedro (Müller et al. 2007).

Some VNSP settlements with evidence of metalworking relied on imported supplies of copper. The settlement at Leceia in the Tagus estuary is a good example, where copper working was first adopted in the Beaker culture phase *c.*2600–2200 BC. Lead isotope analysis indicates that the metal used is unlikely to have come from the few local ore sources, nor does it seem to have originated in the Iberian Pyrite Belt to the south or the metal deposits of south-central or south-east Spain (Müller and Cardoso 2008). The most likely source is the Ossa Morena Zone, at least 100 km east of the major VNSP settlements, where an estimated 650 copper deposits are recorded. The lead isotope data supports an exchange of copper between settlements such as La Pijotilla in the Ossa Morena Zone and the major VNSP settlements in Estremadura during the third millennium BC (Müller and Cardoso 2008: fig. 1). This is supported by research on the arsenical copper used in the VNSP settlements of Zambujal and Vila Nova de São Pedro, which indicates sources in east-central Portugal/ Spanish Estremadura (Montero Ruiz 2010; Müller and Soares 2008).

This model of long-distance supply is supported by new evidence for copper mining for that period from the Ossa Morena Zone. Recent investigations at the mine of Mocissos in the district of Alandroal (Évora), has uncovered evidence of prehistoric copper mining (Fig. 4.1; Goldenberg and Hanning in press). There is evidence for the extraction of mineralized rock along an 800 m long vein, with numerous trenches, shafts, and opencasts of different periods. Stone hammers and other stone implements are exposed in surface deposits of rock spoil. The mineralogy of this spoil suggests there was a substantial gossan on these ore veins, from which the early miners extracted secondary copper minerals (malachite, cuprite, and native copper) produced by the oxidation of primary chalcopyrite.

In 2006 sample excavation and radiocarbon dating of this spoil confirmed that copper mining was undertaken at this location from the later fourth to the first millennium BC. Mining began during the Chalcolithic and continued into the Early Bronze Age, when surface workings associated with stone hammers are in evidence. A smelting crucible from this early phase of mining is significant as it confirms the presence of copper production within the mine area. The mining intensified from the Middle Bronze Age onwards, before ending in the Early Iron Age, periods for which there is pottery occurring in spoil contexts associated with hammerstones (Goldenberg and Hanning in press).

Finally, mention can be made of a discovery in the late nineteenth century at the Quarta Feira copper mine in central Portugal (De Melo et al. 2002). A bronze palstave of Late Bronze Age date was found at a depth of 12 m inside this mine. There are uncorroborated reports of ancient workings, as well as finds of stone hammers, slag, and possibly furnace structures.

SOUTH-EAST AND CENTRAL-EAST SPAIN

It is surprising that no early copper mines have been identified in the provinces of Almeria, Granada, and Murcia. This is a highly mineralized region with an early history of copper metallurgy going back to the fourth millennium BC (Montero Ruiz 1994, 1999). Several important metalworking traditions emerged in this area during the Chalcolithic and Early Bronze Age, associated with the Millaran, Beaker, and Argaric culture groups. This began with the widespread use of arsenical copper and was followed in the early second millennium BC by the production of tin bronze and silver (Montero Ruiz 1993).

Evidence of metalworking is recorded from several Millaran and Argaric culture settlements (e.g. Müller et al. 2004), however no contemporary copper mines are known in those areas. This may be due to the difficulty of recognizing small surface workings, but could also relate to the limited scale of copper production in this region during the Chalcolithic (Montero Ruiz 1994). Grooved stone hammers are reported from Cerro Minado in Huercal Overa, Almeria (in Domergue 1987), however the history of that mine is not known. The identification of a possible Chalcolithic mine (ALS-2 site) in Sierra Alhamilla (Rothenberg et al. 1988) remains to be confirmed. Metal analysis suggests that the mine of Cabezo de Herrerías in Almeria was a likely source of copper and silver metal in this region during the Bronze Age (Montero Ruiz 1994: fig. 16). Stone hammers are recorded from the site, but no definite prehistoric workings (Montero Ruiz 2010: 78, fig. 16). There is also a record of a copper chisel of possible Chalcolithic age found in this mine (Bosch and Luxan 1935).

Recent research has identified a possible early copper mine at Cerro de la Mina in the Lower Segura valley, on the border of Alicante and Murcia. A spoil heap with stone hammers and sherds of Bronze Age and possibly Chalcolithic pottery, have been identified at the site of an early modern copper mine. While definitive evidence remains to be found, it is argued that oxidized copper ores were mined at several sites in the Sierra de Orihuela mountain range during the third and second millennia BC (Brandherm and Maass 2010). Stone hammers are recorded from some other copper mines in the region; for example, at Sotarrenya mine in Alicante (in Domergue 1987).

The location of a possible early copper mine has been identified in eastern Spain, at Loma de la Tejería in Albarracín, Teruel (Fig. 4.1; Almagro Gorgea and Collado Villalba 1981; Montero Ruiz and Rodríguez de la Esperanza 2008). Excavation revealed a small settlement dated by Beaker pottery and other ceramics to the Chalcolithic. A vein of copper ore has been identified in the vicinity of this encampment, though no mine workings of prehistoric date have been identified. The analysis of residues on a stone hammer and stone mortar confirms the processing of copper minerals, while the discovery of a copper prill indicates on-site metallurgy (Montero Ruiz and Rodríguez de la Esperanza 2008). The evidence points to surface extraction of oxidized copper ores over a period of time during the Late Chalcolithic and Early Bronze Age. This seems to have been small-scale activity, probably conducted on a seasonal basis and linked in with animal pastoralism.

Stone hammers are also reported from Torrijos mine in Toledo (in Domergue 1987), however, no archaeological investigations have been undertaken.

The Balearic Islands

Recent research has also raised the possibility of early copper mining on some of the Balearic Islands. The adoption of copper metallurgy coincided with the Beaker culture presence on these islands (Waldren 1998). Copper deposits are recorded on several islands, however, no prehistoric mines have been identified. The potential is indicated by the discovery of some 25 copper smelting sites on the northern side of Mallorca, several of which are now dated to the early second millennium BC. These smelting locations are located high in the mountains away from contemporary settlements, in an area that has small copper deposits (Ramis et al. 2005).

There are also indications of prehistoric copper mining on the island of Minorca (Fig. 4.1; Salvà et al. 2010). This discovery was made at the mine at Mitja Lluna located on the islet of Colom on the north-east side of the island. Stone hammers have been recovered from surface spoil along cliff-top exposures of copper veins (Llull et al. 2012: figs 3–6). Samples from recent excavation produced are radiocarbon dated to the Early Bronze Age (Perelló Mateo et al. 2013).

NORTHERN SPAIN

The oldest known copper mines in Iberia are in the Cantabrian mountains of north-western Spain (Blas Cortina 1999). Copper was extracted in the mines of El Aramo and El Milagro in Asturias, and at La Profunda in León, during the Chalcolithic and Bronze Age, commencing as early as 2600 BC and continuing into the later second millennium BC.

El Aramo

This mine is located in Riosa, about 40 km south of Oviedo in Asturias (Fig. 4.1). First worked during the Chalcolithic and Early Bronze Age, the orebody was also mined by the Romans in the first and second centuries AD. The mine was rediscovered in the nineteenth century and re-opened to extract copper, finally closing in 1956. The scale of production in what is known as the Texeo copper mine was considerable, amounting to some 370 tonnes of copper metal annually between 1947 and 1956 (Loredo et al. 2008: fig. 2).

The oldest part of the mine lies at an elevation of 1,116–1,223 m OD (ordnance datum) on a steep upper slope of the Sierra El Aramo mountains (Fig. 4.6). The ancient workings were first recorded when the mine re-opened

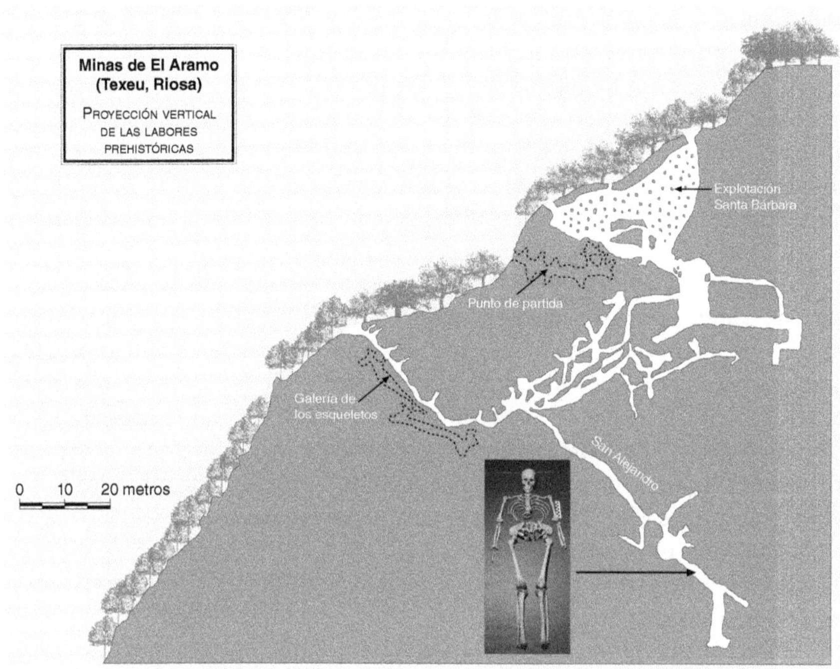

Fig. 4.6. Section across El Aramo copper mine, Asturias, showing prehistoric workings and one of the human skeletons

(*Source*: Miguel De Blas Cortina).

in 1888, with finds of stone hammers, bone tools, and human remains (Dory 1893). The systematic investigation of these remains began in 1987 with a research programme led by Miguel De Blas Cortina (Blas Cortina 1998, 1999, 2005, 2013). While modern mining did disturb the early workings, many of these survived intact along with deposits containing tools and human skeletons.

The host rocks of the mineralization are dolomitized limestones of Namurian age (Loredo et al. 2008). The ore deposit consists of mineralized veins and argillaceous infilled zones within the karstic cavities. The mineral assemblages are present as three successive primary stages, with a well-developed supergene stage. The early association is formed by pyrite and bravoite, with cobalt-nickel arsenides and sulpharsenides and later marcasite. During the intermediate primary stage, major tennantite and sphalerite were deposited in dolomite and/or quartz matrix. The final association is formed by copper-iron sulphides, such as chalcopyrite, talnakite, or bornite (Paniagua et al. 1987; Loredo et al. 2008).

Supergene alteration of this primary ore deposit produced numerous secondary minerals, such as native copper, cuprite, tenorite, azurite, malachite, and erythrite. The prehistoric mining occurred within this oxidized zone, with malachite and azurite forming the bulk of the extracted ore. The mining of this copper ore began on the upper part of the mountain slope where a number of mineralized outcrops were exposed. These first miners also explored a number of natural limestone caves where the green and blue colours of the oxidized copper minerals were visible. Most of this ore was extracted from veins within the dolomite layer. The thickness of these veins ranged from a few centimetres up to 1.5 m, averaging *c*.0.25 m.

An underground survey of the mine reveals a labyrinth of modern and ancient workings. There are also natural tunnels of karstic origin, some of which provided access to the copper ores and also natural ventilation. The prehistoric mine comprised an extensive system of underground tunnels and galleries, which followed inclined mineralized veins that were exposed on the mountain slope. The mine workings have a twisted arboriform structure, with an overall length of 900 m. The deeper parts of the prehistoric mine are 150 m below the surface (Fig. 4.6; Blas Cortina and Suárez Fernández 2010).

The mining progressed upwards, facilitating the disposal of waste rock inside abandoned workings. Some of the old mine tunnels are empty, while others are filled with broken rock debris containing mining tools, animal bone, and charcoal. These underground deposits were extracted in three areas, namely in the lower and upper sections of the Saint Alejandro sector and in the so-called 'Starting Point' gallery. The latter was one of the last parts of the mine to be worked. It is located in the upper part of the mine where an elaborate network of tunnels on a 25–35° inclination removed a one metre thick dolomite layer containing copper ore (Fig. 4.7). This left an intricate network of small vaulted chambers and narrow tunnels, with as many as 18 small rock pillars left at regular intervals (Fig. 4.8). These are not designed for roof support, but were used instead to fix ropes for access and haulage in these steeply inclined workings. There are also small hand-holds and crude steps cut into the wallrock.

The mining techniques varied depending on the nature of the country rock. The softer dolomite was extracted using stone hammers and pick-like tools of wood, red deer antler, goat horn, or cattle bone. Fire-setting was employed with stone hammers where the copper ore occurred within the harder limestone rock (Blas Cortina and Suárez Fernández 2010). Rock spoil in the galleries of the Upper Saint Alejandro section contained charcoal produced in fire-setting. Some form of artificial lighting was also necessary in the interior of the mine. Deposits of charcoal found close to pillars and arches in the 'Starting Point' gallery are probably the remains of lighting torches.

The stone hammers used in this mine come in a range of sizes and shapes, with the majority grooved for hafting. A number of small stone tools of cylindrical shape in the mine may be connected with concentration of ore underground. In this way, the miners tried to limit the amount of rock extract brought to the surface.

Fig. 4.7. Entrance to prehistoric mines at El Aramo, showing rock arch (bottom right) feature

(*Source*: author).

Fig. 4.8. Underground workings at El Aramo

(*Source*: author).

Radiocarbon analysis of 60 samples of charcoal, human bone, and animal remains dates the commencement of mining at El Aramo to around 2500–2400 BC (Fig. 4.9). There was continuous mining during the following centuries in areas such as the San Alejandro gallery. This continued into the Bronze Age, with activity from 1900–1500 BC in the 'Starting Point' gallery. It is not known why the mine was eventually abandoned, but this was probably due to external factors rather than the exhaustion of available copper ore. There may have been some copper mining at El Aramo as late as 1200 BC, however, this was very limited in scale.

The discovery of prehistoric human remains at several locations underground is a particularly important feature of the El Aramo mine. These were first uncovered during the nineteenth-century exploration of the ancient workings, with some recovered during recent archaeological investigation (Fig. 4.6). Radiocarbon dating of the human remains reveals the use of abandoned parts of the mine for mortuary purposes throughout its entire period of operation, *c.*2500–1500 BC. These are believed to be deliberate interments and not the result of underground accidents caused by roof collapse. It is suggested that these bodies were placed in mined-out areas as offerings to gods of the subterranean world in thanks for the extraction of mineral wealth (Blas Cortina 2010).

El Aramo is one of the largest prehistoric copper mines in western Europe, with an extensive layout of underground workings that developed over a millennium or so of continuous copper extraction. Many tonnes of copper ore were extracted and the overall scale of production points to the supply of copper from this mine over a wide area of northern Spain. Given the remote location of the mine it might be expected the copper ore was smelted in the immediate vicinity. Recent excavations conducted close to the mine workings on this mountain slope uncovered evidence of metallurgical processes in a work camp location (Blas Cortina 2013).

El Milagro

This mine is situated on the north-western slopes of the Sobrepeña mountain range in the eastern Asturian municipality of Onis (Fig. 4.1). This is a region along the central and western massifs of the Picos de Europa mountain range that has numerous copper deposits. The ancient workings at El Milagro were discovered in 1850 when the mine was re-opened. It was worked intermittently until 1959, raising a significant amount of copper ore despite problems of underground flooding.

The primary mineralization comprises copper-iron sulphides such as chalcopyrite and bornite, as well as fahlore, with secondary ores in the oxidation zone including malachite, azurite, chrysocolla, and chalcocite. The

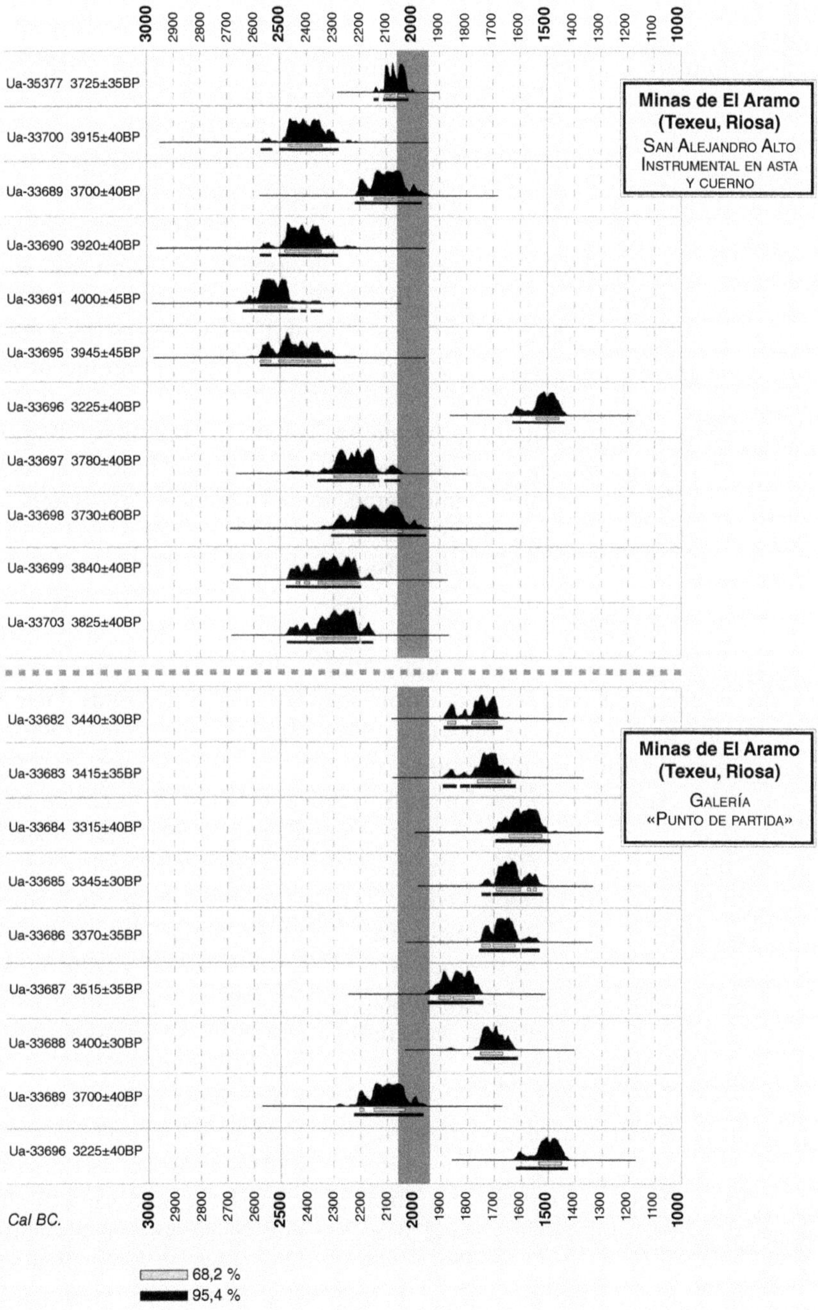

Fig. 4.9. Radiocarbon chronology for El Aramo copper mines (*Source*: Miguel De Blas Cortina).

mineralization occurs within carbonate veins in east–west fault structures within the dolomitized limestone. The veins of copper ore are irregular, ranging from a few centimetres up to 3 m in thickness. They occur within a highly developed karst, where numerous cavities are filled with ferruginous clay containing large amounts of copper carbonate minerals. These secondary copper minerals occur throughout the dolomitic limestone and seem to have been the main target of the prehistoric miners (Blas Cortina 2007–8).

The ancient workings were discovered in the upper part of the mine, known as 'Pozos de Arriba', where the mineral veins dip steeply (50°) in an east–west direction. The prehistoric workings consisted of a series of surface pits, 8–10 m in depth, which led underground to wider and interconnected cavities. Unfortunately, many of these were damaged or removed by mining in the modern era, though some evidence does survive.

The miners used a variety of techniques to extract the ore, depending on whether they were working in the karstic hollows or harder limestone rock. The nineteenth-century miners uncovered arched chambers separated by narrow passages and by rock pillars used to fix ropes for haulage and access. There are indications of fire-setting from the wallrock patterns and the discovery of charcoal deposits. Many of these mines were filled with rock spoil which contain early mining tools, most of which were lost when the early workings were destroyed. These include grooved stone hammers and a range of picks, wedges, and hammers made from shed red deer antler. Only a few of these mining tools survive in museum collections, with the majority buried during the modern mining operations.

The stone hammers consist of well-rounded cobbles of quartzite and sandstone in the 1–5 kg range, mostly gathered from a local river source. The few that survive include an example with a central groove and a lateral facet used to insert a tightening wedge (Blas Cortina 2007–8: fig. 10c). Some larger examples up to 9 kg in weight may have been hung from ropes and swung pendulum-like against the rock face (Huelga-Suarez et al. 2012b: fig. 4). These implements were used in combination with fire-setting in the harder limestone rock, while antler picks were more useful in the softer dolomitized veins and karstic mineralization.

A number of copper and bronze axeheads were also found in the mine area during the nineteenth-century operations. There are records of two flat copper axes, including one with high arsenic content, as well as two palstaves from the mine (Blas Cortina 2007–8: fig. 13). These axes may have been used by the miners to gather and process wood used for mining equipment and fire-setting.

Human remains were also found in this mine during the nineteenth-century operations. Again, most of these are now lost, however, complete or fragments of skulls survive from four individuals. The find context of these human remains is unknown, so it is not possible to tell whether they were deliberate interments or the result of mining accidents. Recent research suggests they are best interpreted as deliberate burials made in compensation for the removal of the subterranean wealth (Blas Cortina 2010; see Chapter 9 for further discussion).

A broad age range for this mine has been established by radiocarbon dating of antler tools preserved in museum collections (Fig. 4.10; Blas Cortina 2007–8). There are eight dates for these implements, which range from the mid-third to the mid-second millennium BC. This is consistent with the typological date range of the copper/bronze axeheads found at El Milagro, and is broadly in agreement with the chronology of the El Aramo mine.

The settlement context of the El Milagro mine is unknown, with no local habitation sites or metallurgical workshops connected with its operation. The copper from this mine was probably exchanged across Asturias where there is a strong distribution of Chalcolithic and Early Bronze Age metalwork (Blas Cortina 1999).

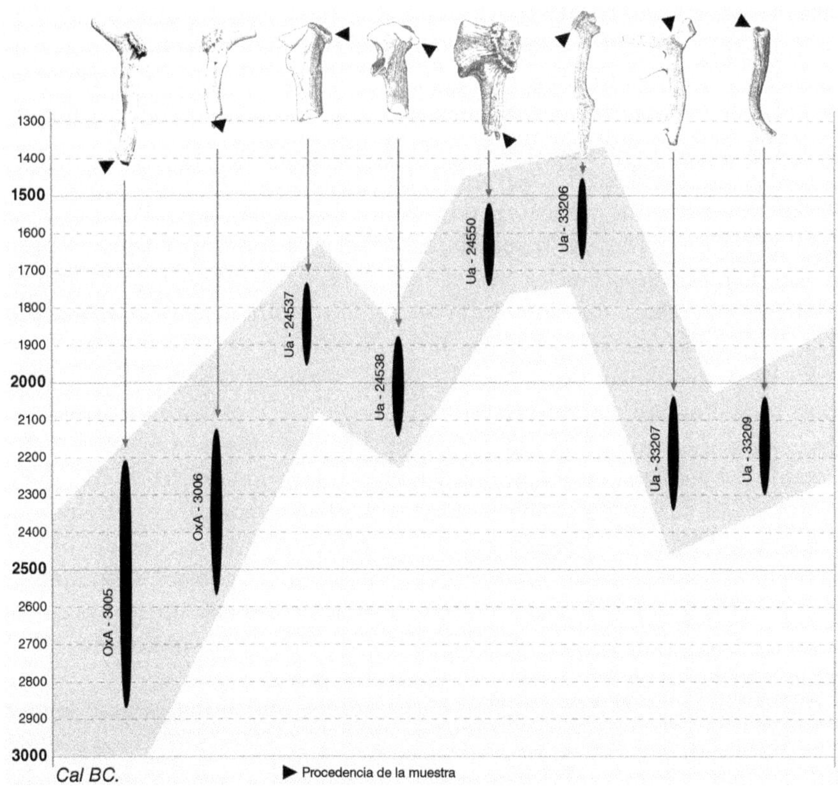

Fig. 4.10. Radiocarbon dating of worked antler implements from El Milagro copper mine, Asturias

(*Source*: Miguel De Blas Cortina).

La Profunda

This mine was discovered in the mid-nineteenth century near La Collada de Cármenes, a mountainous region of northern León (Fig. 4.1). It is located on the southern slope of the Sierra de los Currilliles at an elevation of 1,470 m. There is a large deposit of copper ore within a dolomitized limestone geology. This has been subjected to intense karstification, very similar to El Milagro and El Aramo, where large cavities filled with ferruginous clay carry nodules of copper carbonates. The central area of the deposit, known as La Cuevona ('big cave'), is a pocket of mineralization some 50 m wide and at least 200 m deep (Fig. 4.11).

The prehistoric workings and tools were discovered when the mine was re-opened in 1859 (Matías Rodriguez et al. 2001). As with El Milagro, the miners at that time followed the older workings and in so doing destroyed much of this evidence. An ancient mine gallery, *c.*30 m in length, was discovered within the mineralized dolomite, as well as smaller chambers and tunnels connecting with karstic cavities to a depth of around 36 m (Blas Cortina 2009: fig. 2). The miners extracted small masses of secondary copper minerals, mainly malachite, leaving pillars of rock for roof support. One estimate suggests that up to 18,000 cubic metres of rock with an average copper content of 5 per cent may have been extracted in prehistoric times (Huelga-Suarez et al. 2013).

Though archaeological excavations have not been conducted, there are implements from this prehistoric mine in museum collections. These include a number of grooved stone hammers, as well as a wedge-shaped copper object of uncertain function (Huelga-Suarez et al. 2013: fig. 3). There is also a record of three copper axes, now lost, from the mine (Gago Rabanal 1902; Campos et al. 2007) and evidence for the use of red deer antlers as pick-levers and goat horns as wedges (Blas Cortina and Suárez Fernández 2009). Early records from La Profunda mention fire-setting, however, it is now believed that the blackening of the mine walls there is of minerogenic origin.

Radiocarbon dates are now available for three antlers from the mine (Cortina and Suárez Fernández 2009: fig. 8). These indicate copper mining in the period 2600–2300 BC, though this is unlikely to indicate the life span of the mine. This date range compares closely to the early stages of mining at both El Aramo and El Milagro, pointing to widespread and intensive copper mining in the Cantabrian mountains during the Chalcolithic and Early Bronze Age.

The mine at La Profunda played an important role in the earliest copper production in this part of Spain, *c.*2500 BC. There is no local settlement context for the mine, however, several Chalcolithic settlements are recorded in the wider region. These include the sites of Las Pozas and El Pedroso in Zamora. These small enclosed settlements produced some of the first copper objects in

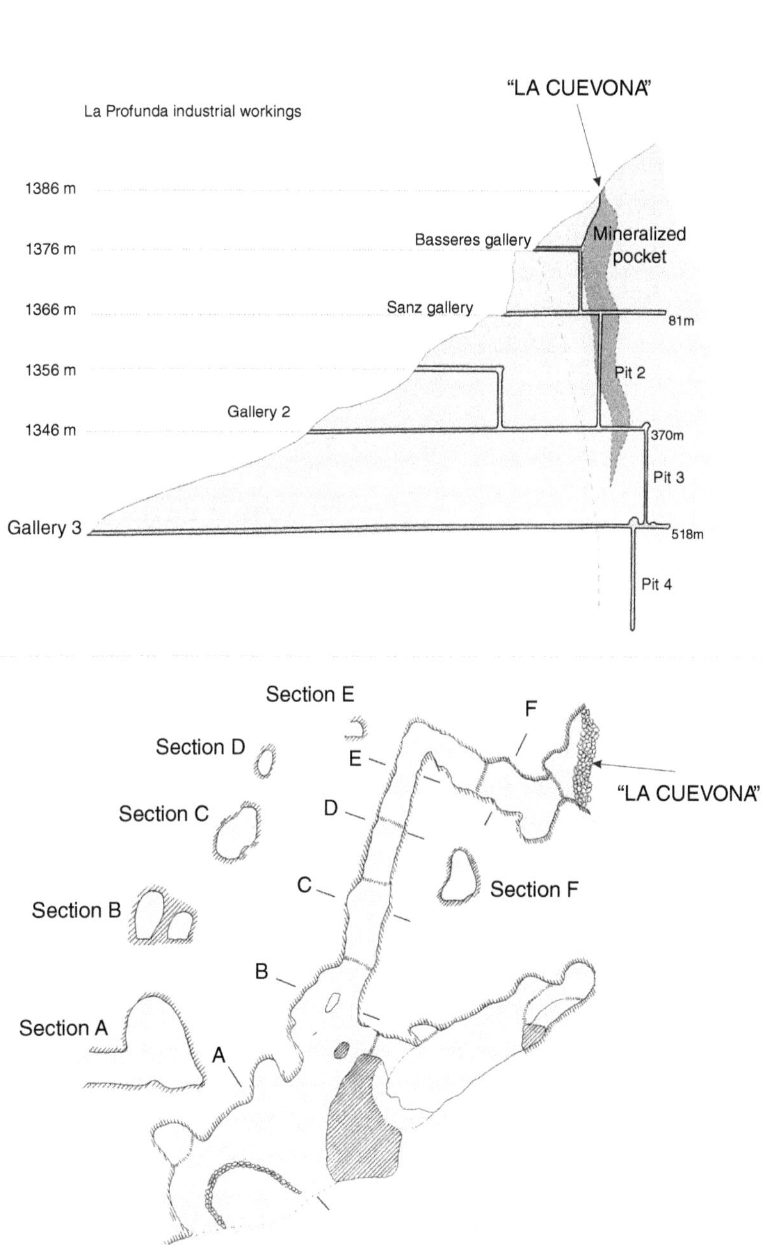

Fig. 4.11. Plan and section of prehistoric workings at La Profunda copper mine, León (*Source*: Miguel De Blas Cortina).

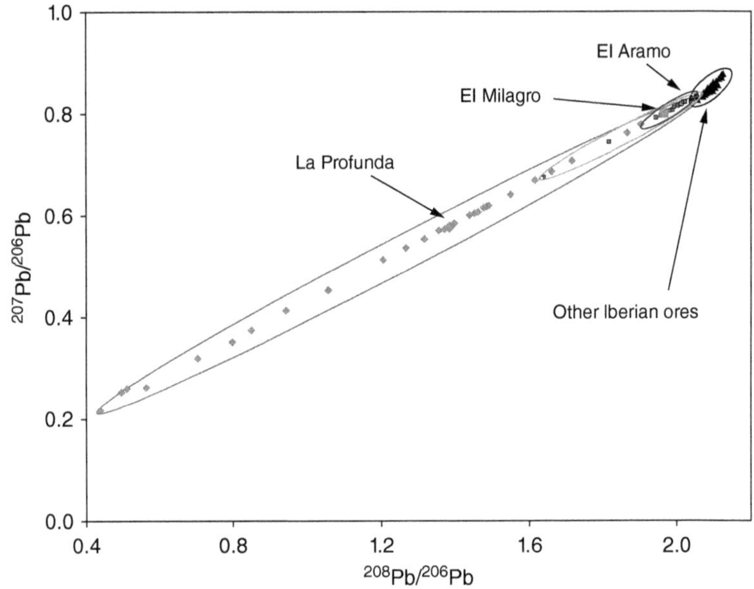

Fig. 4.12. Summary of recent lead isotope analyses for the El Aramo, El Milagro, and La Profunda copper mines in north-west Spain in relation to data for other Spanish copper ores

(*Source*: Miguel De Blas Cortina, from Huelga-Suarez et al. 2013).

this area, with evidence for crucibles, moulds, and slags connected with the production of axeheads, daggers, and awls (Delibes de Castro et al. 2005). There may be other early copper mines in the province of León, including primitive workings known on the southern slopes of Peñas Pintas in Alto Esta (Campos et al. 2007).

Finally, recent research has investigated the lead isotope signature of these prehistoric copper mines in north-west Spain (Huelga-Suarez et al. 2012a, 2012b, 2013). This revealed that the copper ores at El Aramo and El Milagro have a different lead isotope range from other ore deposits in Spain, while that from La Profunda is particularly unique (Fig. 4.12). This offers a possibility in the future of tracing the circulation of copper from these mines within particular metalworking traditions.

Catalonia

Copper deposits are also known in Catalonia and other parts of north-east Iberia. No prehistoric copper mines have been investigated in a region that has a strong distribution of early copper and bronze metalwork (Martín Cólliga et al. 2005: figs 1–2). There are indications of prehistoric copper mining in the Montsant area of Tarragona. They include the discovery of stone hammers at the mine of Solana del Bejo in Tarragona (Martín Cólliga et al. 2005: 215; Vilaseca and Vilaseca 1957). There is no dating evidence for the mine; however, lead isotope analysis suggests a link between the mineralization and a copper axe found at Cova M d'Arboli in the same region (Montero Ruiz personal communication). Other finds in the Montsant area include the recent discovery of stone hammers and primitive workings, as yet undated, at Mina de la Turquesa (Montero Ruiz et al. 2012).

Reference can also be made to the site of Forat de la Tuta in the Riner region, Lerida province, where a small horizontal gallery was previously identified as a copper mine (Serra Vilaro 1920; Montero Ruiz 2010: 74). The site is of particular interest as it contained a human burial accompanied by Early Bronze Age pottery and stone moulds used for casting of flat axeheads. A re-assessment of the evidence from this destroyed site suggests that it probably was not a copper mine, but may have been the burial of a metal-worker in a natural cave (Soriano Llopis 2011).

CONCLUSIONS

The production of copper and associated metalworking was widespread across the Iberian Peninsula during the third and second millennia BC. This is not surprising in one of the most metalliferous parts of Europe, where copper metallurgy was established at such an early date. The control of metal resources was once viewed as an important driver of social complexity in Iberian prehistory. This thinking was particularly influential in understanding the emergence of the Millaran and Argaric cultures of south-east Spain, societies believed to include specialist communities of miners and metallurgists.

Recent research has downplayed the importance of metal in this regard, emphasizing the essentially local character and limited scale of its production and use. Montero Ruiz (1994) and Rovira (2002) argue that copper production (mining and smelting) and metalworking were mostly non-specialist activities throughout the Chalcolithic and Bronze Age, conducted as part of a domestic system of production. The supply of copper was localized, with few major production centres and little evidence for organized trade in metals. This was due in part to the abundance of small copper deposits across the peninsula, which made it difficult to establish any monopoly over supply in societies where the demand for copper was limited. It also explains the basic conservatism of the production processes over several millennia, where direct reduction smelting using crucible technology was adapted to a limited, but steady, supply of copper carbonate minerals from shallow surface workings. Many of the latter were destroyed in the more intensive mining of the Roman to modern era; however, others remain to be discovered at ore deposits of less economic value.

This model of a localized craft activity utilizing the closest sources of copper ore applied to many parts of the Iberian Peninsula during the Chalcolithic and Bronze Age. In those situations, agriculture was the main economic activity, with the significance of metal production balanced with the use of other resources (Bartelheim and Montero Ruiz 2009). It is also likely, however, that this pattern of domestic activity co-existed with copper production on a larger scale in certain regions. This seems to have occurred in the south-west region of the Iberian Pyrite Belt, where there is evidence for the development of a large-scale, specialized copper industry in the Guadalquivir Valley during the third millennium BC (Nocete et al. 2008).

The development of intensive production in certain areas over a long period is a feature of prehistoric copper mining in Europe. This will be considered in the next chapter in relation to the south of France, where, as in Iberia, different scales of mining activity are known.

5

France and the western Alps

The use of copper was first established in the western Alps during the late fifth/early fourth millennia BC. There were several metal-using groups in what is now modern Switzerland during the fourth millennium, including the Cortaillod and Pfyn cultures, followed in the third millennium BC by groups of the Saône-Rhône culture (Strahm 1994). The first direct evidence of copper production, however, only dates from the Late Bronze Age. This is based on the dating of smelting slag heaps in the valley of Oberhalbstein in the canton of Graubünden (Fasnacht 2004). These slags derive from the smelting of chalcopyrite ore derived from pillow lavas of the ophiolite geology in that area (Geiger 1984).

The ability to smelt iron-rich copper ore involved a furnace technology that seems to have been first developed in the eastern Alps (see Chapter 7). No prehistoric mines are known; however, their existence may be inferred from the smelting of local ore at Late Bronze Age sites such as Savognin-Padnal and Marmorera-Stausees in the Oberhalbstein valley. Potential mining sites have been identified (see Schaer 2003), however, these have yet to be investigated in any detail.

EARLY COPPER MINES IN THE LANGUEDOC

There are numerous deposits of copper mineralization in many parts of France. These occur in Brittany, the Pyrenees, the Corbières, on the margins of the Massif Central, the Maures, and the Alps. Research over the past 30 years has identified prehistoric copper mines in several of these areas. Further discoveries are possible in the difficult terrain of the Alps and Pyrenees, and also in areas where early copper mines have not been discovered, such as Brittany where deposits of steam tin and gold are also known.

The oldest metal objects in France are recorded in the Paris Basin, where a small number of sheet copper beads date to the second half of the fourth millennium BC. These include the burial at Vignely (Seine-et-Marne) where a necklace of nine such beads was found with the burial of a five-year old child dated to 3499–3123 BC (Allard et al. 1998). This copper was not sourced locally as there is no mineralization in that region, nor is there any evidence elsewhere in France for mining in that period. The Vignely beads are part of a broader distribution of Neolithic ornaments of fifth and fourth millennia BC date that extends from central Europe to Scandinavia and as far west as eastern France (Ottaway 1973, 1982).

The earliest evidence of copper mining and metallurgy in France comes from the southern region of the Languedoc (Fig. 5.1). There it is possible to separate an early phase of metallurgy, *c*.3100–2500 BC, influenced by Alpine/Italian traditions, from copper production in the second half of the third millennium influenced by Beaker metallurgy with Iberian connections. These Chalcolithic groups produced a range of copper objects made with a distinctive antimonial copper that circulated over a wide area of southern France (Fig. 5.2). The main source of this copper was the small mining district of Cabrières-Péret, in the department of Hérault, central Languedoc.

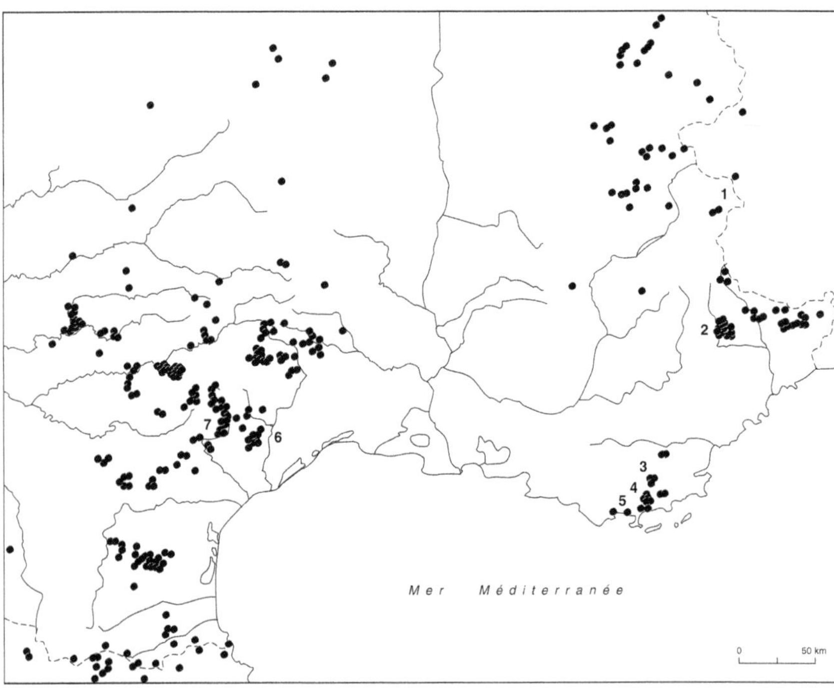

Fig. 5.1. Prehistoric copper mines and copper ore deposits in south-east France. The former include: 1. Saint-Véran; 2. Clue de Roua; 3. Maraval; 4. Le Peirol; 5. Cap Garonne; 6. Cabrières; 7. Bouco-Payrol

(*Source*: Barge et al. 1998b).

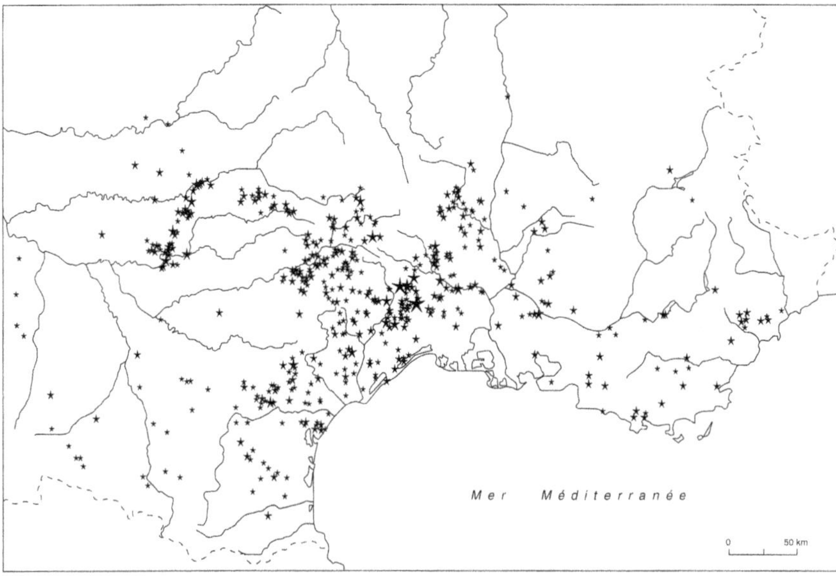

Fig. 5.2. Distribution of metal objects of Chalcolithic and Early Bronze Age date in south-east France. The size of each symbol is proportionate to number of objects from that location

(*Source*: Barge et al. 1998b).

Cabrières

The importance of Cabrières first came to light at the beginning of the twentieth century when stone hammers were found at a number of copper mines (Vasseur 1911). Research over the past three decades has identified a series of copper mines and metal production sites that date from the beginning of the third millennium BC (Ambert et al. 1984; Ambert 1990, 1995, 1996a, 1996b). These occur within a 2–3 km^2 area around the modern town of Cabrières, where three separate areas of early mining have been identified. These include (from north to south) the mines of Pioch Farrus, La Roussignole, and Vallarade, located on small massifs bordering the Le Broum valley (Fig. 5.3).

This is an area of small-scale copper mineralization exposed at several locations in the Devonian geology at the eastern end of the Montagne Noire range. The prehistoric miners worked narrow quartz veins within Hercynian fault structures. These veins contained significant amounts of fahlore and chalcopyrite, as well as oxidized copper ore in the form of azurite and malachite. In addition there are pockets of copper mineralization in karstic cavities within the Devonian dolomites.

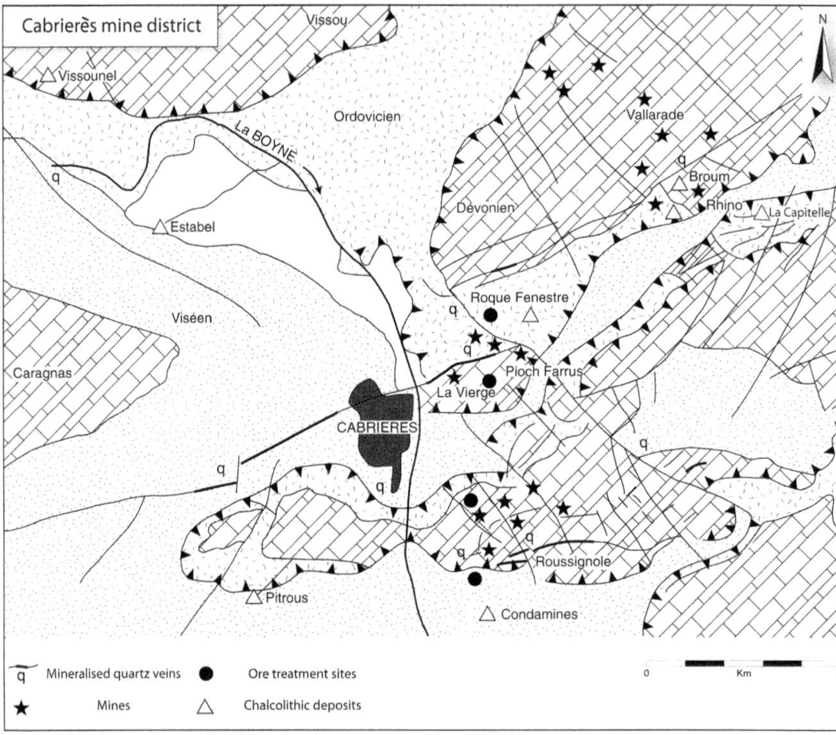

Fig. 5.3. Map of Cabrières mine district showing location of Chalcolithic mines, ore treatment areas, and settlement sites

(*Source*: Paul Ambert with additions).

The principal ore used by the prehistoric miner was a type of fahlore called tetrahedrite, which is naturally rich in copper, antimony, and silver. The smelting of this ore produced a metal with a distinctive chemical pattern containing high antimony and silver content, low lead, and trace arsenic. The high antimony content yielded a metal with technical properties superior to pure copper. This antimonial copper has been identified in over half of the metal tools from the third millennium BC from the Languedoc region. While this was the dominant metal type of that period, other copper ores were also used at Cabrières, including some with higher lead content, as well as small pockets of oxidized mineralization (Ambert et al. 2009).

Pioch Farrus

There may be as many as ten early mines in this area, with excavated examples at Pioch Farrus I and IV, and La Vierge. The early date of these workings is confirmed by the discovery of Chalcolithic pottery in the Pioch Farrus

I working. Similar pottery was found in Pioch Farrus IV, which radiocarbon dating indicates was worked *c.*2400–2200 BC, and subsequently in the Gallo-Roman period (Ambert 1995: fig. 8). The Chalcolithic workings consisted of an irregular, near-vertical trench extracted to a depth of 4–5 m, as well as underground galleries up to 15 m long occurring at depths of 6 m or more (Fig. 5.4; Ambert 1990). The miners followed a mineralized quartz vein, and may have encountered karstic cavities filled with copper ore. Though damaged by later workings, the extant portion of the Chalcolithic mine reveals smooth

Fig. 5.4. Pioch Farrus IV mine, Cabrières, France

(*Source*: author).

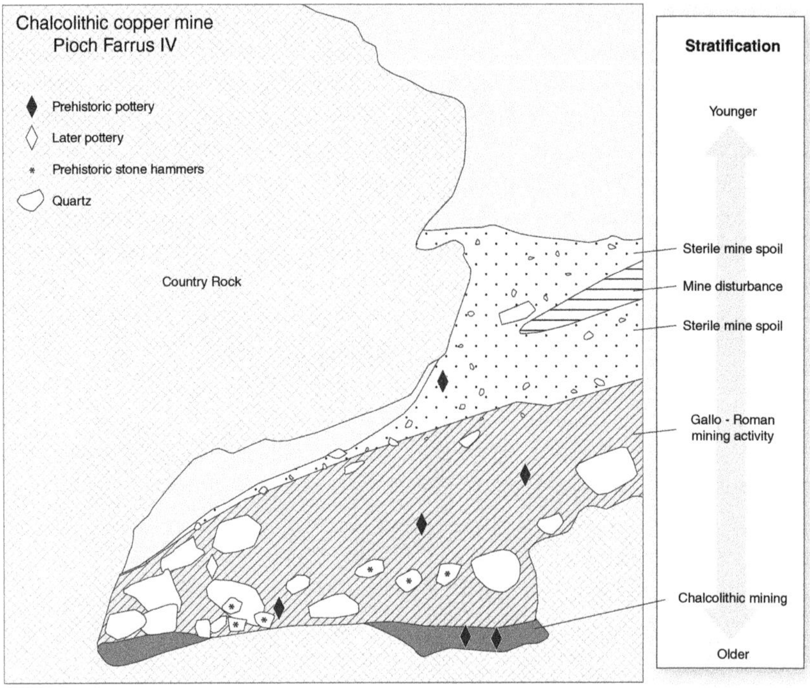

Fig. 5.5. Mine section, Pioch Farrus IV, Cabrières mine
(*Source*: author, re-drawn from Ambert 1996a).

wallrock surfaces with metre-size ovoid depressions indicative of fire-setting. This was confirmed by thermoluminescence measurements of the broken rock from the mine (Castaing et al. 2005), which also contained charcoal residues from the mine faces (Fig. 5.5).

The excavation of a ditch at the site of Pioch Farrus 448 uncovered evidence of copper production connected with the Pioch Farrus mines dating as early as 3100–2800 BC. This feature contained stone mining tools, smelting slags, fragments of crucibles, and copper droplets. There is a small vein containing tetrahedrite and chalcopyrite ore in the vicinity, some of which was found in the ditch (Espérou et al. 1994).

Excavations at a nearby site called Roque Fenestre uncovered further evidence of copper production connected to the Pioch Farrus mines (Espérou 1981; Ambert 1996b). Radiocarbon dates confirm metallurgical activity at this location during the period 2800–2000 BC. A series of four adjoining pits cut into bedrock were identified. These were used at different times in the treatment of copper ore. This involved different stages of crushing and washing of copper ore, using gravity concentration in water-filled pits to separate the metallic minerals. The ore concentrate was then roasted and subsequently smelted in the immediate vicinity of the pits. Chemical analysis of slag and copper prills produced in this manner confirms a link with the mineralization at Pioch Farrus, located only 200 m away. The finds at Roque Fenestre included stone hammers similar to those in the mine, as well as grinding stones and cup-marked stones used in ore and slag processing. Examples of bovine scapula shovels were also discovered (Ambert 1996a: fig. 14).

Excavation of a small mine working at La Vierge recovered stone mining tools and a pottery vessel of the Early Bronze Age (see Chapter 9 for illustration). This small copper mine was discovered in 1979 when a road was widened at the foot of a rock face in the Pioch Farrus mine area. A large pottery vessel was found in a partly infilled mine working measuring 2.3 m by 0.5 m (Ambert 1990: plates 4 and 5). This working followed a mineralized vein containing malachite and sulphidic copper ore. Excavation uncovered sherds of similar pots, as well as stone mining hammers, but no indication of fire-setting.

La Roussignole

These mines are located on the southern massif overlooking Cabrières town, in an area mined for copper during the modern era (Ambert 1990: plate 2). Numerous stone hammers are recorded there, including finds in the so-called Japhet mine, a sub-vertical working where a sherd of prehistoric pottery was found (Ambert and Barge 1990). Evidence of fire-setting is recorded in mine sites 2, 3, and 5 at La Roussignole (Ambert 1990: plate 7; Maass 2005: figs 3 and 4). As with Pioch Farrus IV, this technique seems to have been used to extract mineralized veins in the silicified dolomites.

A significant number of stone hammers are recorded from the early mines in Cabrières and from their associated metal production sites. Some workings have distinctive wallrock markings produced by the use of these implements. The latter consist of irregularly rounded cobbles of basalt, quartz, and quartzite, with varying degrees of use-breakage. These hammerstones range 0.8–2 kg in weight, with larger examples up to 15 kg also recorded. Some were naturally rounded cobbles gathered from the local fluvial drift, while others were fashioned from hard vein rock extracted in the mines. Most were used hand-held, and few have any obvious form of haft modification. There are two main types, namely sub-spheroidal examples used as pounders and some pointed examples used as picks (Ambert et al. 2009). The latter were mainly used in mining, whereas the pounder-type could be used for both rock extraction, connected or not with fire-setting, and for ore beneficiation. The latter required the use of different types of grinding stones and mortars. These include small cup-marked slabs ('galets à cupules') found in the copper production sites in this area (Espérou 1990: fig. 15; Cert 2005).

Vallarade

There was a long history of copper mining on the Vallarade plateau, with prehistoric, Gallo-Roman, and modern workings identified in recent surveys. Following the earliest record of stone hammer finds (Vasseur 1911) indications of ancient mining have been recorded at eight locations in this area (Bouquet et al. 2006: figs 1 and 3). Evidence of Chalcolithic mining dating 2900–2500 BC has been obtained from surface spoil at the Les Neuf-Bouches mine. There is a large vertical trench mine at that location, up to 15 m in depth, where the miners followed quartz veins and karstic mineralization in the Devonian dolomites (Fig. 5.6). There is a similar date range for mine spoil at nearby Petit-Bois where surface treatment of copper ore from the Vallarade 4 mine was undertaken (Ambert 1996a).

Copper mining continued on this massif well into the Bronze Age, with evidence of fire-setting from Vallarade mine 5 and Les Neuf-Bouches dating 1740–1120 BC (Ambert 2002). The technique may not have been used extensively in these mines as the softer dolomite geology could have been worked primarily with stone hammers and bone tools. There is also evidence of some copper mining activity at Les Neuf-Bouches at the end of the Bronze Age, *c.*800 BC (Ambert 2002).

Also noteworthy are several natural karstic caves on the southern side of the Vallarade massif, which contained burials and pottery of the early- to mid-third millennium BC (Fig. 5.3). These include the Rhinoceros caves 1 and 2,

Fig. 5.6. Les Neuf-Bouches trench mine, Cabrières

(*Source*: author).

and the Broum cave (Ambert and Barge 1990: 24–6, fig. 20). The latter has mineralized quartz veins and karstic mineralization, some of which were worked in the Gallo-Roman era. There is no conclusive evidence of Chalcolithic copper mining, however, the cave was certainly occupied during that period.

The mining at Les Neuf-Bouches during the early third millennium BC was connected with copper production at the nearby settlement of La Capitelle du Broum. The site is located on a small terrace in the Broum valley in the eastern part of the Cabrières mine area. This is the oldest known settlement of prehistoric metalworkers in France. Occupation commenced in the period 3100–2900 BC and continued at the site to around 2400 BC, during which time there were several phases of occupation (Ambert et al. 2005, 2009). Pottery from the site is comparable to that found in the cave of Broum and at the nearby settlement of Roquemengarde. The ceramic from these sites indicates a distinct Chalcolithic settlement in the Cabrières region associated with copper production.

The La Capitelle site seems to have been a permanent settlement, with the remains of several substantial house structures found (Fig. 5.7). These had double-faced stone walls, similar in design to buildings of the Fontbouisse culture found to the north of Montpellier (Ambert et al. 2002). Other cultural connections are possible, including influences from the Grands Causses region and groups such as the Treilles culture (Mille and Carozza 2009).

Fig. 5.7. Paul Ambert at the La Capitelle du Broum mining settlement, Cabrières (*Source*: author).

Considerable evidence of copper production was found inside and adjacent to the structures at La Capitelle. All stages of copper production are recorded, from the initial treatment of mined ore to smelting and metal casting (Bourgarit and Mille 2005). Stone hammers and grinding stones used for ore beneficiation were found (Cert 2005). The smelting process was relatively simple, involving a one-stage carbon reduction of copper minerals (tetrahedrite and some oxidized ore) in shallow pit furnaces only 0.4 m in diameter. This produced copper droplets, some of which had to be extracted from a crude slag consisting mostly of partially reduced ore. The slag was finely crushed to extract the metal prills, which were then amalgamated in ceramic crucibles, an example of which was found in the site. Mille and Carozza (2009) also raised the possibility of a primitive matte smelting process at the site, involving two stages of smelting.

There is evidence from La Capitelle for the casting of metal objects, and their finishing using processes of coldworking and annealing. The copper produced in this site was used for a range of objects during the Chalcolithic, including copper awls and needles, chisels, axeheads, daggers, arrow points, and beads (Ambert et al. 2009). Chemical and isotopic analysis confirms that the copper ore used at La Capitelle came from the Cabrières area, with a specific match to the Vallarade and La Roussignole mines (Prange and Ambert 2005). Some of these mines continued on a reduced scale into the Bronze Age. Evidence of fire-setting dated 1740–1120 BC has been identified at Vallarade. There is also evidence of Late Bronze Age mining at Les Neuf-Bouches dated 850–740 BC (Bouquet et al. 2006).

The research at Cabrières has provided much insight into the earliest copper mining and metal production known in France. The beginnings of metallurgy in that area can be dated to around 3100–2800 BC. This is confirmed by the early dates for the mine of Les Neuf-Bouches and the production sites of Pioch Farrus 448, Roque Fenestre, and La Capitelle at Cabrières. The distinctive antimonial copper metal from these mines achieved a wide circulation across south-central France during the early Chalcolithic. This is contemporary with early copper-using groups in the region, such as those represented by the Roquemengarde settlement where some of the earliest metal artefacts in the Languedoc were found (Guilaine 1991). These can be linked chemically to the copper ores from Cabrières, indicating the existence of a metal supply network at this early stage.

The origin of the early Languedoc metallurgy may lie in connections to the east, specifically with the Remedello culture of northern Italy around 3100 BC or with Chalcolithic cultures in Switzerland in the same period. There are indications that the Cabrières copper producers participated in the Beaker network of exchanges from around 2500 BC. This involved connections with Iberia, as reflected in the northward spread of Palmela points in this period, with their distinctive arsenical composition. These include an example from

Cabrières, found in a valley at Les Condamines some 300 m from the Chalcolithic mine at La Roussignole (Ambert et al. 1990: plate 11).

It has been suggested that these Iberian links led to the introduction of the smelting crucible technique to Cabrières (Ambert et al. 2009).

OTHER MINES IN SOUTHERN FRANCE

Other early copper mines have been identified along the south-west margin of the Massif Central. These occur in Cambro-Ordovician geology on the eastern side of the Black Mountains, in the province of Aveyron. A survey in the Montagne Noir and Monts de Lacaune districts identified some 23 potential early copper mines (Barge 1985). There are indications of primitive workings at many of these ore deposits, as well as extraction of copper in Gallo-Roman and modern times. They include the mines at Bouco-Payrol in Aveyron, and in the area of Seronais in the Ariege Pyrenees. The latter mine is considered to date from Roman times, although earlier workings may also occur (Dubois and Guilbaut 1980).

The Bouco-Payrol mines lie near the town of Brusque, about 30 km west of Lodeve (Fig. 5.1). They comprise a series of workings connected to swallow-hole caverns (avens) in karstified limestone (Léchelon 1974, 2001; Léchelon et al. nd).

The deeper workings are Roman period, extending down to nearly 100 m with an interconnecting system of shafts, galleries, and inclines. Closer to the surface there is evidence of narrow tunnels and galleries, with shafts leading to the surface (Barge 1985: fig. 2). Stone hammers were found in these workings, as was pottery of the Early Bronze Age (Barge 1985: fig. 3). There is also evidence for fire-setting, with rounded wallrock surfaces exposed in underground tunnels (Fig. 5.8; Barge 2003: 67). Other mine workings and deposits of rock spoil are recorded in the vicinity. A fragment of human skull impregnated with copper from the surface of one of these mine trenches is radiocarbon dated 2480–2450 BC.

Further evidence of early copper production is provided by the metallurgical workshop discovered at Al Claus in the Aveyron Valley (Tarn-et-Garonne). Excavation uncovered the remains of several timber buildings with evidence of copper metallurgy, including furnace features, tuyères, stone hammers, and mortar stones. One of the pits containing this debris is dated 2448–2175 BC (Carozza et al. 1997). A large amount of pottery was found in this settlement, including some domestic vessels that were re-used as metallurgical crucibles. The analysis of slag residues suggests that some of these pots were re-used as vase furnaces to smelt copper ore. This may have

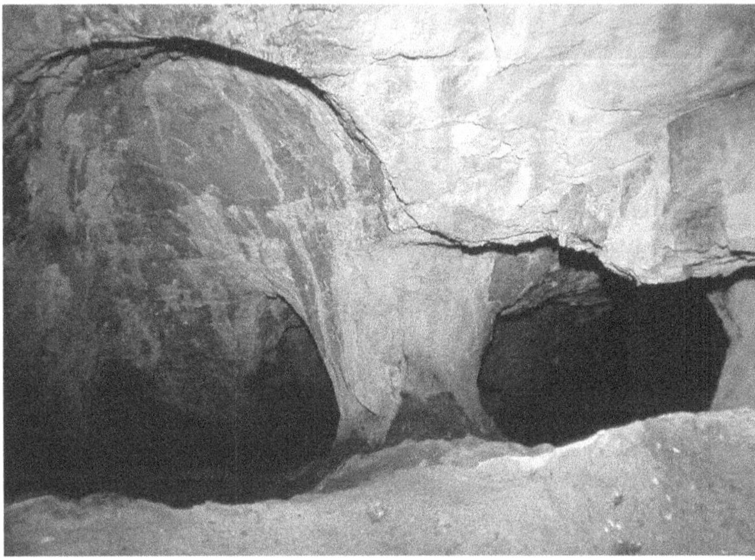

Fig. 5.8. Fire-set wall surface at Bouco-Payrol mine
(*Source* and copyright: Hélène Barge).

involved the use of chalcopyrite ore, as opposed to the fahlore-based production employed at Cabrières (Mille and Carozza 2009: figs 10–11).

Mention can be made of recent investigations at the Roman mine of Les Barrencs (Aude), on the southern slopes of the Montagne Noire in the Languedoc (Mantenant et al. 2012). There are a series of mineralized veins containing copper, lead, silver, and iron. The mining focused on two main veins (Mourral de la Grave and Barrencs de Fournes) 150 m apart, approximately 500 m in length, and of variable thickness. There is a complex network of drifts and stopes connected by vertical shafts down to depths of 100 m or more. Evidence of fire-setting is recorded. Radiocarbon results indicate that mining may have commenced during the La Tène Iron Age, from the fourth century BC onwards, with continued activity to the first century BC into Roman times.

Copper ore deposits are also recorded on the western side of the Pyrenees in the French Basque region. Some of these small ore deposits were mined in the Roman era, but also in earlier times (Beyrie 2005; Kammenthaler and Beyrie 2007).

This was first indicated by palaeoenvironmental studies carried out in the Nives des Aldudes in the south of Lower Navarre (Monna et al. 2004). These results indicate the presence of early metal production in this high altitude Pyrenean valley, with geochemical signals for associated air pollution from the early third millennium BC, the Middle Bronze Age (1500–1300 BC), and the Late Bronze Age (900–800 BC).

Research in the Aspe valley has identified a mine at Causiat worked in the period 2580–2340 BC (Beyrie and Kammenthaler 2008). This is the oldest evidence of copper mining in the western Pyrenees (Fig. 4.1). Located at an altitude of 1,600 m, this orebody contains copper-iron sulphide ore and secondary copper minerals in a Carboniferous geology. The oxidized portion of the mineralization was extracted from the exposed vein using stone hammers and fire-setting. Evidence of the latter may be seen in the smooth, rounded profile of these narrow mine tunnels. The full extent of the mine has not been fully established, with research continuing at the site.

THE FRENCH ALPS

By 2000 BC another important centre of primary copper production had emerged in France. This was based in the Alpine mountain range of the Rhône-Alps and the Provence-Alpes-Côte-d'Azur regions, which border Switzerland and Italy. This environment presented particular challenges in terms of prospection for minerals and the logistics of mining at high altitudes.

Saint-Véran

Important evidence of early copper mining has been uncovered at Saint-Véran in the southern part of the French Alps (Fig. 5.1). This mine is located in the Aigue Blanche Upper valley in the eastern part of the Queyras Massif, close to the Italian border. It is notable for an unusual deposit of rich copper ore that could be mined and smelted with relative ease. The mineralization is of the sedimentary exhalative type and was created by submarine hydrothermal processes during the Mesozoic era. The host rocks are mainly chloritoschists, gabbros, and serpentinites that have a layered structure that was metamorphosed and folded by Alpine tectonic movements.

The copper ore on this mountain occurs within a thin layer of amphibolite, bedded between quartzite and ophiolite formations. This seam of rich copper ore ranged 0.2–0.3 m in thickness, and is tilted vertically along a steep mountain slope at an altitude of 2,250–2,600 m. The mineralization mostly consists of very pure bornite with a copper content of up to 63 per cent. There is no oxidized copper mineralization on the exposure of this ore deposit, however, the presence of secondary minerals cannot be ruled out. Native copper is recorded from the deeper modern mines at Saint-Véran, and it has

Fig. 5.9. General view of Saint-Véran mine
(*Source* and copyright: Hélène Barge).

been suggested that this may have been important in the initial discovery of the mine in the late Neolithic (Bourgarit et al. 2008).

The early workings are recorded along the upper mountain slopes at altitudes of 2400–2600 OD where copper deposits are exposed, making this the highest prehistoric copper mine in Europe (Fig. 5.9). These early mines are mostly infilled, however, one large working known as the 'Tranchée des Anciens' is visible in that area. This narrow trench is about 300 m long by 0.6–3 m wide, and was worked to a depth of around 50 m (Fig. 5.10; Barge 2003). The mine walls have tool marks connected to the use of stone hammers and picks. Investigation of another surface mine close to the 'Tranchée des Anciens' uncovered evidence of fire-setting dated to the Early Bronze Age (Barge and Talon 2012: figs 2 and 3). The remains of fires were discovered, with charcoal evidence to suggest a preference for the burning of local alpine trees (*Pinus Cembra*).

The deeper part of these prehistoric mines can only be accessed through modern cross-cut workings. These reveal that prehistoric mining was conducted at depths of up to 80 m. This mining may even have been conducted during the winter freeze to minimize water seepage in the deeper workings (Rostan and Rossi 2002). The early mine tunnels average 1.2–1.5 m in width, with some as narrow as 0.4 m wide. This is because the early miners removed as little of the country rock as possible around the bornite-rich layer. Some

Fig. 5.10. Tranchées des Anciens mine at Saint-Véran
(*Source* and copyright: Hélène Barge).

mine tunnels have the characteristic smooth profiles of fire-setting, which was used on occasion to extract the harder quartzite rock.

Stone hammers were the main mining tool, made from natural greenstone cobbles and modified for hafting. Goat-horn picks with handles were used in the softer chloritoschist. Other tools found include wooden shovels and torches made using splints of pine. The exploration of these tunnels uncovered wooden timbers used in roof support and in the stacking of rock waste underground.

In 1992 a metallurgical workshop was discovered at the site of La Cabane des Clausis associated with the Early Bronze Age mine at Saint-Véran (Barge 1997; Ploquin et al. 1997; Barge et al. 1998a). The site is located at an altitude of 2,250 m on a natural shelf, 250 m downslope from the prehistoric mines (Fig. 5.9). Excavation uncovered a spread of copper smelting slag, as well as tuyère fragments and part of a stone crucible. No definite furnace structures were found, however, the work surfaces have indications of intense burning and sediments connected with the heat reduction of copper minerals. Stone hammers and grinding stones were used in the beneficiation of bornite ore for smelting. The slag is of a thin platy type, similar to the so-called *plattenschlacke* recorded in later Bronze Age smelting sites in central Europe. This is a high density, homogenous slag with a low copper content, which indicates a controlled smelting process. Smelting experiments carried out using the bornite ore suggest that this may have involved a one-step process using simple pit furnaces and blowpipe techniques.

Radiocarbon dating indicates that the main period of copper mining at Saint-Véran was *c*.2400–1700 BC. Timbers from the underground mines are dated 2200–1700 BC (Barge 2003: 34). The copper production at Clausis is radiocarbon dated 2400–1900 BC, consistent with the finds of Early Bronze Age pottery at the site. Similar dates were obtained for a second smelting site at Saint-Véran, located in a rock shelter at Piniliere just below the Tranchée des Anciens mine. In addition, a smelting site discovered in the Longet valley to the north of Saint-Véran is dated 2210–2130 BC (Bourgarit et al. 2008). Finally, there are radiocarbon dates to indicate some mining on a small scale at Saint-Véran later in the Bronze Age, *c*.1000 BC.

The scale of production at Saint-Véran is impressive. Rostan and Rossi (2002) calculated that an estimated 10,000 m^3 of rock extracted in those mines contained some 2,000 tonnes of bornite, possibly yielding as much as 1,400 tonnes of smelted copper. At an extraction rate of 4 m^3 of rock per month, this level of production would have taken 200 years of continuous production, or possibly twice as long with seasonal exploitation over the known period of mining, *c*.2400–1900 BC.

The background to this copper mining at Saint-Véran may lie in the Chalcolithic of the central Languedoc, in particular the Cabrières production zone. Given the proximity to the Italian border, some influence from that direction is likely. The discovery of rock carvings with depictions of Remedello culture daggers near Saint-Véran is relevant (Rostan and Rossi 2002; see also Rossi et al. 1999). Mille and Carozza (2009) observed that the type of high-yield metallurgical process at Saint-Véran has parallels across the entire Alpine zone during the Bronze Age. This is where the main influences probably came from, as opposed to contacts with Cabrières and other production centres in southern France.

Les Rousses

Another Early Bronze Age mining centre has recently been discovered in the western French Alps to the north of Saint-Véran (Fig. 4.1). A survey in the

Grandes Rousses Massif near Oisans has identified approximately 40 locations where early copper mining may have been conducted at altitudes between 2,000–2,700 m (Bailly-Maitre and Gonon 2008). A series of mineralized quartz veins containing chalcopyrite ore were mined in an area of extensive rock outcrop. Deep trench workings are recorded at the sites of Cavales 4, Cochette, Etendard, and Barbarate. Radiocarbon dating of charcoal samples from spoil dumps near these workings place this mining in the period 2200–1700 BC (Bailly-Maitre and Gonon 2008; Carozza et al. 2010). Fire-setting and stone hammers were used in rock extraction, while the latter implements were employed in ore concentration on level surfaces near the mines.

There are few details as to the organization of this mining on the Rousses Massif. A number of hut structures have been found that may point to seasonal mining activity at this high altitude, however, this remains to be confirmed. No evidence of smelting or metalworking has been recovered at the mines, although there are indications of local metal production in sediment cores taken from nearby Lake Bramant at an altitude of 2,448 m. These reveal a peak in copper content in glacial varves dated 2200–1650 BC. This coincided with a period of warmer climate and reduced glacial activity, which made copper mining at this altitude possible (Guyard et al. 2007). This changed around 1700–1600 BC with a colder climate and ice advances, at around the same time the mining at Les Rousses appears to have ended.

Finally, reference can be made to the region of Provence-Alpes-Côte d'Azur in the south-east corner of France (Fig. 5.1). An examination of copper deposits in the coastal region east of Toulon identified a number of possible early copper mines (Rostan 1994; Barge et al. 1998b: fig. 1). There are records of ancient mine workings at Maraval and also at the mine of Le Peirol where a *galet à cupule* grinding stone was found (Barge et al. 1998b: fig. 2). Further east, there is evidence for the mining of native copper at Clue de Roua (Daluis and Guillaumes), located north of the Alpes-Maritimes (Rostan and Mari 2005). A series of small mineral veins are exposed in sandstone geology along the gorge of the river Var. The veins contain native copper rich in arsenic, as well as secondary copper minerals. There are a large number of mine tunnels, many of which are medieval in date (Rostan and Mari 2005: figs 2–3). There are earlier workings as well, with evidence of fire-setting and the use of crude stone hammers ((Rostan and Mari 2005: fig. 6). Rostan and Mari (2005) have raised the possibility of Chalcolithic mining of native copper based on these stone tools, however, this remains to be confirmed.

DISCUSSION

The origins of copper metallurgy in southern France are generally attributed to external influences. Previously, the emphasis would have been on contacts with Iberia and the role of the Bell Beaker culture in the dissemination of metallurgical knowledge. Influences from that direction are certainly apparent in the Languedoc for the later Chalcolithic period; however, the dates for mining and metallurgy at Cabrières indicate origins much earlier than the Beaker culture.

It is generally accepted that metallurgy did not develop independently in the Languedoc, but spread there from Italy where metallurgy was well established during the fourth millennium BC (see Strahm 2005). Dolfina (2013) argues that the establishment of copper mining and metallurgy in the Languedoc may have been due to several pathways in the later fourth millennium BC. This may have involved contacts with the metal-users of the Rinaldone culture area in west-central Italy and with those of the Remedello culture of northern Italy. The former may have involved a coastal route through Liguria with its early copper mining tradition into Provence. There is also the possibility of maritime contacts from the Rinaldone culture area, extending from Tuscany to Corsica into the French Midi. A separate line of contacts may have been established across the western Alps, possibly connected to the metalworking traditions of the Remedello culture area. That was the world of the Ice-man, the now-famous discovery of a mummified man, dated to around 3100 BC, discovered in the Otztal Alps of northern Italy. The discovery of a copper axe with the body of this Alpine traveller illustrates the potential for long-distance contact with metal across the Alpine zone during the later fourth millennium BC.

The precise origins of the early Languedoc metallurgy remain to be established, as centres such as Cabrières centre lack direct cultural ties with Italy. A further difficulty is the absence of early mining and metallurgy in Provence, a region with abundant copper resources and strategically located between Italy and the Languedoc. The discovery of what seem to be early copper mines in the Toulon region may be significant in this regard.

Cabrières falls into a wider pattern of fahlore metallurgy that was well established in several parts of western Europe during the third millennium BC. While external influences are apparent, the production of copper in that centre developed on a largely independent basis over the course of the third millennium BC (Ambert 1999: fig.3). The spread of Cabrières copper to Atlantic regions of France, along the Garonne corridor is evident by the mid-third millennium BC (Ambert 1999: fig. 5). This was probably stimulated by exchange networks developed as part of the Beaker culture. Ambert has also drawn attention to the prevalence of Beaker metalworking using arsenicated copper in Atlantic France, extending from Normandy and Brittany to the Saintonge region. Numerous flat copper axeheads are known from both Beaker and pre-Beaker contexts in that region, while finds of copper daggers, Palmella points, and other objects can be directly connected with Beaker metalworking (Roussot-Larroque 1998, 2005; Briard and Roussot-Larroque 2002). Ambert (1999) has argued that the arsenical composition of many of these objects in Atlantic France is probably related to the smelting of tennantite ore, from mine sources possibly located in the western Pyrenees.

There is growing evidence of Beaker culture metalworking in Atlantic France using arsenical copper. This includes evidence of copper metallurgy and metal finds from Beaker settlements, for example at Les Florentins (Val-de-Rueil, Eure) in north-west France (Billard 1991) and La République in Talmont-Saint-Hillaire in the Vendée (Ambert 2001). This Beaker metalworking eventually extended into Brittany, possibly the launching point for a spread of copper mining and metallurgical expertise into another part of Europe: Ireland.

6

Northern Europe

Copper objects first circulated in Britain and Ireland around 2500 BC, thus beginning a short-lived Chalcolithic that ended with the rapid adoption of tin-bronze metallurgy after 2100 BC. Both islands have numerous sources of copper; however, these orebodies are not evenly distributed, nor were they all accessible to the prehistoric miner. This is part of the explanation why certain regions developed a strong tradition of copper mining that lasted well into the Bronze Age.

IRELAND

Ireland has long been regarded as a significant producer of metal in the Bronze Age. This reflects the large quantities of Bronze Age metalwork found in a part of Europe with abundant sources of copper. The south-west region of Cork and Kerry was the main centre for early copper production (Fig. 6.1). This began with mining at Ross Island in Killarney, where Beaker culture groups produced arsenical copper during the Chalcolithic and Early Bronze Age (*c.*2400–1900 BC). Farther south, there are seven copper mines now dated to the Early to Middle Bronze Age (*c.*1800–1400 BC) in the peninsulas of west Cork. These are known as Mount Gabriel-type mines, the name coming from the single largest concentration of such workings located on the eastern slopes of this mountain in the Mizen Peninsula (O'Brien 1994, 2003). The recent discovery of trench workings at Derrycarhoon continues the story of Bronze Age copper mining in that area to 1300–1100 BC, after which this activity seems to have ceased (O'Brien 2013).

The study of these mines began during the late eighteenth/ early nineteenth centuries, when mineral prospecting led to the discovery of primitive workings at several locations in south-west Ireland. Described as 'Dane's Workings' in the antiquarian literature, these mines were associated with the use of fire-setting and stone hammers (see quotations from Griffith 1828 and Thomas 1850 (in O'Brien 2003) in Chapter 1). The first systematic research began in

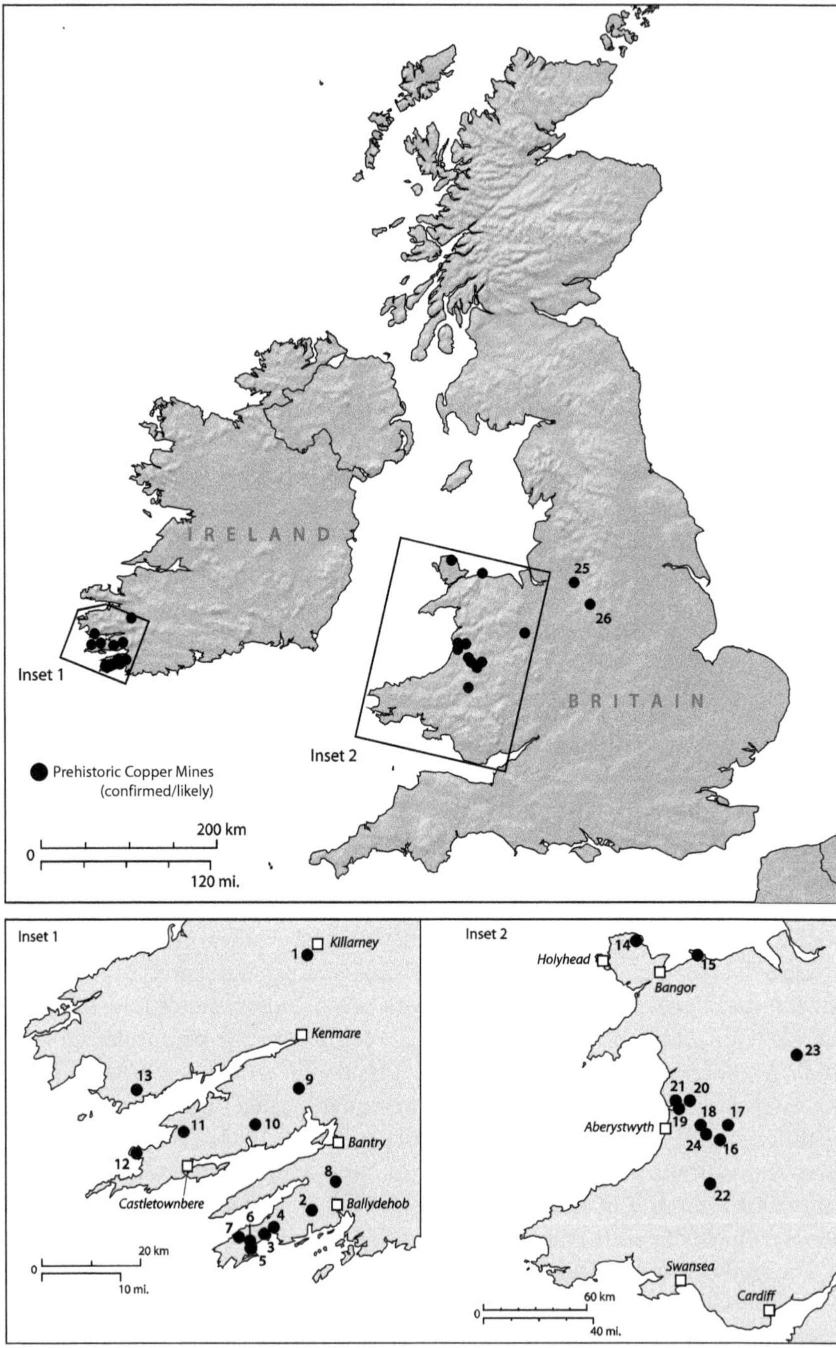

Fig. 6.1. Distribution of prehistoric copper mines in Britain and Ireland

South-west Ireland: 1 Ross Island; 2 Mount Gabriel; 3 Ballyrisode; 4 Toormore; 5 Boulysallagh; 6 Callaros Oughter; 7 Carrigacat; 8 Derrycarhoon; 9 Tooreen; 10 Canshanavoe; 11 Crumpane; 12 Reentrusk; 13 Coad Mountain. Britain: 14 Parys Mountain; 15 Great Orme; 16 Copa Hill, Cwmystwyth; 17 Nantyreira; 18 Nantyrarian; 19 Llancynfelin; 20 Ogof Wyddon; 21 Panteidal; 22 Pen Cerrig y Mwyn; 23 Llanymynech Ogof; 24 Tyn y Fron; 25 Alderley Edge; 26 Ecton.
(*Source*: author).

the 1930s with the discovery of the Mount Gabriel group by the geologist, Tom Duffy. These were subsequently mapped by another geologist, John Jackson, who brought these mines to wider attention when he obtained a Bronze Age date for charcoal taken from mine spoil on the mountain (Jackson 1968). This was the first radiocarbon result obtained for a prehistoric copper mine anywhere in Europe. While doubt was expressed as to the age of the Mount Gabriel mines (Briggs 1983; for reply see Jackson 1984), further research has confirmed their Bronze Age context (O'Brien 1990, 1994, 2003).

Ross Island

Copper metallurgy was introduced to Ireland in the twenty-fifth century BC, coinciding with the arrival of Beaker material culture (Case 1966; Sheridan 1983). The Beaker networks were critical in the transfer of mining expertise in Atlantic Europe, through coastal connections extending from northern Spain and western France to Ireland. The new technology was successfully introduced over a short period, with a prolific output of axeheads, daggers, and halberds over the following 400 years. These copper objects have distinctive arsenic content, connected to the mining of tennantite fahlore from one major source, Ross Island mine in Killarney, county Kerry (Fig. 6.2). The extraction of fahlore copper began there around 2400 BC and continued through the Chalcolithic into the early centuries of the Bronze Age, ending around 1900 BC (O'Brien 1995, 2004).

The early workings at Ross Island consisted of large cave-like openings on mineralized rock exposures (Fig. 6.3). The full extent of the underground mining is uncertain as these workings are no longer accessible due to flooding and roof collapse, problems that prehistoric miners probably also had to contend with. It is likely that the copper ore was extracted to depths of 10–12 m, using simple, but effective, techniques. These included fire-setting, which left distinctive profiles on the mine walls as well as charcoal residues in adjacent spoil heaps (Fig. 6.4). The heat-shattered rock face was pounded with stone cobble hammers, both hafted and hand-held. Thousands of broken examples are recorded close to the mine workings. Other tools included the shoulder-blade bones of cattle used as scoops, with a range of wooden equipment likely to have been used.

A miner's work camp was discovered adjacent to the early workings at Ross Island. This location was used for temporary habitation, with evidence of shelters, food consumption, and the use of pottery (see Chapter 8 for details). The foundation traces of several stake-built huts where the miners sheltered were identified. Food waste in the form of cattle and pig bones, and evidence of flint working, attest to other activities in the life of this mining camp. The animal bones indicate an important agricultural base supporting the mine

Fig. 6.2. Aerial view of Ross Island with location (arrow) of mine (*Source*: author).

Fig. 6.3. Excavation of Chalcolithic/earliest Bronze Age copper mine at Ross Island (*Source*: author).

Fig. 6.4. Chalcolithic/earliest Bronze Age mine workings at Ross Island (*Source*: author).

operation, probably located within the environs of Killarney where copper and bronze axes made with Ross Island metal have been found.

The mine camp at Ross Island was mainly used for activities connected to the production of metal. This began with the crushing and hand-sorting of mineralized rock (ore) using stone hammers and anvils, which was then ground to a coarse granular texture using large basin-shaped querns. This ore concentrate was then smelted in shallow pit furnaces fuelled by charcoal, some of which were stone-lined (see Chapter 8 for illustrations). The process was primitive, but effective, with a large amount of copper metal produced over time. This was due to the large amounts of tennantite ore in the mine, which could be easily extracted and beneficiated using these techniques. This fahlore has a high copper (40–45 per cent) and low iron (<5 per cent) content, with the major impurities (arsenic and sulphur) easily volatilized at relatively low temperatures to be removed as gas. The use of high-grade tennantite with this particular chemistry explains the nature of the smelting processes and the absence of iron-rich slags.

Excavation confirms that copper droplets produced in these pit furnaces were re-melted and converted into small slab ingots, one of which was found in the mine area. These ingots were transported from the mine to settlements around Killarney, where the metal was cast into axeheads and blades in workshop settings. The arsenic content of this metal does not indicate deliberate admixture, but rather the use of a copper ore that is naturally rich in arsenic. The raw tennantite typically contained around 20 per cent arsenic, however, this was reduced by oxidative reduction in smelting, ingot production, and object casting down to a level of 1–5 per cent to produce a distinctive impurity pattern (As > Sb > Ag) in the finished metalwork. Axeheads, daggers, and other objects made with this so-called Type A copper were widely exchanged across Ireland in the period 2400–1900 BC, with some products also reaching Britain (Northover 1982). The currency of this metal correlates closely with the radiocarbon chronology of the Ross Island mine, confirming the latter as the main source in Ireland.

The Ross Island mining was probably organized on a seasonal basis by these miner/farmers. The food waste at the mine camp included cattle bone, which indicates contemporary farm settlement in the Killarney area.

Excavation also uncovered some 400 sherds from at least 20 vessels of Beaker pottery. This ceramic can be directly associated with copper production in the mine during the period 2400–2000 BC. These well-made vessels, decorated with horizontal cord and comb impressions, were used as drinking cups by the miners. They were also employed in a washing process to extract prills of copper metal from the furnace pits. The discovery of this pottery is an important connection with the culture group that introduced copper metallurgy to Ireland at the end of the Neolithic.

On the basis of chronology and ore types, it is certain that Ross Island supplied arsenicated metal to make early copper axes in Ireland. The fact that the earliest axe forms (Castletownroche type) are made of the same type of copper places this mine close to the beginnings of Irish metallurgy. The background to this technology lies not in Britain but in mainland Europe, where the production of arsenicated copper from fahlore sources was part of a wider pattern of metal supply during the fourth and third millennia BC (Strahm 1994). The fahlore copper from Ross Island is likely to have been part of a Beaker metallurgical tradition that extended from Iberia and Atlantic France to Ireland in the mid-third millennium BC. The transmission of this knowledge to Ireland must have occurred along exchange networks established by Beaker culture groups in Atlantic Europe. The technological background to Ross Island may be sought in contemporary mining activity in southern France or northern Spain and in the use of arsenical copper in Atlantic Europe (Ambert 2001). The mining of tetrahedrite fahlore at Cabrières provides an obvious source of knowledge, as do the mines of northern Spain. It may be significant that the stone mining hammers from Ross Island find their closest parallels in the Cantabrian mines of El Aramo and El Milagro (e.g. Blas Cortina and Suárez Fernández 2010: fig. 22).

Mount Gabriel

The decline of Ross Island mine *c.*1900–1800 BC was followed over the following five centuries by a new type of copper mining spread across the peninsulas of west Cork. This involved the extraction of low-grade oxidized ore, principally malachite, from small drift mines located on exposures of sedimentary copper-beds. These mines are concentrated in the hilly interior of the west Cork peninsulas (Fig. 6.1). They occur at altitudes between 60–335 m OD, usually within 7 km of the coast. The single largest concentration occurs on the eastern slopes of Mount Gabriel (481 m OD) in the Mizen Peninsula, Co. Cork, where some 32 workings have been identified (Fig. 6.5). The mining on this mountain was undertaken *c.*1700–1400 BC, and is associated with fire-setting and stone hammer technology (O'Brien 1994).

Mount Gabriel is one of the most intact Bronze Age mining landscapes in Europe, with minimal interference from later mining and excellent waterlogged preservation in a blanket bog environment. The Bronze Age workings dispersed across this mountain occur within a Late Devonian sedimentary geology consisting of thick sequences of purple mudrocks and fine-grained sandstones, interbedded with thin grey-green units of coarser sandstones. The copper mineralization occurs within the grey-green strata of this red-bed sequence. They are mostly small inclined openings driven into vertical rock faces where the green sandstone beds are exposed (Fig. 6.6). The latter contain disseminated copper minerals, made visible when the outcrop is stained green by copper carbonate, malachite. The distribution of the individual workings indicates a careful search for these copper-beds, as well as an empirical understanding of the geological factors controlling their exposure. The miners

Fig. 6.5. View of Mount Gabriel. Mizen Peninsula, Co. Cork, with Mine 3 (inset) (*Source*: author).

Fig. 6.6. Entrance to mines 1 and 2, Mount Gabriel
(*Source*: author).

only extracted rock that might contain copper minerals, moving to other exposures once a particular working was exhausted. Some workings were abandoned after less than 1 m, with the largest excavated example worked to a depth of 11 m. This depended on the concentration of secondary copper minerals present and the difficulties of mine drainage.

The walls of these mines mostly have a smooth concave profile, often with traces of smoke staining. This, together with the discovery of large amounts of roundwood fuel and charcoal within and close to the workings, points to the use of fire-setting in rock extraction. In a daily mining cycle, fires were burnt against the mineralized rock face for several hours causing it to micro-fracture, at which point it was pounded with stone hammers. The miner's task was helped by micro-structures with the rock that allowed fragments to be prised out using fingers and wooden sticks. Experiments have shown that up to five centimetres of rock could be removed in this way before the next firing was necessary.

A single 10 m deep mine on Mount Gabriel might have required up to 200 fires over a period of several months, organized on a 24-hour basis. This process would have consumed huge amounts of wood fuel. It is estimated that the extraction of some 4,000 tonnes of rock from 32 recorded mines on the mountain required anywhere from 4,000–14,000 tonnes of roundwood fuel. The type of fuel used is confirmed by the discovery of a large quantity of branches in one of the waterlogged mines, many with axe tooling marks

(O'Brien 1994: plates 37–9). Tree-ring analysis suggests some form of organized collection to meet these enormous fuel requirements. Oak and hazel were mainly used, however species such as alder, ash, birch, pine, and willow were also collected. Examples of wooden equipment include shovels carved from alder, twisted withies of hazel and willow for stone hammer handles, oak planks used as steps inside the mines, as well as splints of resinous pine used in torches (see Chapter 8 for illustrations).

Once rock was extracted, the next task was to separate the copper minerals to prepare an ore concentrate that could be smelted to metal. This process began with the coarse crushing of rock extract using stone hammers and anvil stones, with continuous hand-sorting of visibly mineralized fragments. Evidence for this survives in the form of low mounds of crushed rock spoil near the mine entrances. These deposits contain large amounts of charcoal and broken stone hammers. Used either hand-held or hafted at the mine face or in surface ore beneficiation, these implements had a brief life, being of similar rock type to that which they were breaking.

The Mount Gabriel mines were short-lived operations, probably undertaken on a seasonal basis due to the demands of the agricultural year and the problems posed by poor weather. Flooding was a serious problem as it hindered fire-setting and may have caused the early abandonment of many workings. To minimize this problem, it is likely that individual mines were worked over a short period, with the extraction cycle at different outcrops overlapping to provide for continuous output during the mining season. The deployment of labour was probably non-specialized and organized around the day-time or diurnal fire-setting cycle. Some individuals were engaged in underground rock extraction and surface ore concentration, while others supported the mining effort by collecting fuel from local woodland or hauling stone cobbles from beach sources up to 4 km away. Food supply was probably organized from local farms where the miners lived. These settlements have not been discovered; however, they can be inferred from the presence of ritual monuments and artefact finds in the general vicinity of the mines.

On present evidence, the mining on Mount Gabriel concluded with the preparation of crushed ore concentrates ready for the smelting furnace. No evidence of smelting has been found on the mountain, probably because these processes were non-slagging and any evidence is concealed by blanket peat growth. The amount of copper ore extracted on Mount Gabriel was small, certainly when compared to the Ross Island mine. Current estimates suggest that these mines may have yielded as little as 15–20 kg of metal a year, still enough to make 40–50 bronze axeheads (O'Brien 1994: table 12).

Copper mining ended on Mount Gabriel sometime around 1400 BC, probably when the supply of copper-bed ore from surface exposures was exhausted. Similar mine workings of the same period, occurring either individually or in small clusters, are known from five other locations in the Mizen Peninsula to the west of Mount Gabriel (O'Brien 2003). One of these is on Ballyrisode Hill, Co. Cork, where, in 1854, a hoard of 12 polished stone axes was discovered in a small fire-set working. This mine is radiocarbon dated 1853–1619 BC. A Mount Gabriel-type stone hammer was discovered at old mine workings at nearby Toormore, in an area where an important votive deposit of Early Bronze Age copper was discovered (see Chapter 9).

Similar mines are also recorded at Boulysallagh in the Goleen area at the western end of the Mizen Peninsula, where 4–5 infilled workings and surface spoil containing stone hammers are visible today. This mine is radiocarbon dated 1883–1691 BC. Other Mount Gabriel-type workings in the same area include Callaros Oughter and Carrigacat, dated 1879–1531 BC and 2009–1693 BC respectively (O'Brien 2003).

This type of copper mining also occurred in the Beara Peninsula of south-west Ireland around the same time (Fig. 6.1). These include an example at Tooreen below the summit of Esk Mountain, near Glengarriff, Co. Cork. This consists of a large cavernous opening with two low drivings at the backwall and exterior spoil deposits (Fig. 6.7). Another example has been discovered at Canshanavoe in the mountains north of Adrigole village, Co. Cork. There are at least two areas of copper-bed extraction at this location, one of which contains three fire-set workings and a large spoil deposit with charcoal dated 1600–1430 BC. Other examples in this peninsula include Crumpane on the mountain ridge overlooking Eyeries, worked in the period 1700–1500 BC, and an undated mine at Reentrusk in the Allihies area (O'Brien 2009).

Fig. 6.7. Bronze Age copper mine at Tooreen mine, Beara Peninsula, Co. Cork (*Source*: author).

In conclusion, these Mount Gabriel-type mines represent a different approach to metal sourcing in the Early to Middle Bronze Age, one that involved extensive exploitation of small ore deposits across a landscape. They present a mining strategy that was well adapted to maximizing returns from low-grade mineralization. There is a proliferation of surface workings, each representing a limited investment of time and resources by small groups working on a seasonal or sporadic basis. This is indicative of the social context in which this type of small-scale mining was undertaken, a subject that will be returned to in Chapter 10.

Derrycarhoon

Until recently it was believed that copper mining in south-west Ireland ceased with the decline of the Mount Gabriel-type mines around 1400 BC. Research at Derrycarhoon, Co. Cork, 8 km to the north-east of Mount Gabriel, has identified copper mines worked several centuries later (O'Brien and Hogan 2012; O'Brien 2013). This site lies in a forest clearing on a ridge where green sandstone beds containing secondary copper minerals are exposed in Late Devonian geology. The re-opening of this mine in 1846 led to the discovery of ancient trench mines filled with considerable thicknesses of peat. Ancient mining tools were also uncovered, including stone hammers, a notched tree trunk ladder, and 'pointed sticks'.

The early mine workings at Derrycarhoon consist of parallel rock-cut trenches, up to 30 m in length and averaging 2–3 m in width, which cross the site in a north-east/south-west direction (Fig. 6.8). These consist of either continuous trenches or a series of contiguous narrow workings, driven vertically along the strike exposure of a copper-bed. The workings were partially backfilled as the mining progressed, and subsequently filled by up to 3 m of peat formed by the slow accumulation of vegetation matter in the flooded interiors.

A portion of one of these mines was excavated in 2011. This exposed a near-vertical, rock-cut trench (Mine 5b-1) measuring 2.75 m long by 1 m wide, with a depth of close to 4 m. The upper part of this mine was filled with compact peat to a depth of 2.4 m, which accumulated over the past 3,000 years (Fig. 6.9). This peat overlay stony sediment containing broken stone hammers with fragments of waterlogged wood dated 1114–921 BC. This material was dumped into the abandoned working when adjacent mine trenches were being worked. It overlay a primary mine sediment consisting of finely broken stone, which contained stone hammers and some items of worked wood, including part of a stone hammer handle, a wedge, and two large branches that had been used as crude ladders. A portion of a red deer antler pick, radiocarbon dated 1386–1132 BC, was recovered in the central floor area. A length of twisted

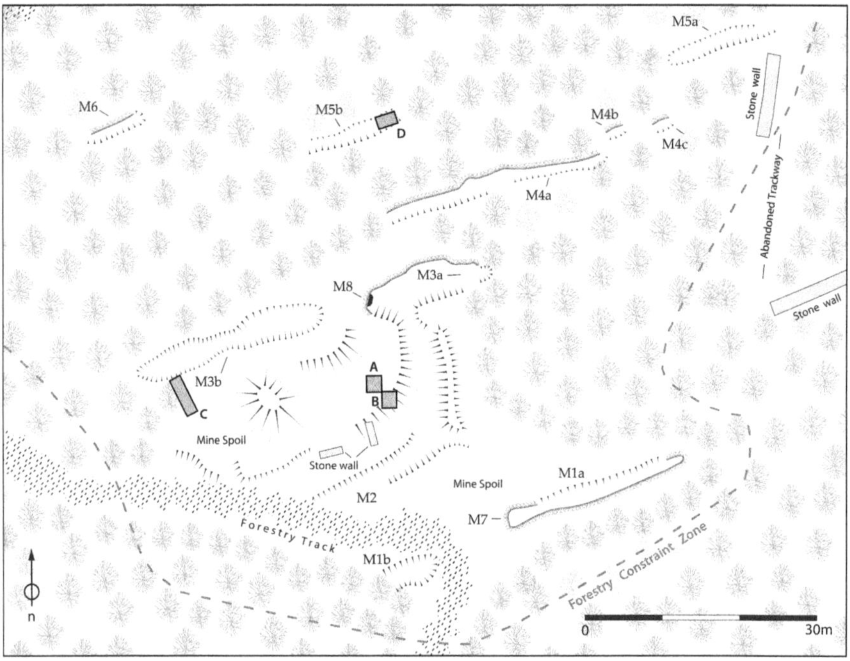

Fig. 6.8. Layout of Derrycarhoon copper mine, Co. Cork, showing Bronze Age trench mines (M1–6) and archaeological excavation (A–D)

(*Source*: author).

hazel withy used as a stone hammer handle, radiocarbon dated 1378–1119 BC, was recovered from an adjacent working. Finally, excavation of surface spoil in the central mine area at Derrycarhoon uncovered numerous stone hammers and wood fragments, with radiocarbon dates of 1300–1100 BC for the latter.

There is no evidence of fire-setting at Derrycarhoon mine, probably because the technique is not suited to the sinking of vertical workings. The miners exploited a closely spaced rock cleavage on a near-vertical bedding direction to remove copper mineralized sandstone using wedges and pick-like tools. A large number of stone hammers are recorded, consisting of well-rounded natural cobbles, 130–213 mm in length and 500–3,500 g in weight, mostly broken through heavy use. These cobbles were sourced in local glacial drift or from beach deposits several kilometres away. The hammers were used hand-held or hafted with twisted hazel withy tied to a wooden handle, examples of which were found in the excavated mine. As with the Mount Gabriel-type mines, little effort was spent on modifying these sandstone cobbles for hafting, as they tended to break easily when striking bedrock of similar lithology. The hammers were also employed in combination with other implements, several

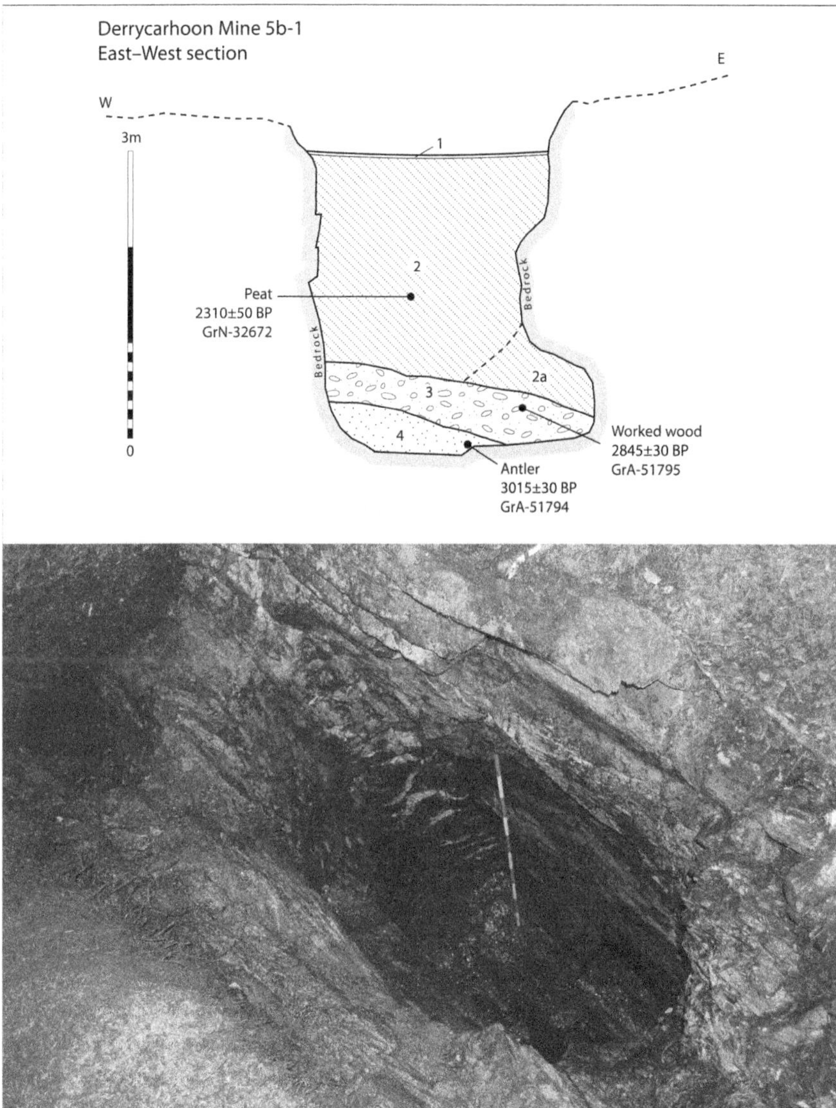

Fig. 6.9. Bronze Age copper mine at Derrycarhoon

(*Source*: author).

of which were found in the mine. These include a short wooden wedge, a number of slender pointed rods of hazel that were used as prise-sticks, and the first evidence for the use of antler picks in the early Irish copper mines.

BRITAIN

> This mine seems to have been wrought in the beginning of times, and before the use of iron was found out, and when mankind knew the use of no tools but stones. I have seen this work open'd and the stone wedges and charcoal taken up with which they split the rocks. Their method seems to be this. They made a great fire of wood in the bottom of their rakes which were always open up on that account, and when the rock was sufficiently hot they cast water upon it, which shiver'd it; and then with stone wedges, which they drove in with other stones, they work'd their way through the hardest rocks, tho' but slowly.
>
> *Lewis Morris 1744 on Twll-y-mwyn mine, Wales; in Timberlake 2003a*

Britain was both a major producer and consumer of metal during the Bronze Age. The earliest use of copper occurred in the Chalcolithic, *c.*2500–2100 BC, when metal was sourced through Beaker culture exchanges with the Continent and with Ireland (Northover 1999; Needham 2002). The mining of local copper deposits began *c.*2100–2000 BC when bronze metallurgy was first developed using the tin resources of south-west England. This led to a period of intensive prospection and mining for copper *c.*2100–1400 BC, which began in mid and north Wales, and spread to the English midlands (Fig. 6.1; Timberlake 2003a, 2009). The supply of copper from these sources declined significantly later in the Bronze Age, when imported metal from the Continent was significant.

The growth of industrial mining in western Britain during the eighteenth and nineteenth centuries led to the discovery of 'ancient' mine workings in several metalliferous districts. Many of these were subsequently destroyed, with occasional reports in the antiquarian and mining literature, where they were variously attributed to the Celts, Romans, or to shadowy Phoenicians. The investigations of Professor Boyd Dawkins (1875) at Alderley Edge near Manchester, and subsequent investigations there by the antiquarians Roeder and Graves (1905), represent the first systematic investigation of an early copper mine in Britain (Fig. 1.5). Those researchers argued that the earliest phase of extraction was Bronze Age, with the use of fire-setting and stone hammers distinguishing these workings from Roman and later mining.

The next landmark came in 1935 when the British Association for the Advancement of Science established a committee to investigate early metal mining in Wales. The secretary was Oliver Davies, then a leading authority on Roman mining in Europe. Davies carried out survey and trial excavation at several copper mines in Wales, notably Parys Mountain, the Great Orme, Cwmystwyth, and Nantyreira, where stone hammers had been discovered (Davies 1937, 1938b, 1939, 1947, 1948).

With no datable artefacts, and without the advantage of radiocarbon dating, he concluded that the use of stone hammers at these sites dated from the Roman or 'Old Celtic' period.

Continued interest in early copper sources in Britain led to further investigation of these sites in the modern era, as well as the discovery of new mines (see Timberlake 2003a for a comprehensive review). This work began in 1976 with exploration at the Great Orme mine on the north Wales coastline, when the first radiocarbon date was obtained for a Bronze Age copper mine in Britain (James 1988). This stimulated further survey and excavation over the next decade, leading to the opening of this mine to the public as a tourism and educational venture.

The investigations at the Great Orme were paralleled by the work of the Early Mines Research Group, established in 1985 by Simon Timberlake and colleagues to investigate early copper mining in Wales (Timberlake 1987, 1988, 1990a, 1990b, 1990c, 1992, 1994). Survey and excavation work at Parys Mountain on Anglesey, north-west Wales, provided conclusive evidence of Early Bronze Age mining. Research was also undertaken in mid-Wales, a metalliferous region with a long mining history, where eight definite and five probable mines of Bronze Age date are known (Timberlake 2009: fig. 7.1). Many of these occur in hill valley and upland settings at altitudes up to 500 m OD. The mineralization mostly comprises surface pockets of chalcopyrite ore within lead-zinc veins that cut across the Ordovician and Silurian geology. The mine workings generally consist of open-cut trenches and opencasts on the exposed mineralization, which can lead into underground tunnels. This involved the use of fire-setting in many instances, as well as stone hammers, bone and antler tools, and wooden equipment of various types. The best known of these mines is located in the mountains east of Aberystwyth in northern Ceredigion (see Crew and Crew 1990, O'Brien 1996, and Timberlake 2003b for a comprehensive review).

Cwmystwyth

This mine is located on Copa Hill in the Ystwyth valley, where mining of lead ore in the early modern era led to the discovery of earlier workings for copper. A large opencast working has been discovered on the brow of this hill (420 m OD) following the outcrop of the copper-rich Comet Lode (Fig. 6.10). This is the only location on the mountain where copper mineralization was accessible to miners in the Bronze Age. These early workings and stone hammers attracted antiquarian attention during the nineteenth century, with the first investigations carried out in 1937 by Oliver Davies. He trenched three of the spoil tips containing stone hammers and concluded that they were connected with Roman lead mining (Davies 1947).

Fig. 6.10. Excavation of Bronze Age copper mine at Cwmystwyth, mid-Wales (*Source*: Simon Timberlake).

Detailed survey and excavation at the site from 1986–99 identified a large Bronze Age mine (Timberlake 1987, 2003b; Timberlake and Switsur 1988). This study revealed that prospection and mining began on Copa Hill *c.*2100–2000 BC, when tin-bronze first began to be used in Britain. The main period of mining was *c.*1950–1750 BC, with the mine eventually abandoned *c.*1600 BC when the accessible ore was exhausted.

The excavation of surface spoil on Copa Hill uncovered broken stone hammers, fire-reddened stone, and fragments of red deer antler. The spoil

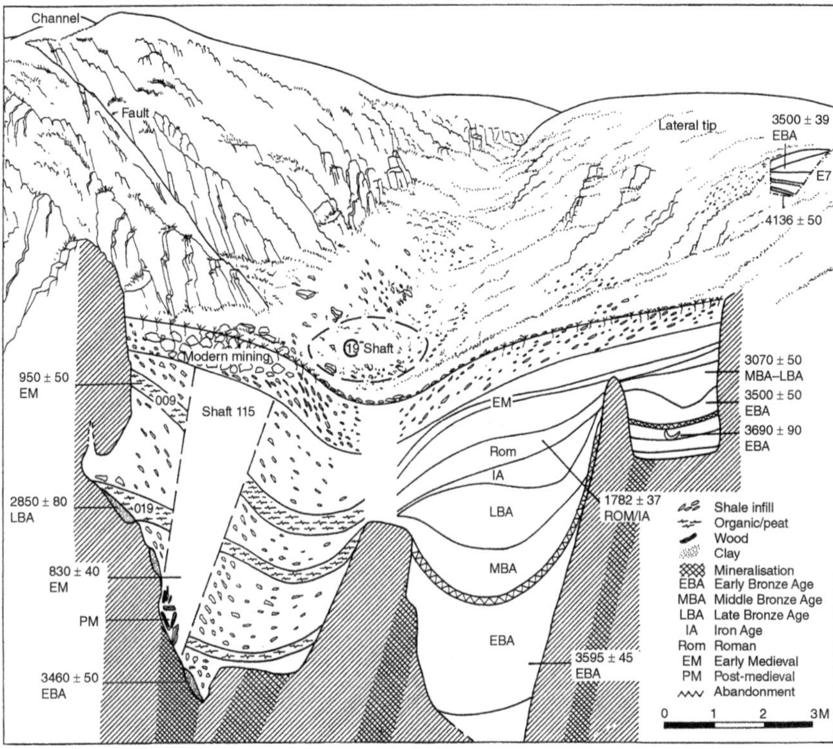

Fig. 6.11. Infill sequence of Bronze Age opencast at Cwmystwyth

(*Source*: Simon Timberlake).

mostly derived from a large opencast, measuring 40 m in length by 5–10 m wide and up to 10 m deep. This large trench is filled to a depth of 10 m by rock spoil and organic deposits (Fig. 6.11). Excavation against the lower north face of the opencast revealed an arched fire-set gallery extending in at least 2 m to follow some of the thinner veins running off the main ore shoot. The wallrock bears the distinctive marks produced when stone hammers were used to pound the fire-weakened rock face.

The Bronze Age miners at Copa Hill extracted a weathered chalcopyrite mineralization that is mixed with lead ore (galena). While it was initially thought these miners were after the chalcopyrite, it is now believed that they were extracting secondary copper minerals in the surface zone of oxidation (Timberlake 2009). The removal of this oxidized mineralization, principally malachite, was thorough and continued down to the limits of the zone of weathering. The scale of this operation is indicated by the size of the resulting opencast, with most of the 3,500 m^3 of rock spoil infilling this opening today believed to be Bronze Age in date.

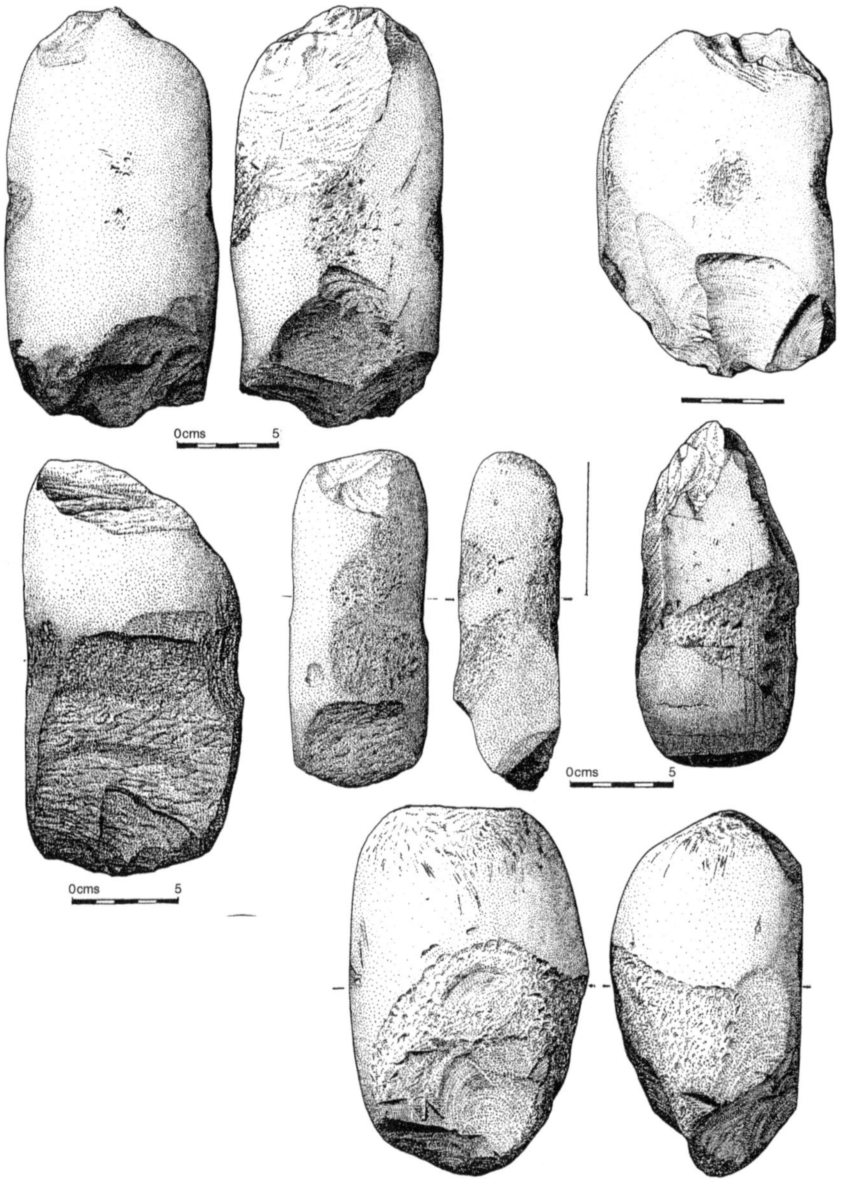

Fig. 6.12. Stone hammers from Cwmystwyth mine
(*Source*: Brenda Craddock).

Excavation uncovered a range of early mining tools at Cwmystwyth, including many items of wood and antler that were preserved in waterlogged conditions (see Chapter 8). These include ropes made of twisted hazel withy, antler picks, basket fragments, mine timbers (stemples), and roundwood fuel used in fire-setting (see Chapter 8 for illustrations). Approximately 1,000 stone hammers were found, the majority broken from heavy use. These implements average 1 kg in weight, with heavier examples up to 7 kg (Fig. 6.12). Most have some trace of hafting, usually in the form of minimal side abrasion or shallow rilling, used to grip handles of twisted hazel withies, examples of which have been found in the mine. The majority of these cobble hammers came from coastal beach sources up to 20 km away. Antler picks were also recovered and the excavator estimates that up to 300 complete red deer antlers may have been used in this mine (Timberlake 2009: 101).

The mining at Cwmystwyth was affected by flooding as early as 1900 BC, due to high rainfall patterns at this altitude and the resulting growth of blanket bog on higher slopes. Excavation carried out at the front of the opencast revealed a sloping, rock-cut channel up to 2 m wide. This feature was created to drain water from the mine, as part of a system that also involved the use of split and hollowed-out wooden logs as launders. A complete example made of alder, approximately 5 m in length, was found in situ in this mine. This remarkable find is one of the earliest examples of mine drainage equipment recorded in Europe. The flow of water may have assisted in the washing of the processed ore, to extract the lighter copper minerals from the lead ore.

Several other Bronze Age copper mines have been investigated in mid-Wales. These include the mine at Nantyreira, located above the 500 m contour on the slopes of Plynlimon, where there is chalcopyrite ore in a lead-silver vein in Upper Ordovician grits. The presence of old open-cut workings associated with stone hammers was recorded in 1858. In 1937 the site was investigated by Oliver Davies who identified an ancient spoil deposit containing stone hammers, charcoal and burnt rock. Davies concluded that these finds were connected to earlier mining dated to the Roman or 'Old Celtic' period (Davies 1938b).

The mine was re-examined by the Early Mines Research Group in 1988, when wallrock traces of fire-setting were identified on the sides of the old open-cut workings. Trial-trenching of the surface spoil revealed broken stone hammers in association with fire-reddened stone and charcoal. The hammers are mostly unmodified cobbles, while a few have side abrasion connected to hafting. Radiocarbon dates for charcoal produced by fire-setting place the early phase of mining at Nantyreira in the Early Bronze Age, 1856–1610 BC. Further evidence of ancient mining in the area comes from the copper mine of Nantyricket, located some 2.5 miles downstream from Nantyreira. An ancient open-cut working has been identified at this location where records of spoil containing stone hammers, charcoal, and burnt rock are also known.

Other copper mines of Early to Middle Bronze Age date have been identified in mid-Wales, all connected to the use of stone hammers (radiocarbon dates after Timberlake 2003a: table 2.1; Timberlake 2009: table 7.1). These include the mines of Tyn-y-fron (2135–1885 BC) in Cwmrheidol, and

Grogwynion in the Ystwyth valley; Nantyrarian (1885–1735 BC) and Twll-y-mwyn (1910–1700 BC) north of the Rheidol river (Timberlake 2006); and Llancynfelin (1745–1645 BC), Pwll Roman, and Erglodd mines (2340–2130 BC) south of the Dovey estuary near Borth Bog. There are historical accounts of stone hammers from mines at Panteidal and Balkan Hill north of the Dovey, with a date of 1890–1630 BC for a mine at Park Lodge (Ogof Wyddon) to the north-east (Timberlake and Mason 1997). In addition to these radiocarbon-dated sites, stone hammers possibly linked to Bronze Age mining are recorded from Llandovery in south Wales (Pickin 1990).

Parys Mountain

This mountain in the north-east corner of the island of Anglesey in North Wales was mined for different metals in the early modern era, and probably also in the Roman period (Fig. 6.1). Mining reached a peak in the late eighteenth century when this was the largest copper mine in the world, with an annual production of some 130,000 tons of copper. These later operations uncovered traces of Bronze Age copper mining near the summit (Sykes 1796).

The geology at Parys Mountain is complex, with several types of mineralization occurring over a 3 km distance associated with volcanic activity in the late Ordovician (480 mya). These include the formation of quartz-chalcopyrite lodes with a high pyrite content, the weathering of which led to iron-rich gossans containing secondary copper minerals. The latter appear to have been the target of the early miners and not the primary chalcopyrite ore. There is evidence now that the copper mineralized veins of the Carreg-y-Doll and North Discovery lodes close to the summit of the mountain were being worked at outcrop as early as 1900 BC (Timberlake 1988, 1990a; Jenkins 1995). Mining may also have been carried out on the nearby coastline, at Pant-y-Gaseg near Bull Bay, west of Amlwch, where stone hammer finds are also recorded.

The discovery of ancient mines at Parys Mountain stems from nineteenth-century records of stone hammers and fire-set drift workings found in this mine. These were initially linked to the many finds of Roman copper ingots from Anglesey. In 1937, Oliver Davies trenched an ancient spoil tip on Parys Mountain that contained stone hammers. In the absence of dateable artefacts, he assigned these to the Roman or 'Old Celtic' period (Davies 1939). This location was re-investigated in 1988 and stone hammers were recovered from an ancient deposit of crushed vein quartz (Timberlake 1990a). Stone hammers have also been found in underground workings. These are small well-rounded cobbles, averaging only 1–2 kg in weight. They mostly derive from local fluvio-glacial drift sources, while some may have come from storm beaches on the north Anglesey coastline. Most were used hand-held, however a small number

Fig. 6.13. Underground workings of Bronze Age date at Parys Mountain, Anglesey (*Source*: David Jenkins).

bear short rills and side abrasion connected to hafting. These implements were found in surface spoil containing charcoal from fire-setting. Radiocarbon dates for this charcoal place the earliest phase of copper mining there in the Early Bronze Age, *c*.2000–1700 BC.

Recent investigations on Parys Mountain identified prehistoric workings at depths of 5–50 m below surface where they intersect with nineteenth-century levels (Fig. 6.13; Jenkins 1995). These may take the form of steeply inclined opencasts that commenced on the surface weathered exposure of a quartz vein. Unfortunately, there are few surface indications of early mining as the entire area is covered today by huge amounts of rock spoil produced in the early modern mine.

Great Orme

This prominent headland of Carboniferous limestone on the north Wales coastline was one of the largest copper mines in Bronze Age Europe, worked more or less continuously from *c.*1800–600 BC (Lewis 1998). There is a long history of mining at this location, beginning in the Bronze Age and continuing with large-scale industrial operations in the nineteenth century. The latter resulted in the discovery of ancient copper mines, initially in 1831, and again in 1849 when miners broke into a large cavern at a depth of approximately 18 m. This contained stone hammers, antler picks, bones, the remains of fires, a fragment of bronze metal, as well as calcite stalactites that hung like tree branches from the ceiling (Hicklen 1863 in Lewis 1994). The primitive nature of these tools led to speculation they were pre-Roman in date, possibly of 'Celtic' origin. Oliver Davies visited the Great Orme in the 1940s during a survey of early copper mines in Wales. He suggested the workings there might be Roman, based on the discovery of a settlement of that period in the vicinity of the mine (Davies 1948). This was not convincing in the absence of Roman pottery and other material culture from the mine itself.

The modern period of investigation began in 1976 when a mining researcher, Duncan James, explored the old workings under Bryniau Poethion. He identified primitive tunnel workings filled with ancient rock debris, which contained stone hammers and bone tools. A charcoal sample from this sediment radiocarbon dated 1410–922 BC provided the first scientific evidence of Bronze Age mining in Britain (James 1988). Detailed underground exploration began in 1987–8, led by the Great Orme Exploration Society who discovered a major complex of Bronze Age mine tunnels in the Pyllau Valley (Lewis 1990, 1994, 1996, 1998). A programme of surface excavation carried out by Gwynedd Archaeological Trust in 1989 around the Vivian's Shaft area provided further radiocarbon dates and information on this important Bronze Age mine (Dutton and Fasham 1994).

The mineralization at the Great Orme consists of a series of sub-parallel veins containing chalcopyrite that cross the mine area in a north–south direction, with major workings located in the Bryniau Poethion and Pyllau Valley areas (Fig. 1.3). The country rock around these copper lodes has been strongly dolomitized, with subsequent alteration, leaving it very soft and easily worked using the most primitive tools. Stone hammers and bone gouges were used to tunnel the softer rock, with the use of fire-setting confined to harder rock. The Bronze Age miners were primarily interested in the oxidized portion of this orebody, where the copper carbonate minerals, azurite and malachite, occur in

Fig. 6.14. Bronze Age trench workings at surface of Great Orme mine, north Wales (*Source*: author).

some abundance. There are also indications that the early miners extracted the oxidized chalcopyrite minerals, partly altered to goethite and malachite (Lewis personal communication).

Mining in the Pyllau Valley began in the Bronze Age with the discovery of mineralization along the numerous scarp exposures of bedded limestone rock. A colourful display of oxidized copper minerals on the exposed vertical veins and dolomitized limestone in the Vivian's Shaft area provided a focus for the initial mining efforts. This began with the hand-picking of malachite ore at surface, leading to the creation of narrow trench workings down to depths of 15 m (Fig. 6.14). These trenches are worked-out mineral channels, from which the ore was removed with great thoroughness using stone hammers with bone picks and scrapers.

Over time this mining progressed to the development of a large opencast in the area north of Vivian's Shaft. This is one of the largest surface copper mines known from the Bronze Age, with an estimated 28,400 tonnes of rock removed. The opencast and adjacent trench workings were taken down to a weakly mineralized level, below which it became necessary to develop a true underground mining system. Openings at the base and sides of the opencast led to a maze of underground tunnels, worked with stone hammers and bone tools (Fig. 6.15). The shape of these underground tunnels was mostly determined by geological controls on the removal of oxidized copper mineralization from the rotted dolomitized limestone and ore veins. The presence of

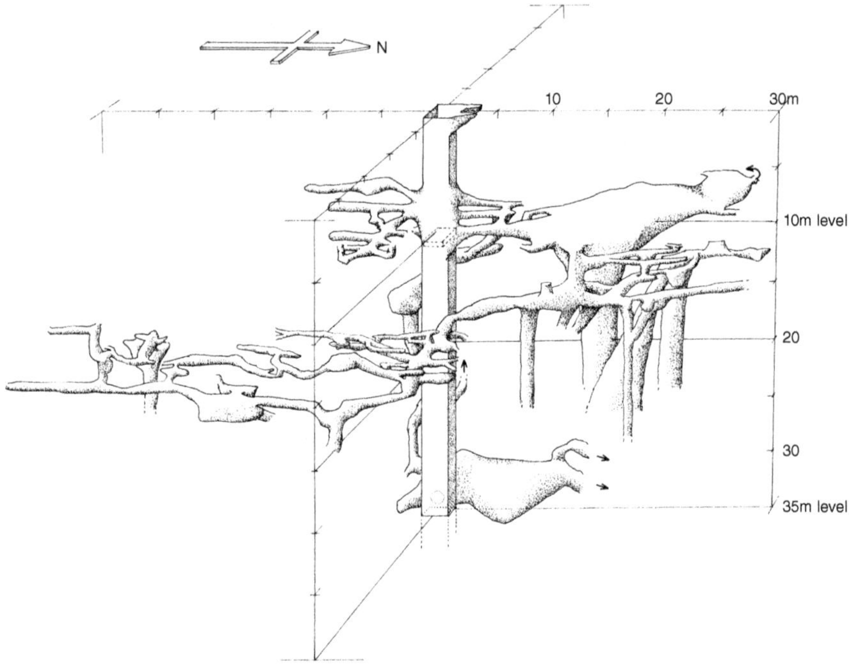

Fig. 6.15. Three-dimensional mapping of Bronze Age mine tunnels at the Great Orme (*Source*: Andy Lewis).

Fig. 6.16. Exploring the Bronze Age workings at the Great Orme mine (*Source*: Andy Lewis).

natural caves and solution hollows in this karst environment enabled the Bronze Age miners to work without any serious threat of flooding. A feature of this underground mine is the small size of the mine tunnels (Fig. 6.16), with some measuring as little as 0.3 m wide by 0.7 m high. Many examples, as well as the surface workings, were systematically backfilled with rock waste as the mining progressed. Some trench workings were deliberately capped with stone blocks, allowing access beneath while the overlying opening was a space to store waste rock.

The discovery of charcoal at depths of up to 70 m indicates some fire-setting, with numerous surface connections providing for ventilation. As stated above, the technique seems to have been confined to hard rock for which stone hammers were also required. Nearly 3,000 stone hammers and fragments have been found in the surface and underground workings, a relatively small number considering the scale and duration of mining. This reflects the long use-life and restricted use of these implements in this geology. The hammer stones are mostly rounded cobbles obtained from glacial drift and local beaches. They range 0.5–13 kg in weight, with some boulder-size examples as heavy as 20 kg. Few hammers were modified for hafting, with most used hand-held in the confined underground workings.

The soft rotted texture of the dolomitized limestone in this mine allowed the effective use of bone tools. A recent study has recorded approximately 16,000 animal bone fragments found in surface and underground workings (James 2011). They mostly derived from cattle, pig, and sheep with a small amount of deer antler. Some of these bones represent food waste; however, many were trimmed for use as points or picks, designed to scoop out copper minerals from the soft rock. The use of these bone points left distinctive marks on the mine walls, as did the pounding action of stone hammers. Many of the bone items are stained green as a result of prolonged contact with copper-rich groundwater.

The scale of copper extraction on the Great Orme is impressive in comparison to other Bronze Age mines in Britain. While the true extent of this complex site remains to be assessed, an estimated 6.5 km of underground tunnels were worked to a maximum depth of 70 m. Total rock extraction in the Bronze Age is estimated to be in excess of 40,000 cubic metres of rock. This partly related to the soft geology of the dolomitized rock, which allowed the Bronze Age miners to tunnel in comparative safety into hard limestone. The extent of these workings relates to the long chronology of mining, with radiocarbon dates indicating that mining commenced around 1900–1800 BC and continued to possibly as late as 600 BC. Much of the mining was concentrated in the Early to Middle Bronze Age, *c.*1700–1500 BC, with the later activity confined to the deeper underground workings.

At least four ore washing sites adjacent to natural springs have been identified on the Great Orme. Initially considered to be of Roman origin, recent research has confirmed Bronze Age activity at those locations. Excavation at two sites revealed dolomitic sands and gravels with traces of copper carbonate minerals, similar to spoil deposits in the Pyllau Valley mine. These ore washing sediments contained stone hammer fragments and copper-stained bones, with a Bronze Age date confirmed by a radiocarbon result of 1877–1496 BC from the site at Ffynnon Rufeinig (Wager 2002).

The Great Orme can also claim the only known Bronze Age smelting site in Britain. The site at Pen Trwyn has produced small bowl furnaces associated with fayalitic slag containing copper prills. One of the furnaces is dated 1675–1500 BC (Chapman 1997). Initial investigations suggest these tiny furnaces may have been for assay purposes only, leaving open the question as to where the bulk of the Great Orme copper ore was smelted during the Bronze Age.

In conclusion, the evidence from the Great Orme points to a steady exploitation of this carbonate orebody over the course of a millennium or so during the Bronze Age. The miners adapted their primitive technology to extract an enriched ore source in a highly organized fashion, within the controls imposed by the carbonate host rock environment. While the scale and duration of this operation partly reflects the unusual geological circumstances, this mine was certainly a major source of metal in the Bronze Age. The copper produced was of high purity, making it difficult to follow its circulation within the wider pool of metal in that period.

In 1990 the Great Orme mine was developed as a tourist attraction, where the public can visit a portion of the extensive surface and underground workings dating to the Bronze Age.

England

Bronze Age workings have been identified at two historic copper mines in the English midlands, namely Alderley Edge near Manchester and at Ecton in the Peak District of Staffordshire.

Alderley Edge

This prominent rock scarp is located 12 miles south of Manchester in the north-east part of the Cheshire Basin (Fig. 6.1). Rising to some 120 m above the surrounding plain, the Edge is formed by bedded sedimentary rocks of Triassic age impregnated with copper and other minerals. These metallic ores occur as stratabound mineral disseminations confined to porous sandstones and conglomerates. They are also concentrated along several of the fault zones that cross Alderley Edge and in joint fissures. Malachite is the most abundant copper mineral, coating sand and pebble grains within the coarser sedimentary rocks, where it also occurs as thin veinlets. The Bronze Age mining targeted these soft malachite-rich sedimentary beds, which were prominently exposed at various locations along Alderley Edge.

Alderley Edge and the nearby Mottram St Andrew mine have a long history of mining, the early stages of which are linked to Bronze Age and Roman operations, with industrial mining from the late seventeenth to the early twentieth centuries. The discovery of ancient copper mines on Alderley Edge dates to large-scale mining in the period 1855–78. This uncovered primitive mine pits at various locations, most notably at Brynlow and along the Engine Vein. The discovery of stone mining hammers and other primitive tools in these workings attracted antiquarian interest. In 1874, William Boyd Dawkins, then Professor of Geology at Manchester University, visited the site and recorded primitive workings and tools (Boyd Dawkins 1875). Records made by a Dr Sainter (1878) reveal that several ancient workings were destroyed in that period. The mine was again surveyed in 1901–05 by the antiquarians Roeder and Graves (1905), who recorded primitive mine pits along the Engine Vein they regarded as worked with stone hammers (Fig. 1.5). By the time mining finally ended in 1919, many of these ancient surface workings and spoil deposits had been destroyed.

The discoveries of stone hammers and primitive mine pits during the nineteenth century confirmed an early phase of copper mining at Alderley Edge. The precise age of these workings was unknown; however, a breakthrough was made in 1992 when an oak shovel recovered from the mine by Dr Sainter was re-discovered and submitted for radiocarbon dating. The resulting date range of 1888–1677 BC places the earliest copper mining on Alderley Edge within the Bronze Age. This was confirmed by investigations in the modern era (Gale D. 1990), which include the discovery of an undisturbed pit working of Early Bronze Age date at the Engine Vein. That mine was worked by fire-setting *c.*1900 BC, after which it gradually infilled before being used *c.*1700 BC as a location for the processing of copper ore (Timberlake and King in Timberlake and Prag 2005).

Bronze Age copper mining at Alderley Edge took the form of surface pitting on mineralized exposures along outcropping veins. The workings at Engine Vein comprise a series of small benched pits connected in places by narrow open-cuts (Fig. 6.17). These were worked to depths of up to 5 m or possibly more. The miners targeted rich pockets of secondary copper minerals (malachite, azurite, and chrysocolla) formed in a zone of supergene enrichment. The action of migrating groundwater created rich pockets of secondary copper minerals along the bedding planes of the sandstone layers close to mineralizing faults, as well as nodules of malachite and azurite within the intervening mudstone horizons (Timberlake 2010: 292). The miners extracted this ore by pounding the soft sandstone and conglomeratic beds with stone hammers. Other primitive mining tools such as antler picks and wooden shovels would

Fig. 6.17. Bronze Age and later trench mining along the Engine Vein, Alderley Edge (*Source*: author).

also have been very effective; however, it is not certain whether fire-setting was ever applied to the extraction of this soft sedimentary rock.

Ecton

Bronze Age copper mines have been identified on Ecton Hill, Staffordshire, a limestone ridge on the east side of the Manifold Valley in the western part of the Peak District in central England (Fig. 6.1). These were discovered in 1997 when an antler pick dated to the Bronze Age was found during the exploration of a cave at the Stone Quarry (Dutchman) mine (Barnatt and Thomas 1998). Subsequent fieldwork led to the discovery of stone hammers in surface spoil deposits at two locations along this ridge. Excavations conducted in 2008–9 at Stone Quarry mine and The Lumb close to the summit of the hill confirmed the existence of early mining (Timberlake 2010: 292). The Stone Quarry mine was reworked in the historic period, however there are some extant spoil deposits containing stone hammers and bone mining tools. The Bronze Age workings on The Lumb were also reworked, but some examples still survive.

The country rock at Ecton consists of a series of tightly folded and steeply dipping beds of limestone with pockets of rich copper mineralization. The surface rocks are highly weathered, with groundwater action leading to the oxidation of the primary chalcopyrite ore to copper carbonate minerals.

Fig. 6.18. Drift extraction of bedded copper mineralization at The Lumb, Ecton mine, Staffordshire

(*Source*: Simon Timberlake).

The extraction of supergene sulphides such as chalcocite and bornite is possible, but it is unlikely that the miners targeted the primary chalcopyrite mineralization. Much of this secondary mineralization is formed in voids within the karst limestone. The ore in The Lumb was formed by dolomitization of a limestone bed, which left pockets of soft iron hydroxide mineral veined with malachite.

As with Alderley Edge and the Great Orme, the miners at Ecton took advantage of solution features in the karst limestone. In the case of The Lumb, they followed a mineralized bed uphill, extracting pockets of malachite from a series of inclined openings, each up to 2 m in depth (Fig. 6.18). The roofs of these small drifts tended to collapse as the mining extended inwards. Rock was extracted using bone and antler tools, as well as hand-held stone hammers sourced from local streams. The former include antler picks and short lengths of cattle long bone made into gouges and points. These were used to scrap malachite from the rotted dolomite and other host rocks, a process that left distinctive marks on some of the early workings in The Lumb. The mineralized rock was then processed by crushing and hand-sorting using stone hammers and anvils, on working floors directly outside the mine.

Radiocarbon dating of worked antler and animal bone confirm that the mine at Ecton was worked *c.*1800–1700 BC. The dating evidence from other British sites reveals a pattern whereby copper mining may have commenced in mid-Wales *c.*2100–2000 BC, at locations such as Cwmystwyth, spreading to the north Wales coastline and the mines of Parys Mountain and the Great Orme soon after 2000 BC. From there copper mining may have spread into north-west central England, reaching Alderley Edge by 1900 BC and Ecton a century or so later (Timberlake and Marshall 2013).

Bronze Age mining in other parts of Britain

There is very little evidence of early copper mining known at present from northern Britain. The absence of early copper mines in Scotland is surprising, given the strong evidence of Bronze Age metalworking in many metalliferous areas. Stone hammers are reported from Bradda Head (Pickin and Worthington 1989) and Langness (Doonan and Eley 2000) on the Isle of Man, however, no early mine workings have been identified.

Also notable is the lack of evidence for Bronze Age copper mining in south-west England, one of the most metalliferous parts of Britain. The copper and tin lodes of Cornwall and Devon have a long history of mining stretching back to Roman and Medieval times. The tin resources of Cornwall are well known, while the copper mining industry in that region was one of the biggest producers in Europe in the nineteenth century. Though there is good circumstantial evidence for alluvial tin extraction in Cornwall during the Bronze Age (Penhallurick 1986), there is no evidence of contemporary copper mining. This partly reflects a lack of fieldwork, and the impact of later mining, though it must be said that, unlike south-west Ireland and Wales, documentary sources from the early modern industry in Cornwall are notably silent on early mine discoveries. The destruction caused by that industrial mining, and the difficulty of dating primitive technology used into medieval times, makes this a challenging area for the fieldworker (for further discussion see papers in Budd and Gale 1997).

The strongest indication of early copper mining in Cornwall comes from lead isotope analysis of five copper objects, forming what Rohl and Needham (1998) termed the 'IMP-LI 2' group of early metal in Britain. These objects have a lead isotope signature and chemistry that is believed to be unique to mineralization in Cornwall. This makes a case for the sourcing of arsenic-nickel (Bell Beaker) and arsenic-only copper in that region (Budd et al. 1999). Bray and Pollard (2012: 60) cite the presence of cobalt in Early Bronze Age metalwork as a possible indication of continued mining of copper in Cornwall into the Early Bronze Age. Both of these theories remain to be substantiated for a region where very little Chalcolithic metalwork has been found (cf. Pearce 1983).

While early mines may eventually be discovered in Cornwall, there are several reasons why this may not have been a significant producer of copper during prehistoric period. In terms of Chalcolithic mining, one factor may have been the strong Cornish tradition of stone axe production, which continued over a long period of the Neolithic, culminating in the use of the Groups 1/1a and 3/3a sources. A comparison may be made with the expansion of flint mining at Grimes Graves during the mid-to-late third millennium BC, at a time when metal was beginning to circulate across southern England. In both cases this may have been a reaction to the shifts in material value posed by the spreading use of metal, or a re-assertion of traditional social values in the face of a threatening innovation (O'Brien 2010). A more compelling reason is that by 2000 BC early metallurgists in Cornwall were in control of another important metal resource, namely tin. With abundant supplies of a metal that was in great demand across Europe, there may have been little incentive to engage in the more laborious processes of copper mining and smelting. Cornish metalworkers could have obtained supplies of copper through a trade in tin conducted with other parts of Europe, including the mines at Ross Island in Ireland.

SCANDINAVIA

While Britain and Ireland emerged as major producers of copper in the Bronze Age, the situation is somewhat different in respect of another part of Europe also rich in copper resources, namely Scandinavia. Metal was already known there by 4000 BC, when copper objects, mainly axeheads, circulated among late hunter-gatherer groups such as the Ertebolle in Denmark and early farming societies such as the TRB (*Tricherrandbecher*) culture (Magnusson Staaf 2002). The copper in question is believed to have come initially from Balkan sources, and later from the east Alpine mines. Copper was also used to a limited extent at that time in northern Scandinavia (Fennoscandia) through contacts established with the Eurasian metallurgical zone, probably through Russian Karelia.

This early use of copper in Scandinavia during the Mesolithic/ Neolithic transition was short-lived, with little evidence for metallurgical processes prior to the Bronze Age. Direct or indirect evidence for mining of local copper ores is also lacking. The treatment of copper ore at the Comb Ware site of Lillberget in Norrbotten, northern Sweden (Halén 1994), cannot be confirmed. Klassen's (2000) suggestion that the Riesebusch copper of Denmark/ northern Germany was produced from sources in central Sweden was subsequently disproved by lead isotope analyses that point to an east Alpine source (Klassen and Stürup 2001).

Much of the early copper circulating in southern Scandinavia during the fourth millennium BC had distinctive arsenic content. A potential source has been identified on the island of Helgoland in the North Sea, located off the coast of Schleswig-Holstein. There are large surface deposits of oxidized copper ore on the island, which were mined in the later medieval period. The possibility of earlier mining was proposed by Werner Lorenzen (1965), based on the accessibility of the ore and the evidence of Neolithic flint extraction on an island where numerous Bronze Age finds are recorded. However, no early copper workings are known and it is not possible to confirm that Helgoland was a source of copper for Chalcolithic or Bronze Age groups in Denmark or northern Germany.

The use of copper in Scandinavia declined significantly in the Middle Neolithic, but was re-established in the late third millennium BC. The circulation of copper gradually increased, accompanied by evidence of local metalworking activity in the closing stages of the Neolithic and beginnings of the Bronze Age. This culminated in the emergence in Denmark and southern Sweden by 1600 BC of a distinctive Nordic style of bronze working based on imported metal supplies (Vandkilde 1996). It included large-scale production of swords and the technology to make complex objects such as lurers and shields. This metallurgy gradually spread into northern Scandinavia in that period, including parts of Norway and Sweden with natural sources of copper.

Bronze Age mining?

It is a long-held view that all of the metal used in the Nordic lands during this period was imported from sources in mainland Europe. This was first expressed by the Swedish archaeologist, Oscar Montelius, to explain the large-scale consumption of bronze in Denmark, a region with no natural sources of copper or tin (Montelius 1872, 1888). There has been a general acceptance of this position (Oldeberg 1960; Thrane 1975; Vandkilde 1996; Kristiansen and Larsson 2005), which over time extended to Fennoscandia where metallurgy was viewed as less developed than in Denmark. The assumption that there are no exploitable sources of copper in Finland, Norway, and Sweden explained, for many, the absence of early copper mines. More recently, a number of researchers have argued for local production of copper in those regions (Janzon 1984; Prescott 2006; Melheim 2012), based on the availability of copper ores and metallurgical expertise. Despite these assertions, no Bronze Age copper mines have been discovered.

In terms of resources, there are numerous copper deposits of varying geological type in Fennoscandia. These include the ore deposits of the Caledonide geology that extends from Norway into northern Sweden, as well as the mining belt of the Fennoscandian Shield from eastern Sweden into Finland. Many of the large copper deposits were mined in the historic era, notably the famous Falun mines of central Sweden and the Kongsberg mine in southern Norway. The latter is notable for the continued use of fire-setting into the nineteenth century, admittedly in circumstances where wood supplies were plentiful.

Many of these orebodies are stratabound sulphide deposits dominated by chalcopyrite ore. Certain environmental factors affect the exposure and geochemistry of this mineralization. Intense glaciation during the Pleistocene means that few orebodies have well-developed oxidation or gossan zones with supergene enrichment of secondary copper minerals. Many of the Swedish copper deposits are covered by extensive deposits of glacial drift, whereas their Norwegian counterparts tend to be better exposed in mountain terrain (Melheim 2012: 324).

It should be noted that many parts of Norway and Sweden with copper ore deposits had a tradition of hard rock extraction during the Neolithic. The search for raw materials may have led to the discovery of copper ores, while expertise in rock quarrying may have facilitated their extraction. However, this was not an inevitable development, as many of the processes in copper mining and smelting are fundamentally different from those in lithic production.

Sweden

The Swedish researcher, Gunborg Janzon, was one of the first to consider the possibility of prehistoric copper mining in Sweden (Janzon 1984, 1988). This was based on an analysis of stone hammer finds in the copper-rich landscape of Norrland in northern Sweden. She was influenced by the discovery of similar implements at the early metalworking settlement of Hallunda in central Sweden, and at the site of Hellerö in Kastervik, Småland, in proximity to a copper lode (Janzon and Noréus 1996; Noréus 2001). The feasibility of processing Swedish copper ores using Bronze Age methods has been demonstrated by experimentation (Bengtsson 1986), however evidence of primary smelting remains elusive.

A recent programme of scientific analysis has affirmed the importance of imported copper supply for the Bronze Age in Sweden (Ling et al. 2013). Lead isotope and chemical analyses were carried out on a selection of bronze items, dated 1600–700 BC, from the copper-bearing districts of Dalsland, Småland, and Värmland in southern Sweden. The results did not match the local ore signatures, or those of other Swedish ores, suggesting that the copper must have been imported from elsewhere. This conclusion is consistent with previous lead isotope analyses of copper and bronze artefacts from Sweden (Klassen and Stürup 2001; Kresten 2005; Schwab et al. 2010). The project is continuing to examine the possibility of local extraction of copper in Sweden, but is also attempting to identify the sources of the metal abroad. Preliminary results indicate the importance of east Alpine sources, such as the mines of North Tyrol in Austria, with a suggestion of supply from as far away as Iberia and Sardinia.

Norway

The possibility of Bronze Age mining in southern Norway was raised by Prescott (2006) arising from the discovery of metalworking in a rock shelter at Skrivarhelleren in the Sogn highlands, an area with known deposits of copper ore. A recent study considered various strands of evidence, including the composition of Bronze Age metalwork in relation to copper deposits in Norway and the significance of stone hammer finds in terms of mining (Melheim 2012). Caution was expressed that compositional patterns of metalwork are not conclusive in relation to mine sources, even if there has been an assumption of supply from Central Europe. Some of this metal could have come from pyritic and fahlore deposits in Norway, even if there is no site evidence of an ability to smelt such ores. This presents a technological barrier to the use of such ore deposits in Norway. It does not rule out the extraction of oxidized mineralization, but again the field evidence is absent.

An interpretation of grooved stone hammers as indicators of early copper mining is central to the Melheim thesis. The problem is that such simple implements can only be linked to metal production when found in the context of a mining or ore beneficiation site. This association has never been proven in Norway, or indeed in Sweden. There may be a general correspondence between distributions of stray stone hammer finds and ore deposits at a regional level, however this is not convincing at a local scale in Norway. The find locations and modified form of the Norwegian hammers does not support a connection with copper mining, even if some examples could have been used in metal fabrication. Where used elsewhere in copper mining, stone hammers tend to occur in large numbers with high levels of use-breakage. It is significant that no examples are recorded in antiquarian or mining records for any copper mine in Norway.

In conclusion, an assessment of the available evidence does not support a connection between stone hammers and copper mining in Norway, while the same is likely for Sweden. The possibility of Bronze Age copper production in Finland based on the use of local ores has also been raised (Huurre 1981; Lehtinen 1996). As with Norway and Sweden, this cannot be confirmed as no mining or smelting sites have been identified. The situation in Fennoscandia illustrates how the presence of copper deposits in a region does not make the exploitation of that mineralization inevitable, even for local groups who were engaged in metalworking. There are other examples of metalliferous regions that may not have developed copper mining to any extent during the Bronze Age, Cornwall and Brittany being good examples. The eventual discovery of a

Bronze Age copper mine in Norway or Sweden will not lessen the significance of imported metal supplies. Trade was an important strategy for metal supply even in the most metalliferous regions. Ireland is a good example, a region with abundant copper mineralization, some of which was extensively mined from the Chalcolithic to the Middle Bronze Age, whereas the bulk of copper/bronze supply in the Late Bronze Age most likely came from the Continent.

This serves to illustrate how the development of an indigenous mining tradition was historically contingent and could be influenced by many environmental and cultural factors. Fennoscandia may be one of those parts of Europe where early copper mining did not develop for such reasons. A reliance on imported metal may be another expression of the strong maritime tradition so apparent in Bronze Age art in Scandinavia. These trade connections may have involved a complex exchange of copper and tin, bronze and gold, Baltic amber and other goods. Some of this metal may have come from Atlantic Europe; however, another important source was the eastern Alps, which emerged as a major centre of copper mining during the second millennium BC.

7

Central and eastern Europe

This survey of prehistoric copper mines in Europe began with the oldest known examples, namely Rudna Glava and Ai Bunar in the Balkans. It is now time to consider some of the largest Bronze Age mines, which were major producers of copper that influenced its supply across large parts of the continent. Much of the focus is on Austria, where the earliest scientific investigations of early copper mines were undertaken in Europe.

THE FIRST METAL WORKERS

The earliest use of copper in central Europe can be linked to a Late Neolithic culture called the Münchshöfen group, best known in south-eastern Bavaria. A small number of copper objects can be associated with this culture group, including axe-hammers and flat axes, awls, beads, and rings. Scientific analysis of these objects reveals that they probably originated in the Balkans, as part of a spread of metal use into central Europe from that area during the second half of the fifth millennium BC (Höppner et al. 2005). This is supported by the material culture of the Münchshöfen group, in particular the ceramic evidence, which finds close typological parallels with metal-using groups in the Carpathian Basin.

It is likely that the same spread of copper use into Austria and southern Germany eventually led to the first attempts to exploit the copper resources of the Alpine region. The evidence comes from the hill-top settlement of Mariahilfbergl near Brixlegg in the Inn Valley of North Tyrol, Austria. Excavation uncovered traces of metallurgical processes in the form of a fireplace with fragments of copper slag, two clay nozzles, and two items of copper metal (Bartelheim et al. 2002, 2003). Radiocarbon analysis indicates a 4500–3640 BC date range, however, the wider cultural context of the site may place these discoveries in the later fifth millennium BC. It is not certain whether smelting took place in this site, though some of the slag-like material suggests the heat treatment of a type of fahlore (tetrahedrite) that is common in the Brixlegg area. Interestingly, chemical and lead isotope analyses of a copper bead and copper strip from the same site context revealed a different chemical composition from that of the slag, one that matches with copper metalwork from Bulgaria and Serbia (Pernicka et al. 1997). These analyses indicate that a possible source of this metal may have been the Majdanpek orebody in Serbia. Though unproven, the suggestion that this mine was first worked in the later fifth millennium BC is not surprising given its proximity to Rudna Glava.

The metallurgical evidence from Mariahilfbergl may be interpreted as experimental smelting of local fahlore by early farming groups in the Inn Valley, who were acquainted with copper through contacts with the Balkan area. It is not known whether this early use of Tyrolean fahlore was more widespread across the Münchshöfen culture group as no mine workings of this period have been identified. What is known is that this earliest metal horizon was followed by a wider use of copper across the Alpine region by the Altheim, Mondsee, and Pfyn cultures of the fourth millennium BC. This is supported by excavations at the Götschenberg hill-top settlement in Salzburg where copper metallurgy from that period is recorded. As with the Brixlegg evidence, no mines of this period have been identified (Goldenberg 1998).

Much of this first copper in central Europe is of the so-called *Singen* type, containing arsenic, antimony, silver, and nickel, and the related *ösenringkupfer* (similar impurities, but free of nickel). The appearance of these metal compositions marks a significant shift in metal supply away from oxidized ores to the processing of fahlores (tennantite-tetrahedrite group minerals). This metallurgical tradition was very strong in the early stages of the Bronze Age, *c.*2300–2000 BC, when tin bronze was initially rare (Kienlin and Stöllner 2009). The discovery of thousands of *ösenringe* ingots in central Europe reveals an extensive use of fahlores from Alpine sources, with mines known in the lower Inn Valley and at other locations in Austria. Other sources may include the Harz and Erzgebirge mountains of central-west Germany and Slovakia, however, no evidence of mining is known in those lower mountain ranges.

The use of *ösenringe* copper *c.*1800–1600 BC overlapped with a new type of copper in central Europe towards the end of the Early Bronze Age. This had a higher purity and was probably smelted from copper-iron sulphide (chalcopyrite) ore. The main source was in the Mitterberg Pongau region in the Austrian Alps, south of Salzburg, where intensive copper production based on chalcopyrite ore is recorded from early in the second millennium BC. There is abundant evidence of Bronze Age copper production in this region and in other parts of the Austrian Alps, in the form of mining sites or else inferred from the presence of smelting sites that used local ores.

THE AUSTRIAN MINES

The importance of the east Alpine copper mines was highlighted by an early study of Bronze Age copper ingots in central Europe (Reinecke 1930). The distribution of these ingots revealed a pattern of large-scale copper supply following major river systems into southern Germany, Austria, and the Czech Republic. This copper probably came from different sources; however, the most important supply emanated from mines in the eastern Alps.

Bronze Age copper mines are recorded at numerous locations in mountainous Alpine terrain of Austria (Fig. 7.1). The focus of this mining was an east–west belt of Palaeozoic sedimentary and low-grade metamorphic rocks crossing the Austrian Alps for some 330 km from the Vienna basin to the Inn Valley. Known as the Greywacke Zone, this geology comprises turbidites, greywackes, schists, and limestones of Ordovician to Devonian Age, and metamorphic rocks of Cambrian/Ordovician age (Lutz et al. 2010). There are rich copper deposits in several regions, mostly vein-style mineralization dominated by chalcopyrite and/or fahlore, with low amounts of secondary minerals.

The largest Bronze Age copper mines are in the inner Salzach valley in north-central Austria. These occur in a mountainous area known as the Mitterberg near Bischofshofen, and also in the St Veit area to the immediate

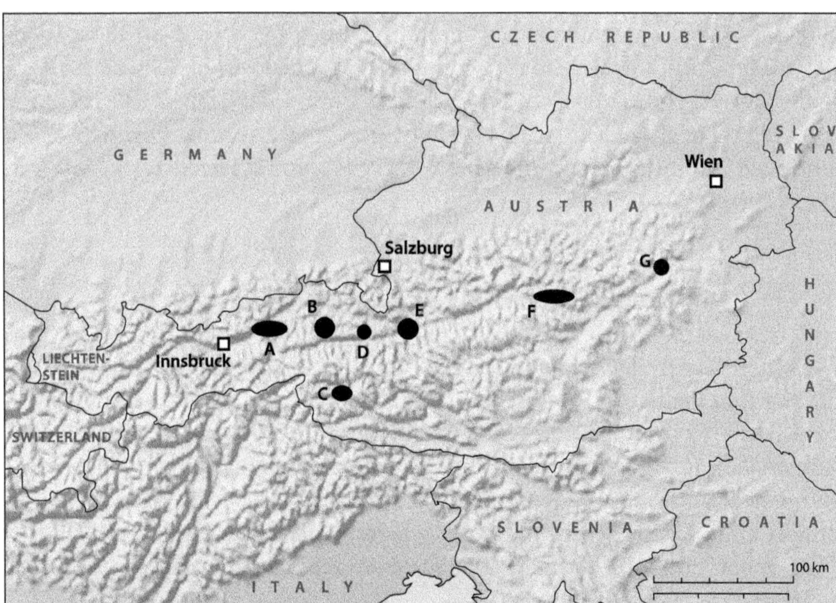

Fig. 7.1. Map of Bronze Age copper mines in Austria. A – Schwaz-Brixlegg, North Tyrol; B – Kelchalm, Kitzbühel, North Tyrol; C – Virgental, Matrei, East Tyrol; D – Saalfelden-Becken, Glemmtal, Salzburg; E – Mitterberg and St Veit, Salzburg; F – Eisenerzer Alps, Styrie; G – Prein, Raxalpe, Lower Austria.

(*Source*: author, adapted from Goldenberg 1998).

south and the Glemmtal area to the west (Eibner 1982; Goldenberg 1998). Farther west, Bronze Age mining was undertaken in the Saalfeldener Becken, also in Salzburg province. To the west there is a major area of Bronze Age copper mining in North Tyrol, including mines near Kitzbühel and a major concentration in the Schwaz-Brixlegg area of the lower Inn Valley.

There are indications of Bronze Age mining in other regions, in the form of smelting activity that was most probably based on local ores. These production sites are recorded at several locations in eastern Austria, notably in the Eisenerzer Alps in Styria. There are also smelting sites recorded in East Tyrol and in Lower Austria, though no actual mines of Bronze Age date are known there. This production also extended across the southern Alps into northern Italy, with Chalcolithic and Late Bronze Age copper smelting recorded in the South Tyrol and the Trentino district to the south (Cierny et al. 1998; Cierny 2008). That was based on the many occurrences of copper ore in this region, though few prehistoric mines are known (see Chapter 3).

Prior to 2000 BC, the production of copper in central Europe may have been based on brief expeditions to fahlore sources in the eastern Alps (and possibly other mountain ranges), undertaken from major settlement zones in the north Alpine foreland. Radiocarbon dating and other evidence indicate that Bronze Age copper mining commenced in Austria around 1800 BC, and intensified during the Middle Bronze Age and the transition to the Late Bronze Age (Stöllner 2009). The emphasis in mining up to 1200 BC was on the chalcopyrite deposits of the Mitterberg, with the fahlore deposits of Schwaz-Brixlegg in northern Tyrol being actively exploited from 1200–700 BC (Goldenberg et al. 2011). It is unclear to what extent this copper production in Austria continued into the Early Iron Age, but there is some evidence to suggest that this occurred.

Bronze Age mining in the Salzburg region

The Mitterberg is the best known area of Bronze Age copper mining in Austria and, for that matter, in Europe. The scale of production achieved there during the Middle Bronze Age was at a level that can arguably be described as industrial, imparting a distinctiveness to an Alpine landscape where large numbers of ancient mine workings and smelting sites occur today. The high altitude and physical setting presented a formidable challenge to the mining of this copper ore, which makes the scale of these operations all the more impressive.

The Mitterberg is a name given to a group of mining districts to the east and west of the upper valley of the river Salzach, in an area between the modern towns of Bischofshofen, Mühlbach, and St Johann, about 50 km south of Salzburg (Fig. 7.2). The evidence of early mining occur at altitudes of 1,200 m

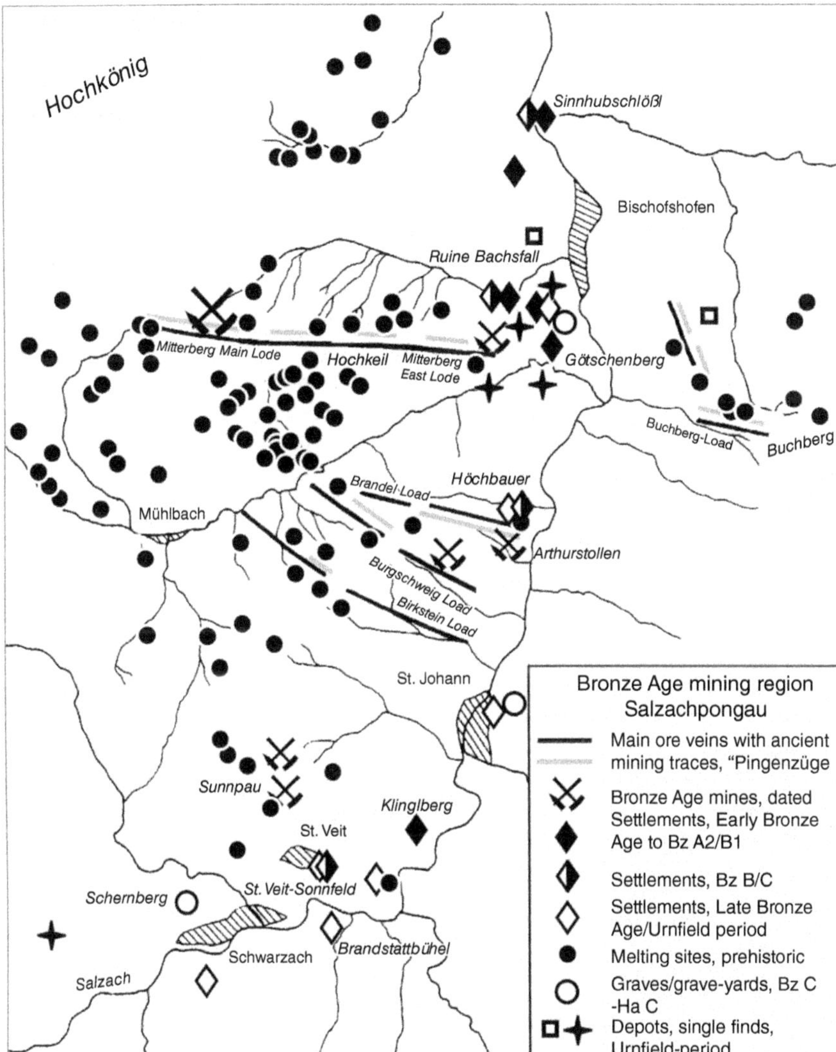

Fig. 7.2. Copper veins and Bronze Age mining in the Mitterberg area, Salzburg (*Source*: Stöllner 2003a).

or so in dramatic Alpine terrain, where mountains such as the Hochkönig (2,941 m) and Hohe Tauern (3,798 m) tower over narrow valleys separated by sharp mountain ridges.

There are three major zones of copper mineralization (Nordrevier, Südrevier, and Ostrevier) in the Mitterberg. These were mined at different times in the modern era to depths down to 700 m, with the last working mine closing in 1977. The Nordrevier copper veins comprise the Main Lode (Mitterberg

Hauptgang) and East Lode (Mitterberg Ostgänge), which cross the north side of the Mühlbach valley in an east–west direction. On the southern side of the valley is the Südrevier orefield, a series of parallel north-west/ south-east veins that include (from north to south) the Brander, Burgschwaig, and the Birkstein lodes. Finally, the copper veins of the Ostrevier occur to the immediate south-west of Bischofshofen, and include the near north–south Buchberg lode and the east–west Winkl lode.

The genesis of this mineralization is complex, with ore deposits differing in geological age and mineral composition. There are occurrences of both concordant and discordant vein-style mineralization, most of which are of hydrothermal origin. These fault-emplaced veins tend to have an east–west strike and also dip steeply to the south. The main veins vary from 1–2 m in thickness, with minor veins branching off (see Pittioni 1951: fig 2). The principal ore is chalcopyrite, with few accessory minerals. The absence of significant amounts of secondary mineralization may be due to erosion in the course of Pleistocene glaciation. The veins, being harder than the country rock, tend to stand out along higher ridges, where the mineral exposures were first discovered in the Bronze Age.

These early mines first came to light in 1828 with the re-opening of some of the Mitterberg mines. Workings of the 'Old Man' (*Alter Mann*) were recorded in the 1840s by a mine manager, Johann Pirchl, with the first systematic study later undertaken by the Viennese scholar, Matthäus Much (Much 1878, 1879, 1893; Stöllner et al. 2011: fig. 2). It was Much who first recognized the prehistoric date of these mines, which encouraged later researchers such as Georg Kyrle (1916) and Oliver Klose (1918). These investigations were taken to a new level in the 1920s with the work of two officials of the Mitterberg Copper Mining Company, the former mine surveyor, Karl Zschocke, and the mine engineer, Ernst Preuschen. Their technical examination of the ancient mines, based on surface surveys and underground access through the modern workings, led to the publication of a landmark study of these mines (Zschocke and Preuschen 1932). Their work is particularly important as most of the deeper Bronze Age mines are no longer accessible.

Preuschen developed an important collaboration with the archaeologist, Richard Pittioni, of the University of Vienna, who was an important figure in the study of early Austrian mines (see Pittioni 1951 for summary). There have been other studies in the modern era, most notably the work of Clemens Eibner in the investigation of smelting sites (Eibner Persy and Eibner 1970; Eibner 1972, 1974, 1979, 1982, 1992; see also Herdits 2003; Herdits and Löcker 2004). Current research centres on the HiMAT project and the work of Thomas Stöllner of the Deutsches Bergbau Museum, Bochum, who has directed new field research and reviewed the Mitterberg mines in relation to the overall pattern of Bronze Age copper mining in Austria (Stöllner 2009; Stöllner et al. 2011).

While many features of this mining landscape remain to be dated, it is possible to present a broad chronology for the development of copper production in the Mitterberg (Stöllner 2009). The earliest evidence comes from the Götschenberg hill-top settlement located at the eastern end of the Mühlbach valley. Excavations in 1979–87 uncovered small-scale habitation with evidence of metal production in the form of crucibles, small amounts of slags, and some copper objects. This activity was connected to the Mondsee culture of the fourth millennium BC, the first significant copper-using population in this part of the Alps. While the source of this copper is not certain, the siting of Götschenberg points to the use of local ores. Clearly, the location was strategic within the Mitterberg as there was also a Late Bronze Age phase of occupation at that site connected to copper mining (Lippert 1992).

The available evidence suggests that sustained copper mining commenced in the Mitterberg in the later stages of the Early Bronze Age. A radiocarbon result of 1940–1680 BC from the Brander Lode (Südrevier) provides evidence of surface workings earlier than the deeper mines of the Middle Bronze Age. The latter probably commenced in the mid-second millennium BC, with evidence from the Arthurstollen dated 1600–1200 BC (Stöllner 2009: fig. 4). There are indications that mining commenced on the Buchberg and Winkl lodes (Ostrevier) in this period (Stöllner 2009: fig. 6). Recent dendro-dating of old wood finds from the Mitterberg Main Lode points to intensive mining of these veins *c.*1400–1050 BC, in the transition from the Middle to the Late Bronze Age (Stöllner 2009: fig. 7). There is some evidence for continued mining at Mitterberg (possibly the Ostgänge lodes) into the early Hallstatt period, however, this was not on the same scale as in the earlier period.

Mining technology

Indications of Bronze Age mining are found all along the major veins of the Mitterberg district (Fig. 7.3). These include mine workings and adjacent spoil dumps, some of which were connected with ore processing. A large number of smelting sites are recorded in the area, many located a short distance down slope from the mines (Pittioni 1951: fig. 3). Recent surveys reveal that a certain amount of ore beneficiation was also undertaken at these smelting sites (Hanning et al. 2012).

The latter are mostly visible in the form of surface depressions known as *pingen*, which are often of considerable size. They include crater-shaped openings (*Trichterpingen*) and trench-like features (*Furchenpingen*) that run parallel to the strike exposure of the veins, and occasionally at an angle following branch and minor veins (Fig. 7.4). These *pingen* are partly infilled today, however geophysical and lidar surveys on the Mitterberg Main Lode and the Brander Lode reveals their extent (Figs 7.5 and 7.6; Stöllner et al. 2011).

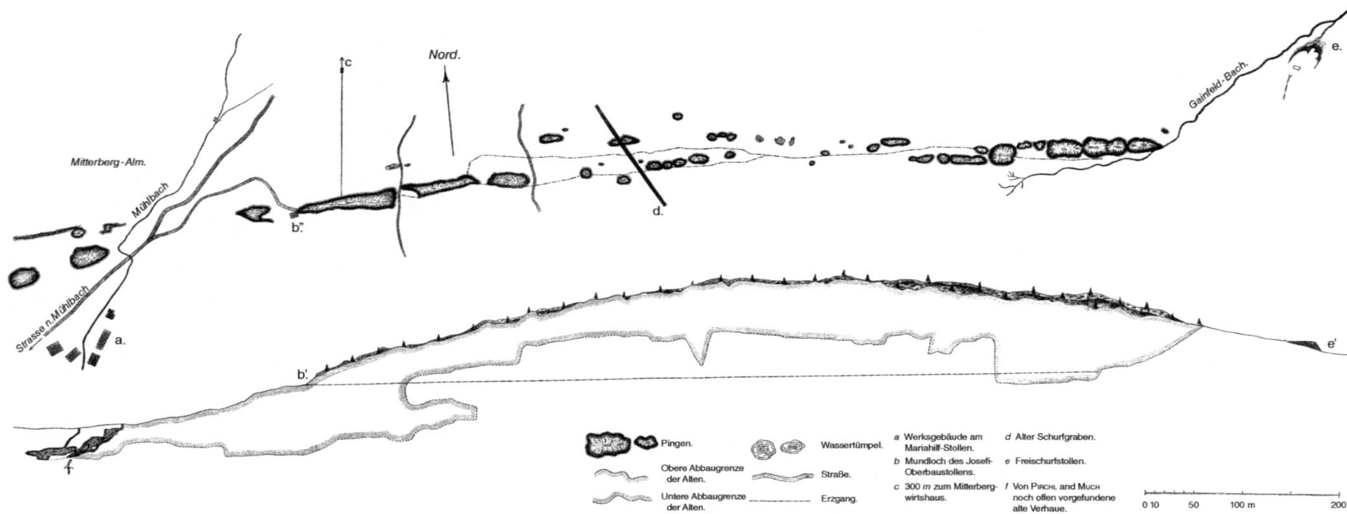

Fig. 7.3. Section across line of Bronze Age *pingen* mines, Mitterberg

(*Source*: Kyrle 1916).

Fig. 7.4. *Furchenpingen* surface mine features in the Mitterberg district

(*Source*: Deutsches Bergbau Museum/Ruhr University Bochum and Thomas Stöllner (left); Georg Kyrle (right)).

This mining commenced on the mineralized outcrop of the veins as small open-cast workings. These continued deeper following the southerly dip of the veins to form inclined tunnels or galleries (*tonnlägiger Einbau*). These were worked at different levels across the mountain ridges where the quartz veins were best exposed. As they progressed deeper, many of these inclined levels were connected to each other in order to facilitate ventilation and mine drainage. Flooding was a significant problem in these workings, with the miners building dams of rock spoil and possibly resorting on occasion to the use of buckets to bail water.

The deepest Bronze Age mines reached an overall length of some 300 m, at a depth of 60–160 m below the surface. Most of these galleries cannot be accessed today, however Zschocke and Preuschen (1932) recorded several examples that were exposed in modern workings. Some of the deeper workings are still exposed in the Arthurstollen, a 5 km long modern tunnel that connects the Mühlbach and Salzach valleys. This cuts through a number of deep galleries of Bronze Age date at depths of 200 m in the Brander Lode (Stöllner et al. 2011: fig. 3; see also Gstrein 1988). The precision by which many of these tunnels were driven and connected underground indicates a

Fig. 7.5. Aerial view of Bronze Age *pingen* mines, Mitterberg district (*Source*: Deutsches Bergbau Museum/Ruhr University Bochum and Thomas Stöllner).

basic understanding of linear and angular survey, suggesting that some form of compass may even have been used (Stöllner et al. 2011: fig. 7).

These veins were mined by removing as little of the country rock as possible. Harder rock was extracted using the fire-setting method together with pick-like tools of bronze, stone, and wood, or these implements on their own where the rock was soft. Bronze picks and hammers are recorded from many workings, as are wooden implements such as tree-trunk ladders, shovels, and troughs. The use of pine chips for illumination in these deep mines is well known, and includes the discovery of a pottery bowl filled with spent examples in the Arthurstollen (Stöllner et al. 2011: fig. 8). The type of tools and the manner in which fire-setting was used in these deep, multi-level tunnels will be examined in the next chapter.

The unique conditions of preservation in the Arthurstollen provide details of the mine timbering that was necessary in these schistose rocks (Stöllner et al. 2011: fig. 6). This involved the transport of roundwood logs from

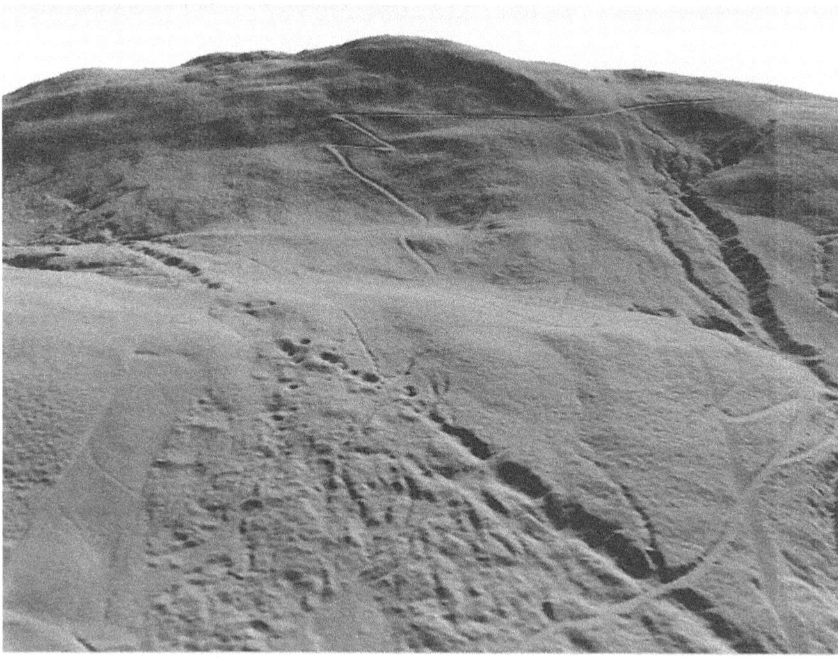

Fig. 7.6. Lidar scan image of Bronze Age *pingen* mines, Mitterberg district

(courtesy Thomas Stöllner. The Lidar Scan was processed by A. Hornschuch, Deutsches Bergbau Museum; data-source: Deutsches Bergbau Museum and Salzburg-GIS).

adjacent forest, which were cut and trimmed at the mine face for use as props and stemples in the timbering over of mine galleries and the construction of working platforms (Fig. 7.7; see also Chapter 8).

The vein material extracted in these mines was initially sorted underground, with barren rock dumped in abandoned tunnels and stored over timbered tunnels. The mineralized portion was brought to the surface to be processed in separating places located on natural or artificial terraces on the upper mountain slopes, generally a short distance from the mine openings. These were also the locations for small encampments where food was prepared and other domestic activities carried out. It is not certain whether hut structures were built in these camps, though some shelter was necessary given the high rainfall and cold at such altitudes (Pittioni 1951: 26).

These separating places are marked by the presence of spoil dumps that can often reach up to two metres in height. This broken rock and finer sediment was produced by several stages of mechanical ore beneficiation designed to prepare an ore concentrate for smelting (Eibner 1982; see Chapter 8). Initially, the mine rock was processed by the hand-sorting of coarse and finely crushed ore, using a combination of stone hammers and anvils. The larger hammerstones (1.5–3 kg) were probably sourced in the drift of the Salzach river valley

Fig. 7.7. Wood stemples from Bronze workings exposed in the Arthurstollen gallery, Mitterberg

(*Source*: Deutsches Bergbau Museum and Thomas Stöllner).

(Fig. 7.8). Some were used unmodified as hand-held implements, while others were side-notched or rilled to allow the attachment of a wooden handle with withy or leather bindings. Smaller hammers were used for fine crushing, while pointed examples served as crude chisels used with wooden mallets (Stöllner et al. 2011: fig. 15).

The next stage involved the use of concave mortars and millstones to reduce the finely crushed chalcopyrite to a coarse powder. This was concentrated

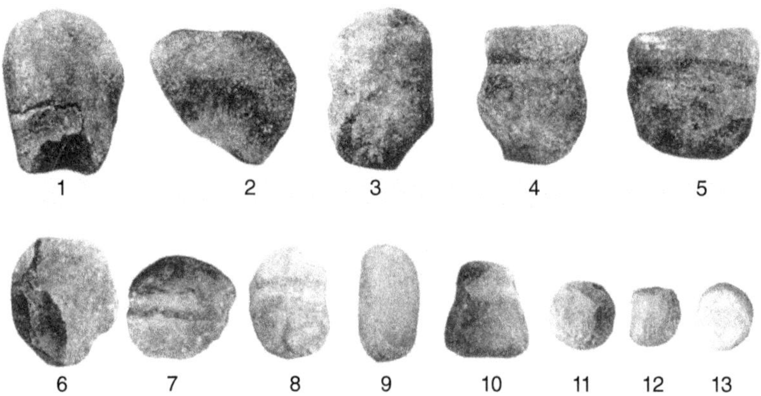

Fig. 7.8. Stone mining hammers from the Höchstollen mines (*Source*: Kyrle 1916).

further by a washing process where the heavier copper minerals were separated in wooden troughs (Stöllner et al. 2011: fig. 20). Wooden knives were used to skim off the lighter gangue material, which included minerals such as quartz, ankerite, dolomite, and siderite. Recent research in the Troiboden area of the Mitterberg points to the use of wooden sluice boxes for ore separation during the Middle Bronze Age (Stöllner et al. 2010).

Smelting of the resulting ore concentrate was generally not undertaken near the mines, but some distance downslope in level clearings with access to water sources (Pittioni 1951: fig. 6). Zschocke and Preuschen (1932) recorded some 138 smelting sites in this area, many of which are now difficult to locate in the forested terrain. These slag heaps are small, often only 15 m in diameter, which is surprising given the scale of Bronze Age copper mining in the area. Recent survey confirms the presence of very low-grade ore and rock mineral in many of these slag heaps, suggesting that a certain amount of primary ore processing took place at these locations (Hanning et al. 2012).

The smelting of this copper ore was a two-stage process, beginning with its roasting in open fires to reduce the sulphur content, followed by smelting in low-shaft furnaces of stone construction. The operation of what is known as the 'Mitterberg process', involving either a two-phase matte smelting or a one-step reduction method, is still debated. It is still uncertain whether copper metal or an intermediary product, matte, was produced at the smelting sites. Whatever process was involved, these furnaces do seem to have been efficient in the production of copper.

The copper-iron sulphide, chalcopyrite, was the main ore extracted in the Mitterberg mines. Some veins contained significant amounts of fahlore (mainly tetrahedrite), however the amount of secondary (oxidized) mineralization seems to have been generally low. The smelting of chalcopyrite yielded a metal with relatively low impurities (Lutz et al. 2010). Recent studies reveal it may eventually be possible to distinguish between different vein sources in the Mitterberg orefield based on their geochemistry and lead isotope signature (Stöllner et al. 2011: figs 17–8).

St Veit

To the south of the Mitterberg there is evidence of early copper production in the area around the modern towns of St Veit and St Johann (Fig. 7.2). This includes a large hill-top settlement called the Klinglberg, strategically placed overlooking the Salzach river. The excavation of this site in 1985–9 revealed a long history of occupation from the late Early Bronze Age, *c.*1800–1400 BC (Shennan 1995). While there is no evidence of smelting at this site, the discovery of raw copper and some ore, as well as large amounts of slag-tempered pottery, reveal a close connection with local copper producers during the Middle Bronze Age.

Scientific analysis of the copper finds suggest that the Klinglberg probably served as a collection and exchange centre for metal produced from a number of mines in the area. These might include the ore veins of the Mitterberg Südrevier, including the Burgstein lode located a few kilometres north of the site, where there is evidence of Bronze Age mining. Other potential sources include a number of copper veins in the Toifengraben area, some 2 km north-west of St Veit. Several early workings have been identified in that area, including funnel and furrow-shaped *pingen* and open tunnels (Gruber 1991; Shennan 1995: fig. 1.3 sites A–G). Prehistoric pottery is recorded from surface spoil dumps (Krause 1991), while a pit in the Auf der Au mine is dated 1700–1500 BC (Krause 2002; Stöllner 2009: fig. 3).

Saalfelden-Becken

This area is located approximately 30 km west of the Mitterberg in the Glemmta valley of the river Saalach in Salzburg province. Early copper mines are recorded at Leogang-Schwarzleo/ Gunzenreit and at Viehofen-Wirtsalm near Zell am See. The deeper workings at the latter site are inaccessible, however, numerous deposits of surface spoil are recorded over a wide area (Preuschen and Pittioni 1956: map 1, figs 3–10). Nineteenth-century mine sources record ancient mining at Viehofen at depths of up to 100 m. This involved the use of multiple inclined shafts to access underground stopes and ventilate fire-setting operations (Fig. 7.9; Kyrle 1916).

The few radiocarbon dates available for Leogang and Viehofen point to mining activity from the Middle to Late Bronze Age, *c.*1600–1150 BC (Stöllner 2009: fig. 8). There are also indications from settlement excavations in the

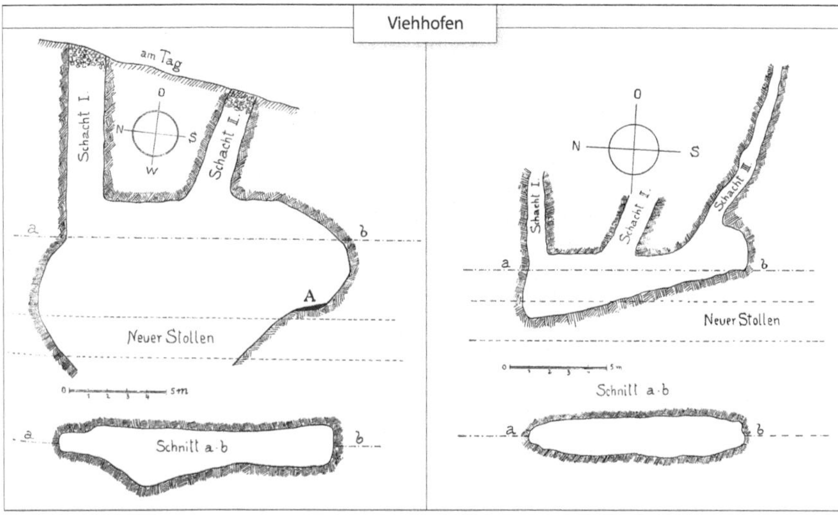

Fig. 7.9. Bronze Age mine workings at Viehofen, showing shafts leading to underground mine levels

(*Source*: Kyrle 1916).

Saalfelden area of copper metallurgy that date from the Early Bronze Age to the Iron Age (Stöllner 2009).

Bronze Age copper mines in the North Tyrol

There is a long tradition of metal mining in this western province of Austria, from prehistoric and medieval to modern times. This includes the famous Schwaz orefield, one of the most important silver mining areas in Europe during the Middle Ages, where there are also significant deposits of copper ore.

Kelchalm

The early mines at Kelchalm are located 10 km south-east of the town of Kitzbühel, approximately 50 km west of the Mitterberg. As with the latter, these mines were first discovered on the resumption of copper mining in the nineteenth century. Surface and underground workings now known to be of Bronze Age date were uncovered at an altitude of 1,700–1,800 m. These are mostly inaccessible today, having either been backfilled by the early miners or collapsed after their abandonment. There are surface indications of this mining (*pingen*), with adjacent spoil deposits up to 4 m in thickness where copper ore was processed. The locations of many small ponds used in these processes are still visible at Kelchalm (Pittioni 1951: plate 2.2). At these high altitudes the miners required habitation facilities, with evidence for the use of pottery and the supply of meat from farms.

Archaeological investigations of the Kelchalm mines began in the mid-twentieth century when Richard Pittioni and Ernst Preuschen began work there. They conducted surface excavations at several locations from 1932–53 (Preuschen and Pittioni 1937, 1954). This included locations where multi-stage treatment of copper ore was undertaken, involving crushing and hand-sorting of mineralized rock using stone hammers, anvils, and grinding stones (Preuschen and Pittioni 1954: fig. 28). There is evidence for the gravity separation of ground ore by washing in wooden troughs (Preuschen and Pittioni 1954: figs 11–13). The latter was facilitated by a series of water channels formed using wooden planking, evidence of which was uncovered due to the excellent anaerobic preservation in these ore separating sites. As well as finds of pottery, stone, and bronze artefacts, numerous wooden objects connected with the mining were discovered, including troughs, drainage launders, and spoon/scoops, as well as domestic objects such as spindles and roofing shingles (Preuschen and Pittioni 1954: figs 36–9; Pichler et al. 2009: fig. 2).

Preuschen and Pittioni (1954) dated the early copper mines at Kelchalm to around 1000 BC based on pottery and metalwork finds. This has recently been revised by dendro-dating of wooden artefacts from those mines, which indicate intensive copper mining during the mid-to-late thirteenth century BC (Pichler et al. 2009: table 1).

Schwaz-Brixlegg

An important group of Bronze Age copper mines and associated production sites has been identified in the Radfeld/ Brixlegg region along the southern side of the Lower Inn Valley. These occur on the eastern margin of the Schwaz orefield, one of the most important copper and silver mining centres in Europe in the late medieval and early modern period (Klaunzer et al. 2010). There are several styles of mineralization in this region (Höppner et al. 2005), the most important being the tetrahedrite-rich fahlore of syngenetic hydrothermal origin contained in the Schwaz Dolomite of Lower Devonian age (Goldenberg et al. 2011: figs 3–4). This copper mineralization contains high concentrations of arsenic, antimony, bismuth, and silver, as well as a distinctive lead isotope signature, which distinguishes it in terms of metal composition from the chalcopyrite ore mined in the Mitterberg, Kitzbühel-Kelchalm, and Viehofen areas.

Unlike the Mitterberg and Kelchalm, many of the Bronze Age workings in this area remain accessible. They include the mines at Maukental,

Moosschrofen, and Kleinkogl near Brixlegg, as well as numerous workings near Schwarz itself, including examples exposed by landslides at Eiblschrofen (Goldenberg et al. 2011: fig. 6). These were first recorded in mining records of the sixteenth century, when they were called 'Pagan Mines'. These mines can extend up to 65 m in horizontal length, with the use of fire-setting creating large dome-shaped chambers connected by narrow tunnels with multiple entrances and pillars left for roof support (Fig. 7.10). There is also evidence

Fig. 7.10. Fire-set mines of Late Bronze Age/ Early Iron Age date at Moosschrofen, Brixlegg

(*Source*: Gert Goldenberg).

for the use of grooved hammerstones, often of amphibolite and related metamorphic rocks, collected in the drift of the Inn Valley.

Following their sporadic use in the fourth and third millennia BC, these fahlore deposits were mined in the Early Bronze Age when they were an important source of *ösenringe* copper. The Buchberg settlement near Wiesing in the middle Inn Valley provides evidence of small-scale copper production around 1800 BC, associated with use of furnaces, tuyères, and small amounts of slag (Martinek and Sydow 2004). Similar results have been obtained from the investigations of Early Bronze Age settlement sites at Brixlegg-Mariahilfbergl (Höppner et al 2005), Kufstein-Tischofer Höhle (Kneussl 1969), and Thaur-Kiechlberg (Töchterle et al. 2009). No mine workings of this period have been identified, however, each of these sites has produced evidence of primary copper smelting of local fahlores from the lower Inn Valley (see Goldenberg et al. 2011).

The copper mining at Brixlegg commenced in the Early Bronze Age and expanded significantly during the Late Bronze Age (1300–800 BC) and Early Iron Age, particularly around 700 BC. These include mines and production sites in the Mauken Valley where there are rich fahlore deposits on an eastern branch of the Schwaz Dolomite. Five of these sites were investigated in 2000–9, as part of the HiMAT research project (Goldenberg and Rieser 2004; Klaunzer et al. 2010; Goldenberg et al. 2011; Schibler et al. 2011). They include the excavation of mine workings ('Mauk B and E'), a mine camp location ('Mauk D'), an ore processing site ('Mauk F'), and a smelting location ('Mauk A') ranging in date from Late Bronze Age to Early Iron Age times (*c.*1200–700 BC).

The mine workings at Mauken are at altitudes of 900–1,200 m OD along the upper valley slopes (Fig. 7.11). The Mauk E mine is a small fire-set working, where several tunnels incline to a depth of 25 m, with multiple entrances, only one of which is open today (Goldenberg et al. 2011: fig 24). Dendro-dating of charcoal indicates that the mine was worked in the period 720–705 BC (Goldenberg et al. 2011: figs 27–9), with later operations dating to the early sixteenth century AD. The mine walls bear the distinctive concave profile of fire-setting with evidence also for the use of wooden lighting splints (Fig. 7.12). The rock extract was sorted underground, with the mineralized portion brought to the surface for further treatment.

The nearby site of Mauk D revealed the remains of a mine camp adjacent to a group of early mine workings on a terrace. This activity dates from the late twelfth century to the ninth century BC. Excavation of a midden uncovered traces of habitation in the form of pottery and animal bone food waste. There is also evidence of ore processing, in the form of grinding stones of gneiss and granite, and numerous stone hammers of well-rounded cobbles gathered from river deposits or glacial drift.

Fig. 7.11. General location of Late Bronze Age/Early Iron Age mines and ore processing sites at Maukental, Radfeld

1 Smelting site Mauk A; 2 Copper mine Mauk E; 3 Mine camp Mauk D; 4–5 further traces of mining (*Source*: Gert Goldenberg).

Fig. 7.12. Fahlore mine Mauk E, Maukental

(photograph and mine plan: Gert Goldenberg).

Farther up the valley there is an ore processing site (Mauk F) in a reclaimed peat bog known as Schwarzenbergmoos. Excavation in 2007–8 uncovered broken rock spoil, stone hammers, and the remains of wooden troughs used in water concentration of copper ore (Fig. 7.13). This site is dendro-dated 900–869 BC (Goldenberg et al. 2011).

Compared to the large number of mines, only one smelting site has been identified in the Mauken Valley. This suggests the use of a central smelting facility, the site of Mauk A located on level ground close to a small stream, not far from the Mauk E mine. Excavations in 2008 uncovered two stone-built furnaces, as well as a two-phase roasting bed (Fig. 7.14). This copper production is dated by ceramic finds and radiocarbon analysis to the later twelfth and eleventh centuries BC. A large deposit of slag sand, up to 1.3 m in thickness, was uncovered, where almost 100 tonnes of crushed smelting slag was processed using water separation in wooden troughs to recover metal prills. Animal bones and fragments of textile preserved by copper salts were found in this slag sand.

Pollen studies and the analysis of wood-charcoal and animal bones from the Mauken sites reveals that these miners lived in a forested environment supported by agriculture zone in the valley below and in mountain pasture. The organization of this copper production and the supply of food to the miners will be considered in Chapter 9.

Fig. 7.13. Wooden washing trough used for ore beneficiation, Mauk F Schwarzenbergmoos (*Source*: Gert Goldenberg).

Fig. 7.14. Roasting bed, Mauk A site
(*Source*: Gert Goldenberg).

There is also potential for the discovery of Bronze Age copper mines in the western part of North Tyrol. There are numerous copper deposits in that region, both fahlores and chalcopyrite, and, while smaller in comparison to Schwaz-Brixlegg, these deposits may have been significant during the Bronze Age. A recent survey identified fire-setting in the mining areas of Innsbruck-Hötting and Navis-Knappenkuchl. A hammerstone was also found in the latter, with examples recorded from another mine, Serfaus-Rotenstein (Martinek and Grutsch 2012).

Other Bronze Age mines in Austria

The Eisenerzer Alps in Obersteiermark (Styria), eastern Austria, was another important centre of copper production at different times in the Bronze Age. Smelting sites and possible mine workings are recorded in this mountain range between the wide valleys of the rivers Palten and Liesing in the south and the narrow valleys of the rivers Enns and Erzbach to the north (Klemm 2003). This area is part of the eastern Greywacke Zone, with numerous deposits of copper ore dominated by chalcopyrite and fahlore mineralization.

The first smelting sites were discovered in this region in 1955 by Ernst Preuschen, who investigated a number of these in later years (Preuschen

1968a). Numerous surveys and excavations were carried out in the 1980s and 90s by Clemens Eibner and Hubert Presslinger (for publications see Klemm 2003: 155–8). Almost 100 smelting sites of probable prehistoric date are recorded, including examples along the valleys of the rivers Palten and Liesing and in many mountain valleys to the north, including those at Gaishorn-Flitzen, Johnsbach, Radmer, Ramsau, and Kalwang-Teichen (Klemm 2003: map 2). The most recent research includes a detailed study of smelting sites in the Ramsau valley, located in the north-east part of the Eisenerzer Alps (Doonan et al. 1996; Klemm 2003, 2004).

No prehistoric copper mines have been identified in this region, however, it is certain that local ores were used in these smelting operations. A number of possible early mines have been identified in the Eisenerzer Ramsau valley (Fig. 7.15), however, this remains to be confirmed. Finally, there is also evidence of Bronze Age copper production at several locations in Lower Austria. Smelting sites of this period, though not well dated, are recorded near Gloggnitz and at Hohen Wand (Stöllner 2009). For the former, there is evidence of ore roasting and smelting at the sites of Prein and Prigglitz-Gasteil adjacent to pingen-type mines (Hampl and Mayerhofer 1958, 1963; Hampl 1976).

Finally, there is evidence of Bronze Age copper production in the East Tyrol region of south-central Austria, an area outside of the Greywacke Zone.

Fig. 7.15. General location of Bronze Age copper smelting sites in the Eisenerzer Ramsau valley, Upper Styria

View of the Erzgebirge ('Ore Mountain') in background, worked for iron ore since the twelfth century AD (*Source*: author).

Prehistoric smelting sites are recorded at Virgental in Matrei, however no associated mines have been found (Preuschen and Pittioni 1953).

Conclusions: Bronze Age mining in the eastern Alps

The development of copper mining in the eastern Alps can be traced from small beginnings in the Late Neolithic to the establishment of industrial-scale operations in the Middle and Late Bronze Age and their subsequent collapse in the Iron Age (Stöllner 2010: fig. 2; Goldenberg et al. 2011). The first use of copper in Austria can be attributed to Late Neolithic groups in the north Alpine forelands, who imported small amounts of metal from south-east Europe in the later fifth millennium BC. This led to an adoption of metallurgical processes by local groups of the Altheim, Mondsee, and Pfyn cultures during the fourth millennium BC, and possibly by Bell Beaker groups in the third millennium BC. This stimulated interest in local copper sources, with an early use of fahlore and possibly native copper and secondary minerals even though no mines are recorded. Finally, there is also evidence of copper production associated with Lagozza and Remedello culture groups of the fourth millennium BC in the southern Alpine region of northern Italy. While evidence is lacking, the existence of Bronze Age copper mines in the Trentino/South Tyrol may be inferred from the extensive smelting activity in that region.

The first significant mining of copper in the Austrian Alps began around 1800 BC, with settlement evidence for small-scale processing of both chalcopyrite and fahlore in the Salzburg and North Tyrol mining districts. This period, *c.*1800–1500 BC, marks the commencement of mining in the Mitterberg. There was a gradual intensification of mining during the Middle Bronze Age (1550–1330 BC) and the early part of the Late Bronze Age in the Mitterberg, Saalfelden, and Kitzbühel orefields, as well as the lower Inn Valley (Schwaz/Brixlegg), with copper mining also underway in Styria and across the Alps in the Trentino. This was made possible by the development of deep mining technology and smelting methods capable of processing chalcopyrite ore on a large scale. There is reason to believe that the advanced furnace technology of the so-called Mitterberg process originated in that district, from where it spread to other mining areas in the Middle to Late Bronze Age.

The development of copper production continued during the Late Bronze Age into the Urnfield period, *c.*1300–800 BC, across all of the Austrian mining districts. This period is marked by large-scale extraction of fahlore in the Schwaz-Brixlegg region of North Tyrol. Copper mining continued in some areas into the Early Iron Age, however, there was a significant decline after 700 BC during the Hallstatt age.

The chronological information now available indicates a more or less continuous development of copper mining in Austria during the Bronze Age. This is true in technological and organizational terms, with evidence of an ability to adapt to different geological settings and styles of mineralization. The extraction of fahlore from carbonate/ dolomitic geology (Schwaz-Brixlegg) during the Early Bronze Age was followed by intensive mining of chalcopyrite in greywacke-schist geology in the Middle Bronze Age (Mitterberg and other sources). Technological advances meant that both styles of mineralization could be exploited by new methods during the Late Bronze Age and Early Iron Age. Fluctuations in the amount and type of copper produced over this 1,500-year period raise important questions as to the economic and social relations of the different districts that formed part of this east Alpine mining tradition.

The preservation of wooden mine equipment in flooded workings is an important feature of these mines. Early research in the Mitterberg and Kelchalm uncovered examples of buckets, ladles, launders, lighting chips, mallets, shovels, tool handles, tree-trunk ladders, and troughs, amongst other items. The schistose geology of the Mitterberg and Kelchalm mines required extensive use of support timbering, as exposed today in the Arthur-stollen tunnel. The bedrock structure in those mines encouraged the use of bronze picks, which was an important innovation in these Austrian mines. Their use contrasts with the use of fire-setting in the dolomitic geology of the Mauken mines in Brixlegg. There are relatively few finds of stone hammers in those workings, and it seems likely these implements were mostly used in ore beneficiation (Goldenberg personal communication).

In conclusion, the production of copper in the eastern Alps developed from a small-scale activity associated with seasonal pastoralism in the Early Bronze Age to a largely mining-based economy in the Middle to Late Bronze Age. There is evidence of permanent mining communities in the Alpine zone by 1600 BC. This is reflected in the establishment of hill-forts associated with the production and supply of copper, sites like the Götschenberg, or at Klinglberg in St Veit (Shennan 1995). The scale of the Austrian copper mines is impressive, with evidence of an industrial level of copper production by 1500 BC employing an advanced shaft furnace technology. Estimates from the Mitterberg suggest that as much as 18,000 tonnes of raw copper were extracted there during the Bronze Age. This copper was exchanged over a wide area, as evidenced by discoveries of large quantities of copper ingots along river valleys in the north Alpine lowland leading into southern Bavaria, Lower Austria, and Moravia. By the Late Bronze Age, the circulation of east Alpine copper extended across Europe, reaching as far as Scandinavia, Britain, and Ireland.

THE GERMAN MOUNTAIN RANGES

While the eastern Alps was the major source of copper during the Bronze Age, there are numerous other ore deposits in central Europe that may have been exploited at that time. These include the ores of the Harz Mountains and the Saalfeld-Kamsdorf district in Thuringia, the Erzgebirge mountains on the border of Bohemia and Saxony, and the extensive *Kupferschiefer* of central Germany. Many of these were important sources of copper and other metals in the historic period, with famous mines such as Rammelsberg and Freiberg.

Some of these orefields can be excluded as potential sources of copper during the Bronze Age due to the geological occurrence and low concentration of the mineralization. This would include the low-grade ore of the Kupferschiefer and many ores in the Harz Mountains. It is possible that some of these Thuringian deposits originally had significant zones of secondary mineralization that are now removed (Niederschlag and Pernicka 2002; Niederschlag et al. 2003).

There has been much speculation that central German ore deposits were a source of copper (and tin) during the Bronze Age. This view is encouraged by the widespread occurrence of this mineralization in an area that was heavily settled during the Bronze Age, with major metal-using groups such as the well known Unetice culture (Niederschlag et al. 2003: fig. 1). The widespread use of bronze is evidenced in grave and hoards finds, and in the large number of copper ingots in circulation during the Early Bronze Age (Niederschlag et al. 2003: fig. 9). This suggests some involvement in primary copper production, including the possibility that copper and tin deposits within the Unetice culture area were exploited.

Much of the focus in this debate has centred on the Erzgebirge ('Ore Mountain') on the border of Saxony and Bohemia. The Erzgebirge is famous for tin extraction in the medieval and modern eras, when copper mining was also undertaken on a smaller scale. Researchers have long speculated on the possibility that this area was a major source of copper and tin during the Bronze Age (Niederschlag et al. 2003: 61). The ore deposits there could be considered optimum for the development of early bronze production, given the proximity of tin placers to copper mineralization, occasionally even within the same deposit. There is also the discovery of metallurgical finds of Bronze Age date, such as moulds and bronze casting residues, in the vicinity of many ore deposits.

Despite these promising indications, modern field surveys have not identified any prehistoric copper mines, while evidence of early tin streaming is also scarce (see Bartelheim et al. 1998). In the absence of field evidence, this issue has been explored through scientific studies that compare copper and tin ores of the Erzgebirge with metal used in the Bronze Age. This approach was first attempted in the mid-twentieth century, when Otto and Witter used chemical analyses to argue that this region was a major source of copper. This view was not widely accepted, with others arguing that the eastern Alps, the Slovakian Ore Mountains, and even the Transylvanian mountains, were a more likely source of early copper in central Europe (Neuninger et al. 1957; Neuninger and Pittioni 1963).

The significance of the Erzgebirge ores has been examined in the modern era using new approaches, including lead isotope analysis (Niederschlag and Pernicka 2002; Niederschlag et al. 2003). This research combined a review of these copper sources in terms of prehistoric mining (Niederschlag et al. 2003: table 4) with a programme of lead isotope analysis. This revealed no correlation between Early Bronze Age metalwork in south-east Germany and these copper ores in the Erzgebirge. The ingot results, in particular, do not support an Erzgebirge copper source for this metal, while evidence of contemporary tin extraction in that area is also lacking. Some possibilities for early copper mining in the Harz Mountains were identified, however this remains to be confirmed by further research.

SLOVAKIA

Other ore deposits in east-central Europe must be considered as potential sources of copper in the Chalcolithic and Bronze Age. Though research is lacking, there is evidence of early copper mining in central Slovakia where pottery of Chalcolithic and Late Bronze Age date has been found in mine contexts with stone hammers. These include the copper mine at Spania Dolina-Piesky, located 7 km north of Banská Bystrica at the southern end of the Lower Tatras mountains in central Slovakia (Tocík and Zebrák 1989). The mineralization consists of quartz-sulphide veins with primary chalcopyrite and fahlore (tetrahedrite) ore, as well as secondary copper minerals.

Intensive mining at Spania Dolina-Piesky during the sixteenth and seventeenth centuries has destroyed or concealed earlier mine workings (Tocík and Zebrák 1989: figs 9.1–9.2). Evidence for the latter rests on the discovery of grooved hammerstones in surface spoil containing finds of prehistoric pottery. The first archaeological investigations were carried out in 1971–2, led by A. Tocík, with later investigations conducted by Pavel Zebrák. This revealed traces of early mining at an altitude of 800 m along the ridge between the valleys of Piesky and Richtárová (Zebrák 1995). An estimated 200 stone hammers are recorded from spoil deposits in the mine, all coming from secondary contexts. These are well-rounded quartzite cobbles of local origin, of varying shapes and weighing from less than 1 kg up to 2.5 kg (Tocík and

Zebrák 1989: fig. 9.3). Other finds from the mine include a cake of raw copper, as well as pottery dated to the Bronze Age.

Other indications of primary copper production are provided by the discovery of a smelting site at Slovenské Pravno dated to the later fourth millennium BC by pottery of the Lengyel culture. Similar pottery is recorded from the Spania Dolina mine located 35 km east of this site. Finally, stone hammers are recorded in the copper mine of Spanie Pole in the Rimavská Sobota region of south-central Slovakia where prehistoric pottery has also been found (Tocík and Zebrák 1989).

EURASIA

It is obvious that an ability to conduct copper mining in the prehistoric period was influenced in the first instance by the geological occurrence of the raw material. We have seen how widespread copper mineralization is in Western Europe, and yet how localized the sources can be at a regional level, leading to disparities in raw material supply. The situation is more extreme on the eastern side of the Continent, where there are few copper deposits over the vast distance of the Great Eurasian Steppe extending from the Carpathian Mountains to the Urals, the latter generally taken as the boundary with Asia. Yet, it is there, on the Russian border with Kazakhstan, that the largest Bronze Age copper mine of all is to be found.

Kargaly

This enormous mining and metallurgical centre is located on the open steppe landscape of the southern Urals, in the Orenburg district of the Russian Federation. The area is highly mineralized, with copper deposits spread over approximately 500 square kilometres in a north-west/ south-east direction along the valley of the river Kargalka, one of the tributaries of the river Ural. The ore occurs as rich pods or scattered lenses of secondary copper minerals (mostly cuprite, malachite, and azurite) in bedded sandstone geology.

The early copper mines at Kargaly were first recorded in the eighteenth century, coming to archaeological attention in 1989 when a major programme of research was conducted by Evgenij Chernykh of the Russian Academy of Sciences (Chernykh 1992, 1994, 1998a, 2002, 2003). Over the next decade the mining landscape at Kargaly was recorded in detail, with the discovery of as many as 20 settlements connected to the Late Bronze Age mine. These include the settlement site of Gorny, located directly within the mining complex (Chernykh 1998b, 2004a).

There are two distinct phases of early mining at Kargaly. The first dates from the late fourth millennium/ early third millennium BC, and is associated with the Yamnaya-Poltavka culture of the Early Bronze Age. These nomadic stockbreeders occupied vast areas of Eurasia and are known archaeologically from thousands of burial mounds (kurgans), many of which contain copper from Kargaly. Three kurgan cemeteries are recorded within the mine area, one of which was excavated to reveal the burial of a young man, accompanied by a copper axe and an axe mould, dated to 2900–2700 BC (Chernykh 2004b).

There may have been a break in mining at Kargaly from 2500–1900 BC, only to the resume at a greater level of intensity in the Late Bronze Age. The peak of production *c.*1900–1300 BC was connected to steppe herders of the Srubnaya (timber grave) culture. This group of settled pastoralists were widely distributed from the southern Urals to the Black Sea coast. Excavation of their settlements reveals they were strongly involved in copper mining in the Ural sandstones during the later second millennium BC.

Fig. 7.16. Aerial view of part of Bronze Age mine landscape, Kargaly, Russia (*Source*: Evengij Chernykh).

The Bronze Age operations at Kargaly were distributed across three major mineralized zones, with eleven separate mining locations over an area of around 140 square kilometres (Chernykh 1994: fig. 2). The scale of the Bronze Age operations is staggering, with an estimated 35,000 surface features connected to mining over these two millennia (Fig. 7.16). This has left a lunar-type surface with thousands of infilled mine shafts, drift mines, and prospecting pits, and a labyrinth of underground tunnels (Fig. 7.17; Chernykh 2002). The presence of rock spoil around many of these surface depressions indicates that they are backfilled shafts, while others represent the collapse of underground mine cavities.

The copper ore lies close to the surface, scattered through beds of red sandstone down to depths of approximately 90 m. This bedrock is overlain by up to 12 m of alluvium, with some mineralized exposures along the sides of ravines and river courses. These outcrops were streaked with the bright colours of oxidized copper minerals, which is how the ore was discovered 5,000 years ago.

The miners initially worked the exposed mineralization using drifts and opencasts. Having established the physical occurrence of the ore, they

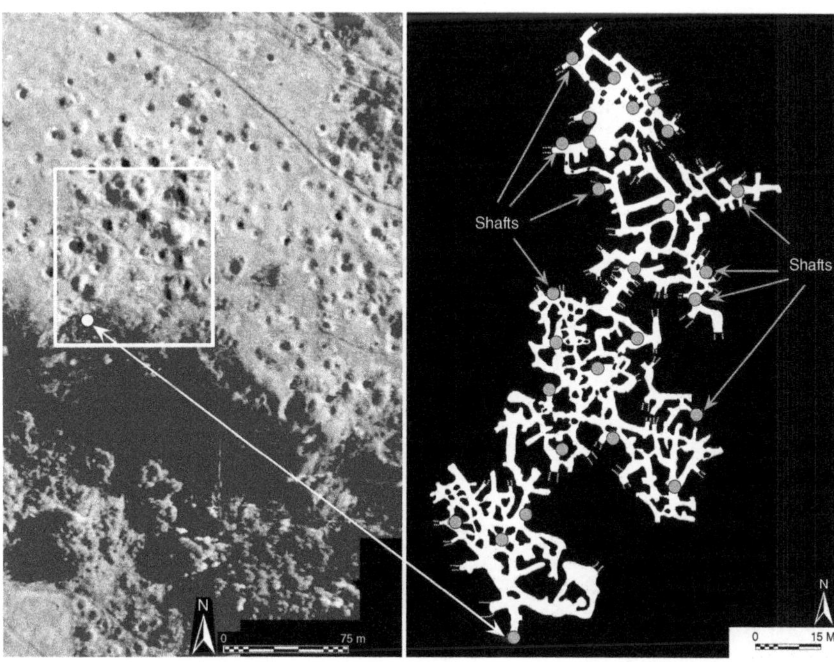

Fig. 7.17. Bronze Age mine shafts at Kargaly with detail of underground tunnel systems

(*Source*: Evengij Chernykh).

developed a new strategy by digging into the surface alluvium in search of the cupriferous sandstone beds. They sank vertical or inclined shafts in close proximity to a depth of 10–12 m into the bedrock (Fig. 7.18). Many of the vertical shafts are circular in section, with some having rock-cut handholds (Fig. 7.19). On reaching bedrock they proceeded to tunnel outwards to follow the ore in all directions, often down to depths of 40 m.

This was not an easy process due to the scattered nature of the copper ore and the dangers of tunnelling in this soft sandstone geology. If they failed to locate mineralization after a few metres of tunnelling, the miners abandoned the working to commence a new surface shaft a short distance away. Where copper ore was discovered these tunnel systems could become extensive, developing from narrow passageways into larger stopes up to 20 m high where rich pockets of ore were removed. Many mine tunnels connected with each other, to create a labyrinth of underground workings believed to extend several hundred kilometres across the entire orefield (Chernykh 2002: fig. 2.8). There is some access to these underground workings today, however, many examples have collapsed with others extensively re-worked in the modern era.

The mining methods were surprisingly basic for such intensive operations conducted over nearly 2,000 years. This may be explained by the soft nature of the wallrock that allowed the use of pick-like tools and hammerstones without any need for fire-setting (Fig. 7.20). The latter was probably not an

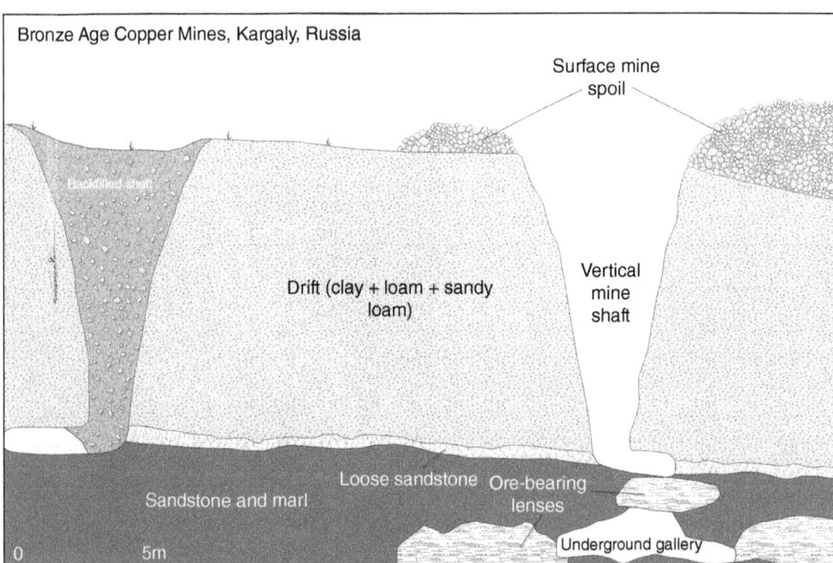

Fig. 7.18. Schematic section showing geological setting of Bronze Age shafts, Kargaly mine section

(*Source*: author, re-drawn from original provided by Evengij Chernykh).

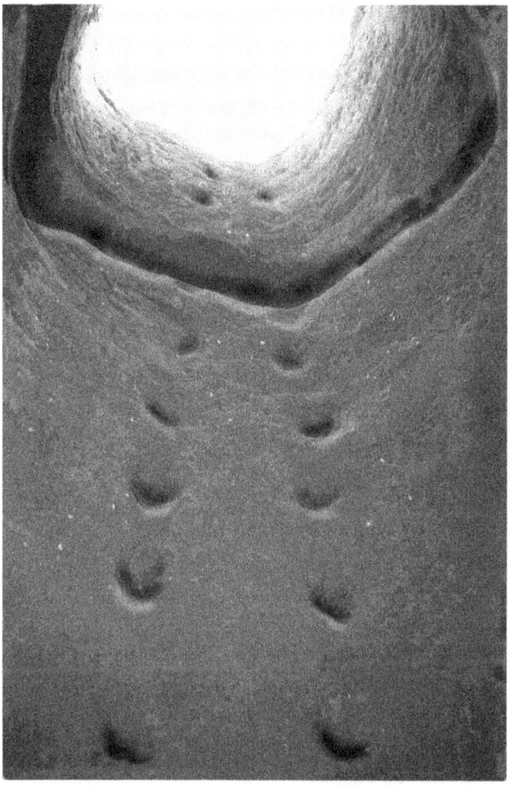

Fig. 7.19. Bronze Age mine shaft at Kargaly showing hand-holds (*Source*: Evengij Chernykh).

option given the limited wood supply in the open steppe lands. The miners probably used bronze picks, but mainly employed short picks made of split cattle bones, together with cattle scapula shovels and stone hammers (Rovira and Navarrete 2005: fig. 12). These tools have left distinct wallrock traces and their effectiveness is confirmed by experimental studies (Chernykh 2004a: figs 9.9–9.14).

An important feature of the Kargaly research is the investigation in 1992–2002 of a mining settlement of Gorny. This site is located on a low hill directly within the mining complex. This area is covered with mine workings, spoil dumps, and smelting sites of Bronze Age date (Chernykh 2004a: fig. B1). The excavation uncovered an early phase of pit dwellings that were occupied on a seasonal basis, followed by more permanent houses that combined residential space with smelting facilities and ritual areas. There is much evidence of domestic activities in the form of pottery and animal bone food waste. The excavation of part of the settlement uncovered some 120,000 sherds of pottery from an estimated 5,000–6,000 vessels. An estimated 2.5

Fig. 7.20. Stone mining hammers from Kargaly

(*Source*: Evengij Chernykh).

million animal bones were recovered in this excavation, mostly domesticated animals dominated by cattle. Many of these were used to make mining tools. Large amounts of processed copper ore were also found in the settlement, including an estimated 4,000 finds of raw copper, ingots, and finished metal-work. Some 190 casting moulds for the production of axes, daggers, and other implements were found, including examples used to cast copper picks and shovels for the mines (Chernykh 2003).

The excavation also yielded a large number of stone hammers, used along with anvil stones in the concentration of copper ore (Chernykh 2004a: figs 6.1–6.10). This was achieved by repeated crushing and hand-sorting of the green/blue copper minerals. Though furnace structures are lacking, the smelting of this oxidized ore appears to have been a fairly primitive, direct reduction, process that yielded only small amounts of slag. A certain amount of copper ore was smelted at Kargaly for the production of mining tools; however, most was exchanged over an area as far as the Volga, where it was smelted in settlements of the Srubnaya culture that had access to charcoal fuel.

There are certain aspects of the Gorny settlement that suggest unusual rituals and beliefs connected to mining and copper production. Excavation underneath one of the house structures uncovered a passage that seems to imitate a mining tunnel. This was filled with pieces of rich copper ore in a manner that cannot be easily explained. The discovery of approximately 2.5 million fragments of animal bone within an excavated settlement area of around 1,000 square metres is unusual. A number of excavated pits contained what seem to be sacrificial deposits of animal bone, while stone objects interpreted as phalluses were deliberately placed in the ground.

Mining resumed at Kargaly in the eighteenth and nineteenth centuries, when an estimated five million tonnes of copper ore was raised to produce an estimated 130,000 tonnes of copper metal. Some of this was produced by deeper mining, which extended the prehistoric shafts and drifts downwards to depths of about 90 m. This also involved extensive re-working of spoil heaps left by the Bronze Age miners. The scale of Bronze Age mining was similarly enormous, with estimates of 55,000–120,000 tonnes of copper for this period (Chernykh 1998a, 2004b). This output is even more remarkable when compared to the total amount of Bronze Age metalwork recorded in Eurasia.

The impact of the Kargaly mining on woodland resources was considerable in terms of the supply of charcoal for smelting. The scale of production at Kargaly and its long-term environmental consequences will be considered further in Chapter 9.

8

Technology and work practices

The production of copper in prehistory was a complex undertaking, involving a sequence of inter-related activities that required specialist knowledge and informed organization. The operation of early copper mines demanded a long-term commitment that had important implications for the societies in question. There are obvious parallels with modern mining in terms of the multi-stage nature of the process, from the initial search for copper ores to their extraction, treatment, and eventual conversion to metal. This explains why early mining is often characterized as an 'industry', with its own specialists and techniques.

Given their professional background, it is not surprising that the earliest researchers in this field focused on the geological setting, technology, and engineering aspects of prehistoric copper mines. Modern research also deals with these aspects, but is more concerned with understanding the wider societal context of this activity. It is now recognized that any reconstruction of the *chaîne opératoire* of early copper mining cannot be based solely on technique and process, but needs to incorporate the dynamic social context of this activity (Fig. 1.9). It also needs to recognize that human choice played a significant role in determining the process of mining. While some areas are not amenable to archaeological enquiry, a reconstruction of task structure and work routines should capture many aspects of decision making in these mines. This will be revisited in the next chapter to explore the distinctive nature of prehistoric mining communities and how they interacted with the external world. Before doing so, it is necessary to examine the process of mining in terms of the techniques employed and the work environment of these miners.

THE SEARCH FOR COPPER

The opening chapter of this book considered the concept of ore, as it would have applied in prehistory. The modern definition is essentially an economic one that relates the cost of mining a mineral deposit to perceived financial returns or strategic interests. The effort of extraction must be set against the concentration or 'grade' of metal present in the rock, which will determine whether it can be profitably extracted. While modern cost–benefit concepts of ore are inappropriate in pre-monetary societies, there is no doubt that early mining groups applied their own value thresholds below which it was not profitable or technically feasible to extract metal. That said, the lengths these miners went to in search of even small amounts of copper were extraordinary, confirming the high value they placed on this metal.

Whereas modern ore prospecting employs a range of scientific methods, such as geophysical and geochemical survey, the search for copper in prehistory was more basic. Bedrock exposure was essential, which explains why most prehistoric copper mines are found in mountainous areas. Kargaly in southern Russia is an obvious exception, where Bronze Age miners were able to use limited surface exposures to infer the presence of copper mineralization covered by surface drift. In most cases the discovery of copper ore involved a careful search in areas of favourable geology, looking for rock outcrops stained with the bright colours of oxidized copper minerals.

In some instances, these discoveries were connected to an earlier history of Neolithic settlement in the same area. Early farmers collected hard rocks and exotic minerals to make tools and ornaments. This interest, and their close familiarity with the landscape, particularly upland areas where seasonal herding was practiced, led on occasion to the discovery of copper minerals. It is often stated that these minerals were first appreciated as sources of pigment, with some evidence for this coming from Copper Age settlements in the Balkans. Their significance as potential sources of metal became apparent as copper objects and metallurgical knowledge spread into those regions. A decision to actively search for these ores may have followed a local adoption of metallurgy. In some instances this did not happen as early metalworkers could trade for copper and engage in recycling. Where local sources were lacking, prospecting and mining expeditions may have been conducted outside of home territories, which raises the issue of resource ownership and access.

Geological setting

The opening chapter outlined how copper mineralization occurs in geological formations of different rock type and age. Many ore deposits were not formed by a single geological process but by separate mineralization events, which partly explains their heterogeneity. There is a broad distinction between epigenetic ore deposits, that formed later than the enclosing rock, and syngenetic mineralization, that developed more or less at the same time. The former includes vein-type deposits that cut across the country rock at differing angles, while the latter are commonly found as stratabound deposits laid in conformity with the host geology. Mineralization was created by complex geological processes, from volcanic activity deep in the earth to hydrothermal sedimentary processes at the surface. Some ore deposits only contain copper, but most are polymineralic to some extent, where metals such as lead, iron, and zinc are also present. Many of these metallic minerals were discarded during mining as part of the waste rock (gangue): however, minor or trace amounts of these metals will survive in the smelted copper and, in some instances, affect its properties.

With such variety, copper ores can be very complex in their geological setting, mineralogy, and chemistry. Prehistoric mining was heavily influenced by these geological controls and by the surface exposure of an ore deposit. Mineralized strata and veins were often contorted where the country rock was folded or fractured through faulting. The mine record reveals an empirical understanding of geological controls on the mineralization, which was only gained by many hours at the mine face. The miners learned to adapt to variability within an orebody, only selecting those copper ores that could be extracted and processed with the prevailing technology.

Prehistoric miners used different geological indicators to detect copper deposits. In the Mediterranean regions they were assisted by the presence of an iron hat (gossan), which could range from a few metres to as much as 40 m in thickness. These leached formations, made up mostly of iron oxides, had bright red and yellow colours, with small amounts of native copper, copper oxides, and copper sulphates also present. Gossans such as those in Cyprus or south-west Spain often overlie zones of secondary enrichment, where supergene mineralization in the upper part of the orebody has higher copper contents than the primary sulphide ore (Fig. 1.2). Gossans do not generally occur in the glaciated landscapes of temperate Europe where ore deposits mostly have a thin oxidation zone. There are some exceptions to this, including gossans with supergene mineralization at Parys Mountain, north Wales, which was mined for copper during the Early Bronze Age.

The extent to which these landscapes are different today from the prehistoric period must be considered in relation to the discovery of copper sources. The passage of Pleistocene ice laid down a mantle of fluvio-glacial till that conceals many ore deposits today, particularly those in lowland settings. Early miners had to contend with greater tree cover in some areas, though in other situations there could have been greater bedrock exposure where peat growth was not initiated.

A gradual accumulation of prospecting experience would have enabled miners to intuitively recognize the rock strata more likely to contain copper. Over time, they came to recognize and develop their own understanding of various geological controls surrounding this mineralization. Colour was the most important indicator of copper, with rock surfaces often brightly stained green or blue from oxidized minerals such as malachite and azurite. Many copper ores, such as the fahlore minerals and copper-iron sulphides, have a shiny appearance. The weight of mineralized rock fragments can also be indicative of metal content. These miners mostly relied on visual examination

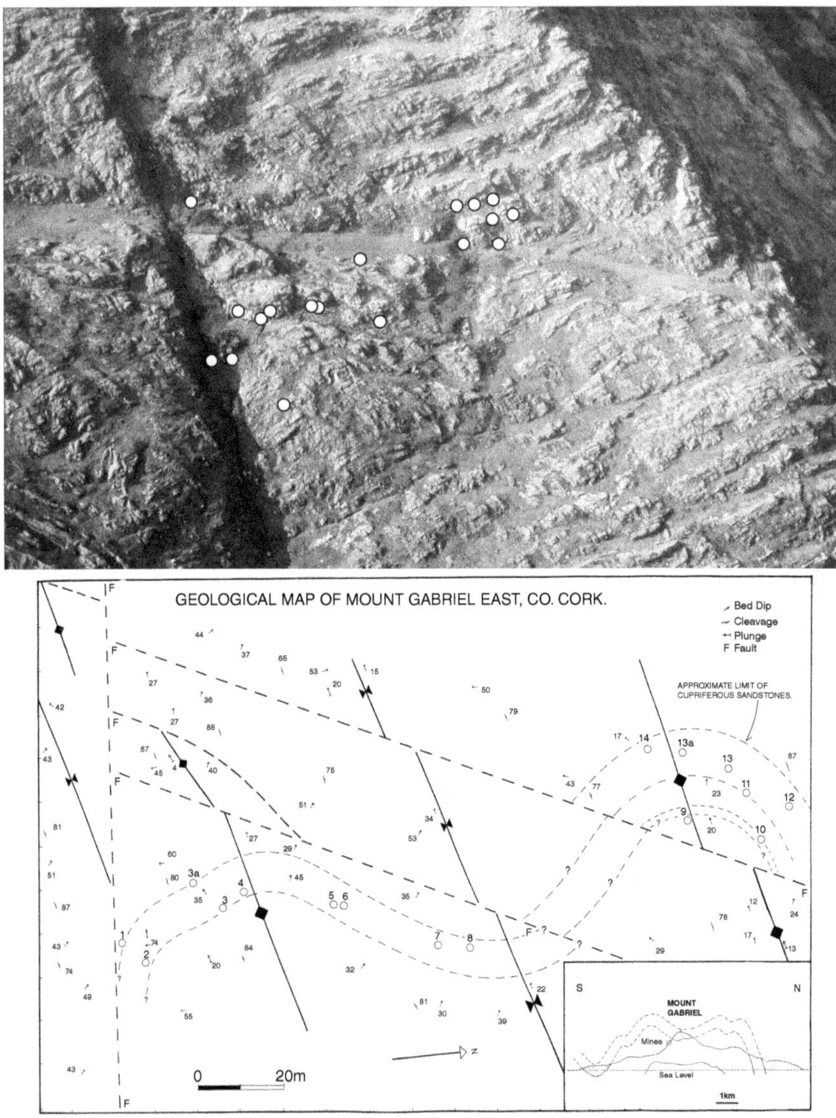

Fig. 8.1. Geological setting of Bronze Age copper mines, mid-slope on Mount Gabriel, south-west Ireland

The prospecting strategy of the miners involved a search for surface indications of copper minerals in a grey-green sandstone unit bedded within a purple mudrock sequence.
(*Source*: author and Michael O'Sullivan).

and experience to assess the amount of metal present in a piece of rock. It is not known whether this was taken further to use properties such as smell or flame colour to detect copper minerals in heated rock samples.

Mount Gabriel in southern Ireland provides a good example of the prospecting expertise that developed during the Bronze Age. The distribution of mines on this mountain indicates a careful and systematic examination of the rock outcrop in an effort to locate copper mineralization. The miners were helped by the presence of a Copper Horizon in this area, where copper-beds occur within red-bed sequences at a particular horizon in the Late Devonian sedimentary geology (Fig. 8.1). The mineralization was only found in coarser lithologies of greenish grey sandstone. The miners searched for these thin concordant bands within a mainly purple mudrock siltstone sequence. Once located, the miners walked along the strike outcrop of the sandstone beds searching for bright green staining, where small amounts of malachite were formed by surface oxidation.

A similar approach relying on the exposure of copper-stained outcrops of rock is evident in the discovery of other Bronze Age copper mines in Europe. In a few examples the miners were able to infer the existence of sub-surface mineralization by following the continuation of bedded deposits from limited surface exposures. This occurred in the case of the Kargaly mines, where the mining strategy was similar to that used in early flint mines, such as Grimes Graves in England.

Finally, other prospecting methods such as dowsing are unlikely to have been significant in the discovery of copper deposits. High levels of copper and other metals will affect the growth of certain types of vegetation. In some environments this can be a useful way for archaeologists to recognize deposits of mine spoil and smelting slag. There is no evidence, however, that geobotanical indicators were significant in the discovery of copper deposits during the prehistoric period.

THE APPROACH TO MINING

The opening chapter of this book considered how the selection of a particular copper deposit for mining during the prehistoric period was determined by several factors. These include the geological environment and topographic setting, the nature and concentration of the mineralization, and the technology available to exploit this ore. It is important to consider the decision-making aspect of the mining *chaîne d'opératoire*. Several 'push' factors can be considered in relation to the initial decision to commence mining. These include a perceived need or desire for copper, along with some understanding of the likely benefits that would arise from participation in mining. Access to suitable mineralization in terms of exposure and resource ownership was obviously important. The availability of sufficient labour and specialist knowledge, as well as access to supplies of equipment and service materials, are also relevant.

The type of mining carried out was determined to a large extent by the geological setting and surface exposure of the ores. Mining almost always commenced directly on the mineralized exposures, following the ore underground to a depth that was determined by a number of factors. We can distinguish between mining carried out on sources of concentrated mineralization, such as quartz-sulphide veins, and those situations where the ore is dispersed through a particular rock stratum. In all cases, the miners came to understand the various geological controls, whether it was the dolomitization process on the Great Orme or the colour and lithology of copper-beds on Mount Gabriel. In metallurgical terms, they also learned how to adapt to different types of copper mineral within an orebody.

There is an important distinction between mining in hard siliceous rocks and in softer, more friable types (Lewis 1996). The mining of soft calcareous rocks tends to produce more extensive workings, particularly where the geology was affected by weathering and chemical alteration. This influenced the type of tools used, such as the use of bone gouges and pick-like implements. Fire-setting was required for harder geology, for example sedimentary rocks such as sandstones, mudrocks, and slates (Alderley Edge, Cwmystwyth, Mount Gabriel, and the Mitterberg), or igneous geology (Parys Mountain). These rock types vary in ground water chemistry, which affects the hardness of the county rock and the type and grade of mineralization.

Bronze Age miners worked in many different geological environments where the ease of extraction depended not just on rock type, but on the degree of surface weathering, the influence of microstructures like cleavage, and the oxidation of sulphide ores. Mineralized strata and lodes often occur in a contorted state where the host rock has been folded or fractured through faulting. The miners had to adapt to these different geological circumstances. With karst geology they could take advantage of natural tunnels and caves to access copper ore. This occurred at the Great Orme, the Bouco-Payrol, and Grotte du Broum mines in France, El Aramo and El Milagro in Asturias, and the Grotta della Monaca in Calabria.

Mine workings

The mines themselves could take different forms depending on the geological setting and concentration of the mineralization, the duration of mining, and the techniques used. As they were mostly adapted to local geology and topography, it is not possible to create a typology of the mine workings characteristic of different periods of mining. That said, early copper mines have several features in common. Even though these miners had an ability to tunnel at will, in most cases there was an economy of effort where only rock containing copper minerals was extracted. For this reason, many workings are essentially emptied ore channels, be they mineralized veins or bedded deposits. Mine development in the modern sense of exploratory or service shafts and adits was unknown. An exception is the Kargaly mine, where vertical access shafts were dug through a 12 m overburden of alluvium to access the mineralized rock (Fig. 7.18). Another example is the systematic arrangement of inclined galleries in the Mitterberg mines, to facilitate the drainage and ventilation of deep underground mining (Fig. 8.4).

While they may be irregular in appearance, there is a degree of planning behind many of these early mine workings. Most large mines were not carefully planned from the outset, but developed in an organic way over centuries of more or less continuous working. This progress could be limited by geological factors and operational difficulties (flooding etc.), as well as fluctuations in the supply of labour and other necessary resources.

Surface workings were heavily influenced by the topographic expression of the orebody, which might determine whether they were horizontal, inclined, or vertical. Some took the form of small surface pits following the horizontal exposure (strike) of the orebody. A good example would be the many irregular hollows along the Brynlow and Engine Vein deposits at Alderley Edge in England (Fig. 8.2). Some surface pits were so large they could be described as opencast workings. These include a 45 m long by 8–20 m pit following the Comet Lode in Cwmystwyth mine. Another impressive example in Wales is the Great Orme, where an estimated 28,500 tonnes of rock was removed from a large open pit (Fig. 6.14). This overlay a complex network of underground tunnels, the form and extent of which reflects the occurrence of copper minerals in a soft rock environment heavily affected by karstic processes and dolomitization. Bronze Age mine workings were identified to vertical depths of 70 m, extending laterally for some 240 m over an estimated area of 24,000 m^2 (Fig. 6.15).

The form of many surface workings was determined by geological controls where the mineralization was confined to particular formations of rock. This type of stratabound mining was common and includes the so-called drift mines of the Mount Gabriel and Ecton (The Lumb) type (Figs 6.6 and 6.18). Trench mines are another variant of this mining. This is where the strike exposure of a vertical or steeply inclined mineralized vein or bed was mined to create a long, narrow working. There are examples at Ai Bunar, Alderley Edge, Chinflon, Cuchillares, Derrycarhoon, and Vallarade Les Neuf-Boches at Cabrières. Some examples can be very large, for example the 'Tranchee de Anciens' at St Véran, or the 'Blue Hole' on Ross Island. Mine trenches could continue over long distances as a series of short trenches; for example, at Ai

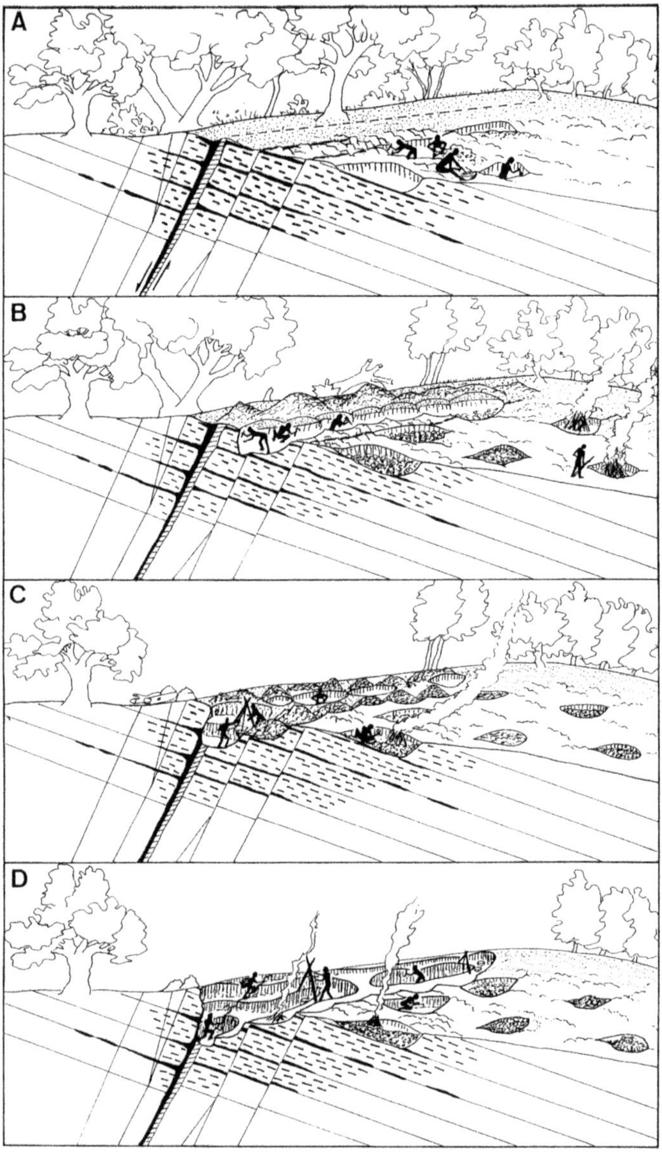

Fig. 8.2. Reconstruction of Bronze Age copper mining at Alderley Edge, Cheshire (drawing Brenda Craddock; source: Timberlake and Prag 2005).

Bunar where trenches measuring 15–110 m long followed the exposed veins for a distance of 1.5 km (Fig. 2.7). The *furchenpingen* of the Mitterberg are even more impressive (Fig. 7.4), while the same strategy is evident on a smaller scale at Derrycarhoon mine in Ireland (Fig. 6.8).

Vertical shafts were less common, but include the huge number recorded at Kargaly (Fig. 7.17). Many workings have an irregular profile, such as those at Rudna Glava where a series of steeply inclined tunnels follow the twisting shape of the ore veins to depths of 25 m (Fig. 2.3). Some workings have a cave-like appearance, such as those in the Western Mine at Ross Island. There are records of cave-like workings, now destroyed, in the Rio Tinto mines, which extracted supergene ore from beneath the gossan. Reference has already been made to the mining of copper minerals in karst environments, such as the example at Bouco-Payrol in the south of France where mine workings inter-connect with natural cave tunnels (Fig. 5.8).

Many surface workings represent more or less the entire extent of the prehistoric copper mine. In other cases, these openings served to access extensive underground workings of varying shape and size. The Great Orme is a good example, where trench mining and opencast extraction on the surface overlay a labyrinth of narrow underground tunnels connected to larger extraction chambers or 'stopes' (Fig. 6.15). The Kargaly mines are another example where a combination of workings (shafts, tunnels, and stopes) was employed.

Some operations involved a proliferation of small surface workings, moving from one mineralized outcrop to the next. The Mount Gabriel-type mines in Ireland are a good example, where a proliferation of small workings targeted small, mineralized exposures across a wider landscape. The relative success of this mining effort is obvious on Mount Gabriel itself, where some 30 workings ranging from less than one metre up to 12 m in depth are recorded (Fig. 8.1). This size variation reflects the varying concentration of copper minerals at each outcrop.

While many prehistoric copper mines were worked by 'flitting' from one mineralized exposure to the next, larger workings reflect a concentration of mining effort on a single rich deposit over a period of centuries or more. These include the Great Orme where underground tunnels extend over some six kilometres to a depth of up to 70 m. Another example is the mine at El Aramo in northern Spain, where steeply inclined tunnels followed the ore veins to depths of 150 m below the surface.

METHODS OF ROCK EXTRACTION

The ability to mine metal ore from the Earth's surface has a background in thousands of years of rock extraction during the Stone Age. Many methods of rock extraction in the earliest copper mines can be paralleled in contemporary hard rock quarries and flint mines of the Neolithic (see Weisgerber 1987). This is evident in the mining of bedded mineral deposits, as in a comparison of the

Kargaly and Grimes Graves operations, and in the extraction of minerals in karstic caves such as the Grotta della Monaca in Calabria. Many of the basic techniques in stone quarrying, flint mining, and early copper mining combined human muscle power with the use of stone, bone, and wooden tools. There were, however, important differences, not so much in relation to the use of metal implements, but in the manner that rock was extracted. The most important was the use of fire-setting and stone hammers to extract copper ore in a comminuted form, as opposed to the detachment of blocks of rock for lithic reduction. There were also differences in the heterogeneous nature of the rock extract, where copper minerals were contained within a matrix of barren rock (gangue).

Fire-setting

This is a rock extraction method of great antiquity, found in early mining contexts the world over. The technique continued in some regions into historic times, including Norway where it was used as late as the nineteenth century (Timberlake 1990c; Willies 1994; Weisgerber and Willies 2001). The technique involved the burning of wood-fuelled fires against a rock face, causing this to weaken and micro-fracture (Fig. 8.3). Some rock might be removed by direct thermal exfoliation, but mostly through pounding the fire-cracked face with stone hammers, and prising out the ore with fingers and pointed implements.

The effectiveness of this method depended very much on the country rock. It was particularly suitable for hard siliceous rocks and less so for calcareous geology. Experimental studies confirm that the effects were variable, depending on the heterogeneity of the orebody and the geological setting (O'Brien 1994: table 10). Different rock types react in different ways to heat treatment, depending on their chemical and mineral composition, texture, thermal expansion ratio, and internal structure (Willies 1994: 2–4). The thermal shock effect may have been increased in some instances by sudden cooling with water. While water quenching is often cited in relation to prehistoric fire-setting it is difficult to prove and is unlikely to have been widely used.

The application of fire-setting is generally recognized by the smooth concave profiles that resulted on the mine walls, as well as the presence of fuel residues inside the workings and in surface spoil. The technology is not well suited to vertical workings, but is particularly useful in the driving of inclined mine tunnels of the type found in the Mitterberg or on Mount Gabriel. The use of this method could also be constrained by fuel supply, as large amounts of wood had to be gathered from local woodland in an organized manner. This explains why it was not widely used in early copper mines in south-west Spain or in the treeless steppe of Kargaly.

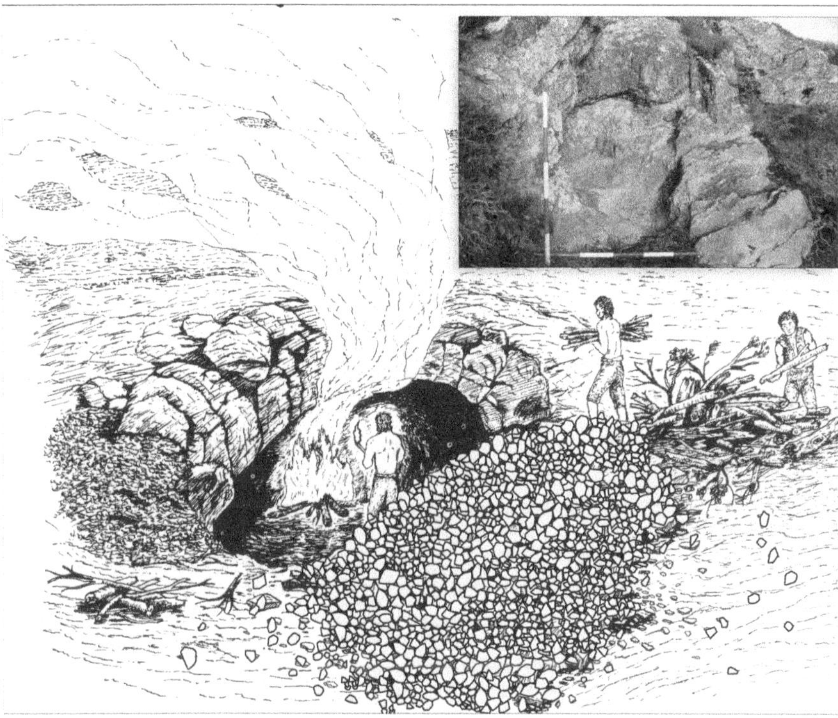

Fig. 8.3. Artist reconstruction of fire-setting in a Mount Gabriel-type mine, Ireland, with example (inset) of a Bronze Age fire-trial

(*Source*: author).

The use of fire-setting was particularly sophisticated in the mines of the Mitterberg (Zschocke and Preuschen 1932; see also Pittioni 1951). Fires were lit directly on the mineralized outcrop once the latter was fully exposed on the mountain slope (Fig. 8.4). The fire-setting progressed on an inclined plane, with support timbering required after a short distance underground. This arrangement served to divide the mine tunnel into different sections to facilitate the flow of air. Barren rock was stowed on top of the support timbering, which also provided a platform to direct fire on the tunnel roof. As the mining progressed new tunnels were opened at a higher or lower level to the first, using the same fire-setting technique to extract ore. The different levels were eventually connected to facilitate mine drainage.

Water would have been a problem for fire-setting in inclined workings, particularly in high altitude mines with higher rainfall. In the case of the Mitterberg this problem was also addressed by building dams at the tunnel entrances, with hand-bailing possibly used as a last resort. Pittioni (1951) noted that the discovery of flooded Bronze Age workings was a particular hazard for miners in the Mitterberg in the modern era.

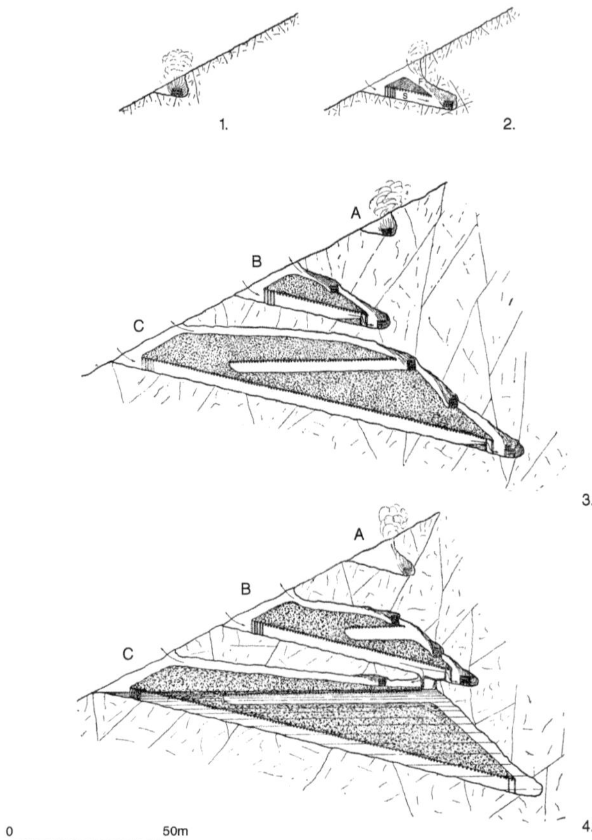

Fig. 8.4. Stages in the application of fire-setting in Bronze Age inclined workings at the Mitterberg

(*Source*: author, original from Zschocke and Preuschen 1932; see also Pittioni 1951).

Stone hammers

The use of stone hammers was essential in the effective use of fire-setting in prehistoric copper mines (Fig. 8.5; see Gale 1995). Sometimes referred to as mauls, these were rounded cobbles of natural origin, gathered from glacial deposits, river, or beach sources. In some cases the sources were local; however, in other instances they had to be transported some distances. In the Mount Gabriel mine cobbles had to be hauled from beach sources 4 km away, or up to 25 km in the case of Cwmystwyth in Wales. Along with the collection of wood fuel, this was a laborious task that may have required its own work teams. It is also possible that the miners themselves would have cached a sufficient number of cobbles at the start of each mining season.

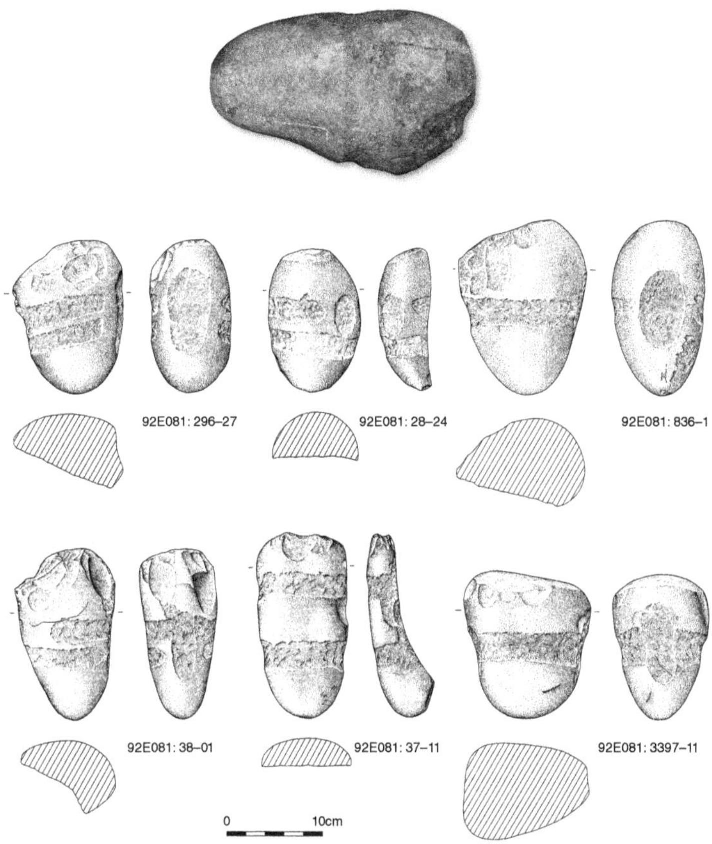

Fig. 8.5. Selection of stone hammers of Chalcolithic date from Ross Island copper mine, Ireland.

Examples of haft modification include single and double rills with lateral facets for insertion of tightening wedges.
(*Source*: author).

These hammers were used either hand-held or hafted to break rock at the mine face or in ore concentration at the surface. Depending on the required use, they can weigh 0.5–10 kg, with most examples in the 1–3 kg range. Many cobbles were modified for hafting prior to use, to allow either a twisted withy, leather strap, or a rigid wood handle to be attached (Fig. 8.6). This ranged from minimal side abrasion, such as that applied on Mount Gabriel and many of the Welsh examples, to complete or partial lateral rilling or deeper grooving (Fig. 6.12). The former usually indicates a brief use-life, whereas more modified examples, such as those used at Alderley Edge, El Aramo, Ross Island, and Rudna Glava, lasted for longer periods. This often reflects a distinction between the use of hammerstones of similar lithology to the rock they were

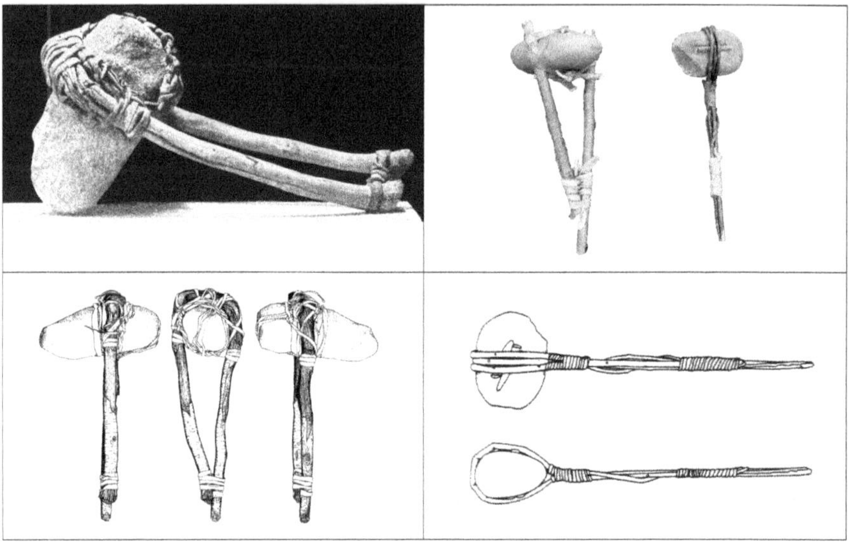

Fig. 8.6. Examples of methods used to haft stone mining hammers

(*Source*: Brenda Craddock) with photograph of Chuquicamata hammer, Chile (top left) taken from Kyrle (1916).

breaking, and those of a harder type used against softer lithologies. Most examples tended to break easily, particularly when used in heavy-duty rock breaking. This is why they are often found broken in large numbers in surface spoil, which represents an important archaeological indicator of this activity. Some were used in the coarse crushing of rock extract using block-on-block techniques, and subsequently in fine crushing along with mortars and grinding stones.

Other mining implements

The miners would have exploited any structural weakness in the country rock, such as veining, bedding, cleavage, joints, or other fracture planes, using pick-like implements and fingers to prise out loose material. Wooden implements, including pointed sticks, picks, and short wedges were discovered on Mount Gabriel, though whether these were all used in rock extraction is unclear (Fig. 8.7).

There is evidence for the use of antler picks in Chalcolithic mines at Rudna Glava and Ai Bunar in the Balkans. They are also recorded in the mines of El Aramo, El Milagro, and La Profunda in northern Spain. The mine at El Milagro has also produced antler pick levers and hammers with transverse perforations, which are not known in other early European copper mines

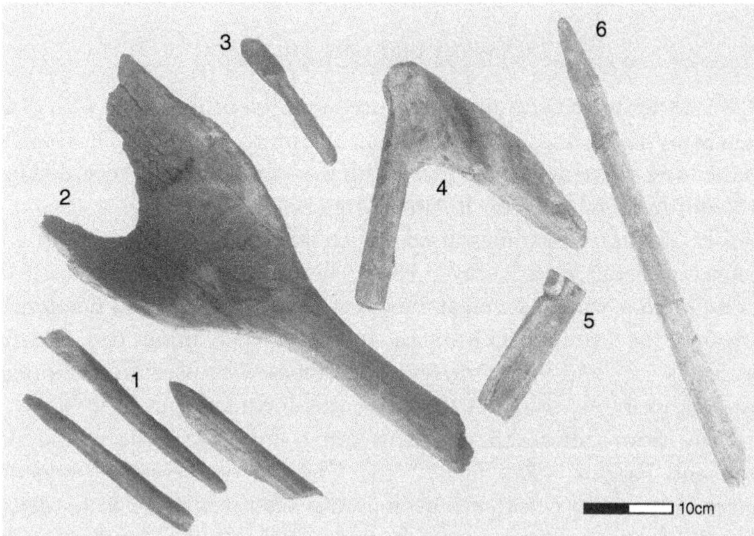

Fig. 8.7. Selection of Bronze Age mining implements found in excavation of waterlogged Mine 3, Mount Gabriel, south-west Ireland

They include: 1 Pine lighting chips; 2 Short-handled alder shovel; 3 Carved peg of alder (unknown function); 4. Hazel pick; 5 Pine wedge; 6 Pointed prise-stick. A number of oak planks and a hazel withy (not illustrated) were also discovered in this mine.
(*Source*: O'Brien 1994).

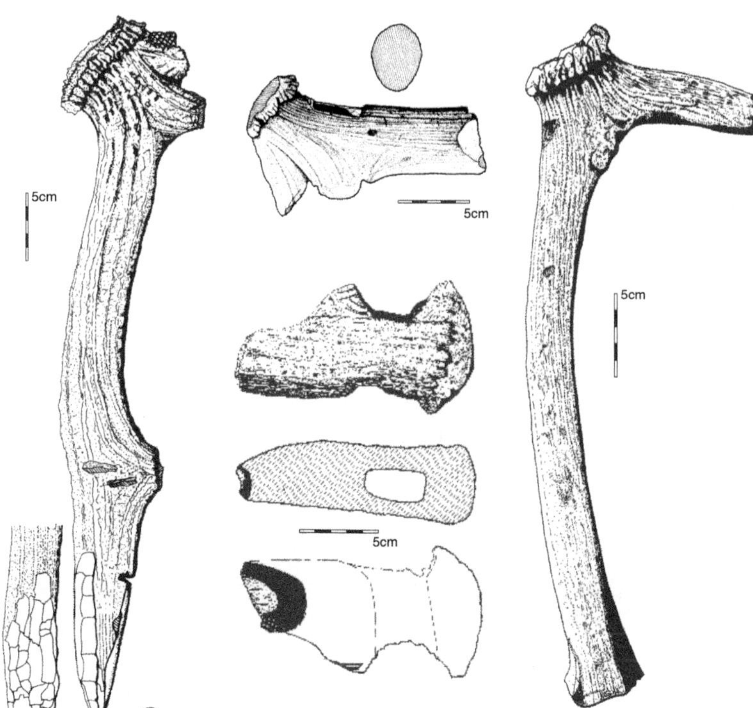

Fig. 8.8. Selection of red deer antler tools of Chalcolithic and Early Bronze Age date found in the El Milagro mine, Asturias

They include (left to right): antler chisel, two antler hammers and an antler pick.
(*Source*: Miguel de Blas Cortina).

(Fig. 8.8). Antler tools continued to be used into the Bronze Age, with discoveries at Cwmystwyth and Ecton in Britain, and more recently at Derrycarhoon in Ireland. The use of goat-horn picks with wooden handles is recorded in the Saint-Véran mine in the French Alps (Barge 2003).

In other situations the miners used pointed animal bones to scrape or gouge out copper minerals from decayed rock. These simple bone points left their mark on the mine walls, as seen at the Great Orme (Fig. 8.9) and Ecton mines, at El Aramo, the Grotta della Monaca, and the Kargaly mines (e.g. Chernykh 2004a: plates 7.21 and 9.9). Different bone types were used from a range of animals; for example, in the Great Orme this included split limb bones and ribs mostly from cattle and pig, with some sheep, goat, deer, and horse (Dutton and Fasham 1994: figs 12–13). The use of fire-setting and stone hammers in mines of calcareous geology was often restricted to harder, less weathered portions of country rock, particularly as the depth of mining increased.

With a few exceptions, metal tools were generally not used in the extraction of rock in these mines. Two shaft-holed copper implements were found in the Ai Bunar mine in Bulgaria. They include an axe-adze and a hammer-axe, the latter of a type similar to the gads used in historic mining (Fig. 2.8). Whether these implements were actually used in mining, or in some ancillary activity, is

Fig. 8.9. Wallrock marks from use of bone picks and scrapers (inset) in Bronze Age mining on the Great Orme, Wales

(*Source*: Andy Lewis).

Fig. 8.10. Bronze picks discovered in the Mitterberg mines, Austria (*Source*: Kyrle 1916).

uncertain. There are records of bronze mining picks in the Mitterberg (Fig. 8.10) and also in the Kargaly mines where bronze shovels are also reported (e.g. Chernykh 2002: plate 5.8). A shaft-hole hammer of bronze from the Mitterberg mines (Pittioni 1951: plate 4.2) is unique in the early mine record in Europe. The significance of tiny fragments of bronze discovered in the Great Orme mine is uncertain. Lewis (1996) has suggested

Fig. 8.11. Artist reconstruction of underground copper mining at Ross Island, Ireland (*Source*: author and David Hill).

that bronze tools as well as fire-setting might have been required for extraction of harder rocks in the less weathered depths of that mine.

The most important use of metal tools was in the collection and working of wood for different purposes, including fire-setting and the preparation of mining equipment. Metal axes and blades were rarely discarded in the mines, however there are some finds. Copper axes were discovered in ancient workings in Portugal, with both copper and bronze examples recorded from the El Milagro and La Profunda mines in Spain. There is also the evidence of tooling marks on mine timbers, roundwood fuel, and items of equipment, as seen on Mount Gabriel. The use of metal axes can also be inferred from the absence of broken stone axeheads in these mines. These include a broken example found in Ross Island mine and a cache of 12 examples from Ballyrisode mine, also in south-west Ireland (O'Brien 2003: fig. 11).

In conclusion, a remarkably similar range of techniques was used to mine copper during the prehistoric period. These represent the same cultural responses to the demands of rock extraction and ore treatment in broadly similar geological environments. The earliest mines were located on mineralized outcrops that only required basic mining techniques (Fig. 8.11). Deeper mining developed over time as experience accumulated, leading to the development of more advanced mining methods. This included the first use of bronze mining tools; however, these do not seem to have been widely used and rarely feature in the archaeological record.

The practical application of these tools in different geological environments may be considered in the following two case-studies.

The Great Orme: mining in soft rock

The geological setting of this ore deposit was the most important influence on the overall form and size of the prehistoric mine (Lewis 1996). Various structural controls can be identified where the primary mineralization was directed along a series of north–south fractures and joints, and to a lesser extent along east–west fractures (Fig. 1.3). There were also lithological controls where the mineralization is confined to dolomitized limestones, particularly those with a dominance of coarse-grained calcite crystals.

Oxidation of the primary chalcopyrite produced significant amounts of hydroxycarbonate minerals, dominated by the brightly coloured malachite and azurite, as well as hydrated iron oxides and other secondary minerals. Their formation was related to rotting of the dolomites, especially in the immediate vicinity of the fractures, a process that extended in certain parts of the mine to a depth of 130 m. The extent and shape of the mine workings relates directly to the removal of this ore in the zone of rotted dolomites, centred around the mineralized fractures and, to a lesser extent, in the adjacent mudstones.

Lewis (1996) has proposed that copper mining began in the Pyllau valley around 2000 BC with the extraction of these secondary copper minerals from the rotted dolomite. This brightly coloured ore was easily identified on the surface exposure of the north–south veins. The extraction of these minerals using bone gouges and other pointed tools, with some use of stone hammers, created a series of parallel trench workings (Fig. 6.14). The removal of copper minerals from the rotted dolomite would have been fairly rapid; however, the less weathered rock was more difficult, requiring greater use of stone hammers and possibly fire-setting. These harder rocks may have initially been left, and were only removed at a later stage of mining in the Bronze Age.

As mining progressed, it was discovered that ore-bearing rock could be followed from the sides or base of the trench workings. This led to the development of a large opencast. Lewis (1996) has discussed whether this was formed as a single operation, or arose from a series of interconnected workings that were extended to make the open pit. As mining progressed, openings at the base and sides of the opencast led to a maze of underground tunnels (Fig. 6.15), which incorporated natural solution hollows in the karst environment.

The weathering of the copper mineralized rock in this mine became less pronounced with increasing depth, reducing the effectiveness of bone tools. This is confirmed by the complete absence of bone tools in workings where stone hammers and fire-setting charcoal are found. The later stages of mining in the Bronze Age are represented by the deeper tunnels and by an extension of the underground mine northwards into the hillside beneath Bryniau Poethion. Spoil from these later operations was probably backfilled into the earlier opencast. The supply of copper minerals steadily declined once the limits of the ore-bearing dolomite were reached, resulting in the eventual abandonment of the mine early in the first millennium BC (Lewis 1996).

Mount Gabriel: mining in hard rock

The geological setting of the mineralization on this mountain influenced the overall approach to mining during the Bronze Age. The copper-beds are essentially stratiform, i.e. commonly parallel to the stratification of the enclosing sedimentary rocks. They occur as concordant grey-green bands with gentle bedding of 20–30° westwards, which outcrop on the face of joint-controlled escarpments. This bedding attitude determined the typically sub-horizontal trend of these mines, driven as they were into the scarped exposure of the mineralized units (Fig. 8.12). The limited continuity of disseminated mineralization in a thin sandstone bed determined to a great extent the final shape of these mines. They range from shallow surface workings of less than one metre in depth to larger underground mines of 11 m or more. Their overall size was determined by the depth of the water table during the mining season, and by the concentration of secondary copper minerals present in the surface zone of oxidation.

The mining strategy was particularly suited to this type of small-scale mineralization. It involved the use of drift workings where separate showing of copper minerals were pitted along the strike exposure of these copper-beds. The intensive nature of this prospecting is demonstrated by the occurrence of 14 workings across a 200 m long strike of a copper bed (Fig. 8.1). There was an obvious economy of rock extraction in the shallow confined profile of these mines, where only the grey-green sandstone hosting copper minerals was removed. This required intensive fire-setting and the pounding action of hammerstones to extract even small amounts of rock. They were assisted by microstructures that helped weaken the hard siliceous country rock, such as joint sets, cleavage, and minor quartz veining.

While the copper-beds on Mount Gabriel are bedded in conformity with the sedimentary sequence on this mountain, they occur in a contorted state within minor fold structures that control outcrop exposure. The configuration of these copper-beds will vary depending on their structural position within

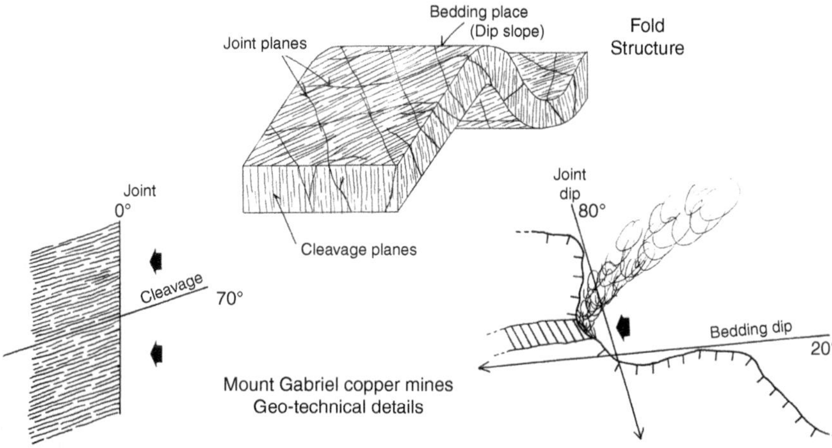

Fig. 8.12. Geo-engineering controls on Bronze Age mining on Mount Gabriel, Ireland

Example of Mine 5/6 showing green copper bed within a purple mudrock sequence. The primary exposure sites on the eastern slopes of Mount Gabriel are scarp faces that are glacially accentuated. These scarps are defined by A/C joint planes that strike 350°–020° and dip steeply (070°–090°) to the east. Cleavage planes strike 070° and dip steeply to the south (070°–090°). These planes are more closely spaced in mudrocks (pressure solution cleavage X mm) and more widely spaced in sandstones (fracture cleavage X cm). Drivings made into east-facing joint-controlled scarps will have side walls defined by cleavage planes, while the hanging and footwalls will be defined by gently dipping, bedding planes (020°–030° south-west).
(*Source*: author).

these minor folds. This is apparent at several workings; for example, Mine 3 on the mountain was driven into a tight S-shaped fold (Fig. 8.12), however the anticlinal portion was soon abandoned to mine deeper into the core of the syncline. The first 5–6 m of extraction was done in conformity with the bedding of these rocks, at which point the miners decided to tunnel across the bedding to an eventual depth of around 11 m. This transition from stratabound drift mining to tunnelling was an important development in early hard rock mining. These mines were driven parallel to an axial planar rock cleavage. In one instance (Mine 3b) the miners erred in driving against this natural lamination (Fig. 8.3 inset), a mistake they quickly recognized and so abandoned the working.

THE MINING ENVIRONMENT

While surface mining was considerably less complicated, deeper workings presented practical difficulties and potential hazards that prehistoric miners learned to deal with through experience and the use of suitable equipment. It is necessary to go beyond the static archaeological record to imagine a scene of bustling activity and noise at the mine face, in the underground passages, and the surface operations. There was constant movement of miners, their equipment, and fire-setting fuel, as well as the haulage of rock extract out of the mine (Fig. 8.13).

It was usual to remove rock extracted in the mine to the surface for further treatment. In some deeper mines there is evidence of underground sorting and

Fig. 8.13. Underground mining environment in the Great Orme

Andy Lewis retrieving a hammerstone in the Great Orme mine, north Wales. Such underground excavation by archaeologists provides an insight into the practical realities and overall experience of daily work in these Bronze Age copper mines.
(*Source*: A. Lewis).

stowing of waste rock in abandoned tunnels. This occurred in the Great Orme, partly for support reasons, but also to reduce the effort of bringing large amounts of barren rock to the surface from a considerable depth. It was done in an organized fashion, to maximize backfilling of waste rock underground while maintaining access (Lewis 1996). This also occurred in the Mitterberg mines where it was connected to the maintenance of airways for underground fire-setting (Fig. 8.4). The movement of materials and rock extract required containers of some sort, probably in the form of leather sacks or wooden baskets. There is little evidence for such containers from these mines, however the remains of woven baskets may have been found at Cwmystwyth (Fig. 8.14). Wooden troughs were used in ore separation in the Austrian mines (see The Treatment Of Copper Ore section) and similar containers may have been employed in the haulage of rock extract. The use of wooden shovels is recorded in the Mitterberg mines (Pittioni 1951: plate 7.1), at Alderley Edge (Timberlake and Prag 2005: fig. 1.1) and Mount Gabriel (Fig. 8.7). There is evidence from Ross Island for the use of cattle shoulder-blade bones as scoops and cattle rib bones as spatulae in ore separation (O'Brien 2004: plates 75–7).

While many prehistoric copper mines were surface operations, some provision for access had to be made in deeper workings. This includes the cutting of footwall notches in steeply inclined workings, as recorded in the Mount Gabriel and Chinflon mines. Rock-cut hand-holds were used in the vertical mine shafts at Kargaly (Fig. 7.19). Vertical workings required some form of rope, however, there are no recorded examples from these mines (twisted withies found at Mount Gabriel and Cwmystwyth are probably associated with either basketry or stone hammer handles). Notched tree-trunk ladders are recorded in the Mitterberg mines, at Chinflon, and at Derrycarhoon. At the latter mine short lengths of roundwood with chopped offset branches were also used for this purpose.

Ventilation was an issue for deeper mines, particularly where fire-setting was carried out. The underground workings in the Great Orme had numerous connections to the surface for this purpose. There is some evidence from that mine for stacking of rock waste to control air-flow for fires lit at depth, a practice also recorded in the Mitterberg (Fig. 8.4). In many operations there may have been a daily work routine, where the miners vacated underground workings while the fires were lit, using this time on the surface to concentrate ore, collect service materials like wood, prepare mine equipment, or simply rest and consume food. In some instances the mine fires may have been lit at night, allowing time for smoke to clear before mining resumed at daybreak.

Lighting was a considerable problem in the case of deeper workings. The use of chips of resinous wood, such as pine, was common, as they burn with a brighter flame. Charred splints of pine, used individually or bundled as

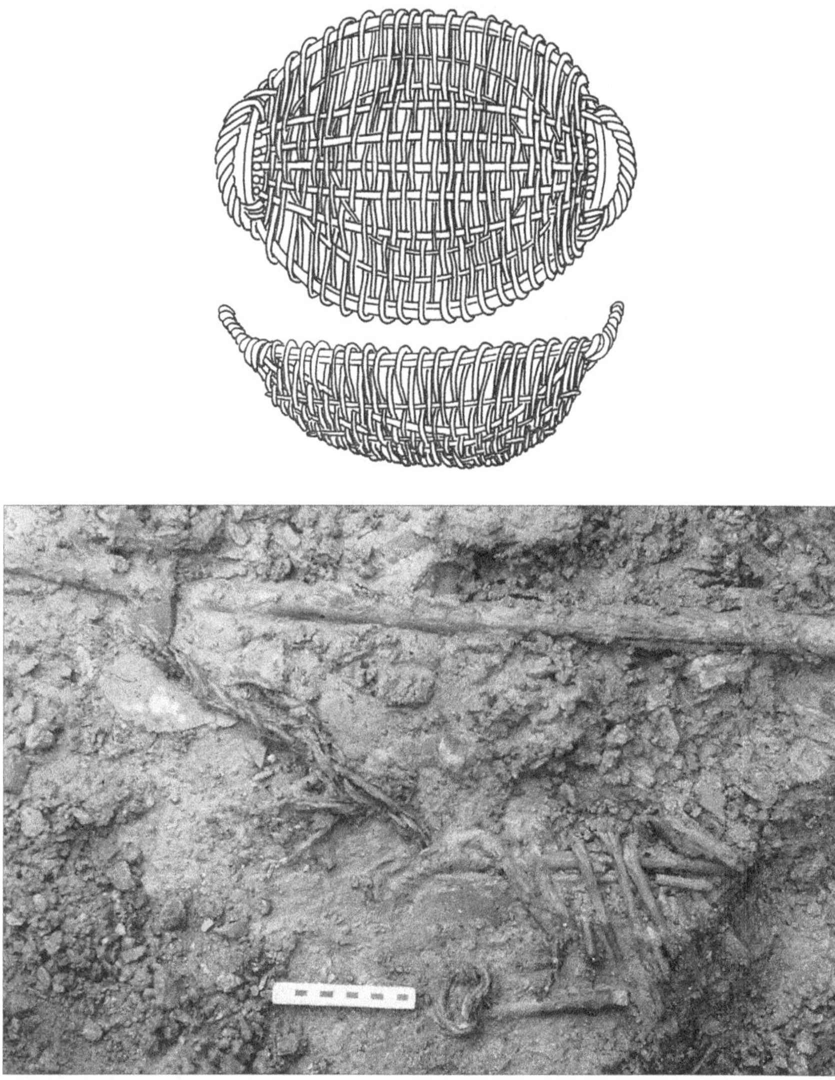

Fig. 8.14. Fragments of woven basketry (with reconstruction) excavated at Copa Hill mine, Cwmystwyth, Wales

(*Source*: Simon Timberlake; drawing: Brenda Craddock).

torches, were found at the Mitterberg, Saint-Véran, and Mount Gabriel. There is little evidence for animal fat lamps, with the possible exception of a small clay example found at Chinflon mine.

Like their counterparts in historic times, Bronze Age miners had much to fear from roof collapse and flooding. That said, there are few recorded instances of the former, mainly because the mining process did not disturb the rock strata to the same extent as modern explosives. Fire-setting tends to create workings with smooth load-bearing walls, which explains why support timbering was not widely used. Roundwood stemples and other split timbers were required in some mines, with the best-known examples coming from the Arthurstollen workings in the Mitterberg (Fig. 8.15). There is evidence for the stacking of waste rock and support timbers in the Saint-Véran mines. Pillars or arches of rock for roof support are recorded in many mines. The numerous small pillars in some of the workings at El Aramo mine in northern Spain may have been used to fix ropes for access or haulage purposes in those steeply inclined tunnels (Fig. 8.16).

There is little evidence of fatalities in prehistoric copper mines (the significance of human remains will be examined in the following chapter). This is probably due to the nature of the mining processes, but could be due to the recovery of bodies for burial elsewhere. It may also be because archaeologists have not excavated collapsed workings. The discovery of human leg bones at Ross Island and Chinflon may indicate accidents similar to the fate of the 'Copper Man', a mummified corpse from Chuquicamata prehistoric mine in northern Chile, associated with hafted stone hammers, wooden tools, and

Fig. 8.15. Wooden stemples used in roof support in Bronze Age workings exposed in the Arthurstollen gallery, Mitterberg

(*Source*: Deutsches Bergbau Museum and Thomas Stöllner, with inset drawing by Georg Kyrle 1916).

Fig. 8.16. Arches of rock at entrance to prehistoric copper mines, El Aramo, Spain (*Source*: author).

baskets (Bird 1975). There is an early record of two undated skeletons found with stone hammers in a collapsed mine at Peñaflor, Seville province (Hunt Ortiz 2003). Mining was always a dangerous activity, and some evidence of protection rituals and superstitions might be expected.

Though less of an issue for mining in arid regions of the Mediterranean, water was certainly a significant problem in the mines of temperate Europe. This was particularly the case where fire-setting was employed in upland areas of higher rainfall. Many workings are permanently flooded today; however, the extent of this problem during the prehistoric period is not clear, as local rainfall patterns were probably different than today. The miners may have confined their operations to the drier months, working day and night to minimize flooding, using animal skins or other containers where hand-bailing was necessary. Evidence for drainage equipment comes from excavations at Cwmystwyth where a wooden pipe (launder) was uncovered leading out from a Bronze Age working (Fig. 8.17). There are similar finds from the ore processing sites at Kelchalm (Pittioni 1951: fig. 5.1). In some situations, the miners had to contend with more serious flooding; for example, at Ross Island where mining in permeable limestone was conducted close to the lakeshore.

Fig. 8.17. Wooden launder used for drainage in the Bronze Age copper mine at Copa Hill, Cwmystwyth

(*Source*: Simon Timberlake).

THE TREATMENT OF COPPER ORE

Following the extraction of mineralized rock, beneficiation was the next stage in the mining operation. This refers to a series of filtering processes by which the copper minerals are separated from a matrix of barren rock (gangue), to produce an ore concentrate suitable for smelting. Modern mines achieve this by exploiting some physical or other property of the metal ore through water flotation or chemical treatment. While prehistoric methods were more basic, there is evidence of an organized, multi-stage approach to the concentration of copper ore.

Evidence of ore beneficiation in early copper mines generally takes the form of deposits of crushed rock spoil in the vicinity of the mine workings. Where intact, the size of these spoil dumps provides an indication of the extent of the underground workings. The ore was separated from its gangue matrix by a combination of crushing and hand-sorting, guided by the colour, lustre, and density of the copper-bearing minerals. It is common to find stone tools used in this process, as they were generally discarded in situ when broken or heavily worn. They include stone hammers and anvil slabs, mortars, and grinding stones (Fig. 8.18). The mineralized rock was repeatedly crushed and hand-

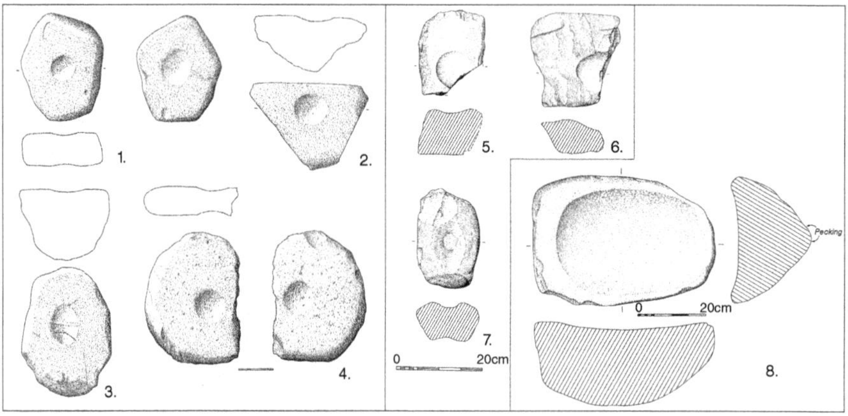

Fig. 8.18. Grinding stones used in ore concentration from Cabrières mine, France (1–4) and Ross Island, Ireland (5–8)

(*Source*: Ambert et al. 2009 and author).

sorted into different ore grades, before being ground and further concentrated, in some instances by water separation. While hand-sorting is laborious, it was an effective beneficiation technique in early mining contexts where labour costs were low relative to the value of the metal product. The efficiency of this approach can be seen in the sparcity of copper minerals in many prehistoric mine dumps.

Water flotation may have been an option for many copper ores; however, it is difficult to obtain archaeological evidence for this when carried out adjacent to a river or stream. The washing of ore concentrate and crushed slag is well known from the Bronze Age mines of Austria (see Eibner 1982). Wooden troughs used for this purpose were excavated at Kelchalm (Pittioni 1951: plate 4.1), Mitterberg/ Troiboden (Stöllner et al. 2011: fig. 20) and the Mauk mines near Brixlegg (see Figs 7.13 and 8.19).

There is evidence of water separation of copper prills from furnace sediment at Ross Island mine, where Beaker pottery vessels were used for that purpose (O'Brien 2004: plates 50–1). The practice is also recorded for the processing of fahlore at Cabrières in southern France in the third millennium BC (see Beneficiation of Fahlores section). Where water separation was not feasible, as in arid environments, finely ground ore may have been winnowed. Merkel (1985) suggests that this was done in the Late Bronze Age/Iron Age copper mines of Timna, Israel. There is no evidence the technique was used in early European mines.

The following case-studies illustrate the different treatment of copper ore in these prehistoric mines.

Fig. 8.19. Wooden troughs used in ore washing, Mitterberg mines, Austria (*Source*: Georg Kyrle 1916).

Beneficiation of oxidized mineralization

The Mount Gabriel mines in south-west Ireland provide a good example of the recovery of low-grade ore. Sampling of surface spoil in these mines proved problematic as the residual mineralization does not reflect the grade and mineralogy of ore extracted by the Bronze Age miners. The mineralization on this mountain does not conform to modern perception of what constitutes a viable ore because of the low metal concentration in these cupriferous sandstones. It was even argued at one point that these mines represent extensive trials that failed to find suitably rich ore (Crew 1994; but see O'Brien 1997). This is not supported by the scale of mining on Mount Gabriel over at least two centuries, nor by the sorted nature of the surface spoil.

The miners on Mount Gabriel were able to extract small quantities of secondary copper minerals by careful beneficiation of the mineralized rock using stone hammers and anvil slabs. The mineralization is low grade (less than 0.5 per cent wt% Cu), fine-grained (0.1 mm to 2 mm in diameter) and disseminated within a silica-rich matrix. A regional study of this mineralization indicates that the highest grades in outcrop range 0.5 per cent to 6 per cent wt% Cu, usually occurring over distances of under one metre (Snodin 1972: table 3). The ore minerals are chemically simple in terms of their metals, though their mineralogy is more complex. They include such copper sulphides as djurleite-chalcocite, digenite, covelline, and blaubleibender covelline, together with minor copper-iron sulphides such as bornite, altered bornite (idaite), and trace chalcopyrite. This sulphide component occurs in both disseminated and stringer form, but is considerably less abundant than the oxidized mineralization, which is dominated by malachite with minor azurite and cuprite (Ixer 1994).

There is evidence of a multi-stage approach to the concentration of this low-grade mineralization. The archaeological record consists of small dumps of broken rock debris containing fire-setting charcoal, located directly outside the mine entrances (Fig. 8.20). The comminuted nature of this rock spoil, and the presence of broken stone hammers, indicates that the first stage of ore concentration involved the crushing of the mineralized rock extract with continuous hand-sorting of malachite-stained fragments. Further processing of this ore was undertaken elsewhere in the mine camp, where excavation uncovered stone slabs surrounded by finely crushed spoil and broken stone

Fig. 8.20. Spoil heap adjacent to Mine 3 on Mount Gabriel, Ireland, showing (inset) stone anvil and hammer stones used in beneficiation of copper minerals

(*Source*: author).

hammers (Fig. 8.26). The final stages of ore concentration produced a finely crushed malachite-rich concentrate for smelting, which contained a small amount of sulphide minerals. There is no evidence for the grinding of this ore, nor any indication that it was washed to concentrate the copper minerals. Very little mineralization is visible in these spoil heaps today, which points to an efficiency in the recovery of rock fragments with malachite staining.

Similar methods of mechanical reduction and hand-sorting are recorded at other early mines where oxidized copper ore was extracted. Examples include the Great Orme mine in Wales (Lewis 1996), Alderley Edge and Ecton mines in England (Timberlake and Prag 2005; Timberlake 2013), Chinflon and similar mines in Huelva, Spain (Rothenberg and Blanco-Freijeiro 1981), the Kargaly mines in Russia (Chernykh 2003), among others. Stone hammers and anvil stones were widely used in this process, often accompanied by the use of grinding stones to produce a fine concentrate for smelting.

Beneficiation of fahlores

Fahlore (*fahlerz* from the German 'pale ore') is the name commonly given to the tennantite–tetrahedrite group of minerals, a series of copper sulphosalts containing varying amounts of antimony, arsenic, bismuth, and silver (see Ixer and Pattrick 2003 for details). The smelting of fahlores with arsenic (tennantite range) or antimony (tetrahedrite range) as significant impurities will yield copper with superior technical properties than the purer form of the metal.

The mining of tetrahedrite-rich fahlore is recorded in the Cabrières area of the central Languedoc. These veins contained significant amounts of tetrahedrite and chalcopyrite, as well as oxidized copper ore in the form of azurite and malachite. A number of production sites are recorded close to these mines at Cabrières (Espérou 1981; Ambert 1990, 1996a). Excavations at Roque-Fenestre uncovered four adjacent pits used in the treatment of copper ore. This involved different stages of crushing and washing of copper ore, using gravity concentration in water-filled pits to separate the metallic minerals. These were then roasted and subsequently smelted in the immediate vicinity of the pits. The tools used include stone hammers similar to those in the mine, as well as grinding stones and cup-marked stones used in ore and slag processing (Fig. 8.18; Cert 2005).

Evidence of copper production in a ditch at Pioch Farrus (site 448) included stone mining tools, smelting slags, fragments of crucibles, and copper droplets. All stages of copper production are recorded in the La Capitelle settlement, from the initial treatment of mined ore, to smelting and metal casting. Stone hammers and grinding stones used for ore beneficiation were found. The smelting process involved carbon reduction of copper minerals (tetrahedrite and some oxidized ore) in shallow pit furnaces. This produced copper droplets that had to be extracted from a crude viscous slag consisting mostly of partially reduced ore.

The Ross Island mine in southern Ireland provides evidence for the processing of tennantite-rich ore. This began with hand-crushing and separation of mineralized rock brought up from adjacent mine workings. Rock extract was initially sorted outside the mine entrances to discard any obviously barren rock, as well as charcoal residue from the fire-setting. An advantage of the latter, particularly when used with stone hammers, is the comminuted nature of the extracted rock. This reduced the amount of crushing necessary in ore treatment and rendered the rock fragments more friable. The sorted ore was taken to a work camp location, to be finely crushed using stone hammers and anvil slabs (O'Brien 2004: plates 70–1), with hand-picking of visibly mineralized fragments to provide a concentrate for smelting. The miners were assisted by the shiny silvery appearance of the tennantite, which occurred in thick or thin seams largely free of gangue or mixed with the yellow chalcopyrite or purple bornite ore. The latter were discarded in favour of the more easily smelted tennantite, which explains why these minerals were commonly found in the prehistoric spoil.

The ore concentrate was further reduced using large grinding slabs, examples of which were found in the mine camp (Fig. 8.18; O'Brien 2004: figs 165–6). This was done as part of the final preparation for smelting and is confirmed by the presence of granular copper ore around several pit furnaces. Bone spatulae were used in the sorting of this finely crushed and ground copper ore, which may have been washed in the adjacent lake to separate heavier metallic ore from the limestone gangue. There is no archaeological evidence for such gravity concentration, except to say that the miners did employ a washing process to recover roasted ore and/or metal droplets from pit furnaces in this site (O'Brien 2004: 224–34).

Recent investigations in the Mauken Valley, part of the Schwaz-Brixlegg orefield of North Tyrol Austria, provide detail on the treatment of fahlore during the Late Bronze Age (Goldenberg et al. 2011). Evidence of ore crushing and hand-sorting was discovered in a work camp (Mauk A) near mine workings dating from the late twelfth century to the ninth century BC. Another ore processing site (Mauk F) of the same period was identified farther down the valley, where stone hammers and sorted spoil was found, as well as wooden troughs used in water concentration of copper ore (Fig. 7.13).

Beneficiation of copper-iron sulphides

A similar range of processes was applied to the beneficiation of chalcopyrite ore in the Bronze Age mines of the Mitterberg and Kelchalm in Austria (Eibner 1982). Broken rock from the mine was sorted underground, with

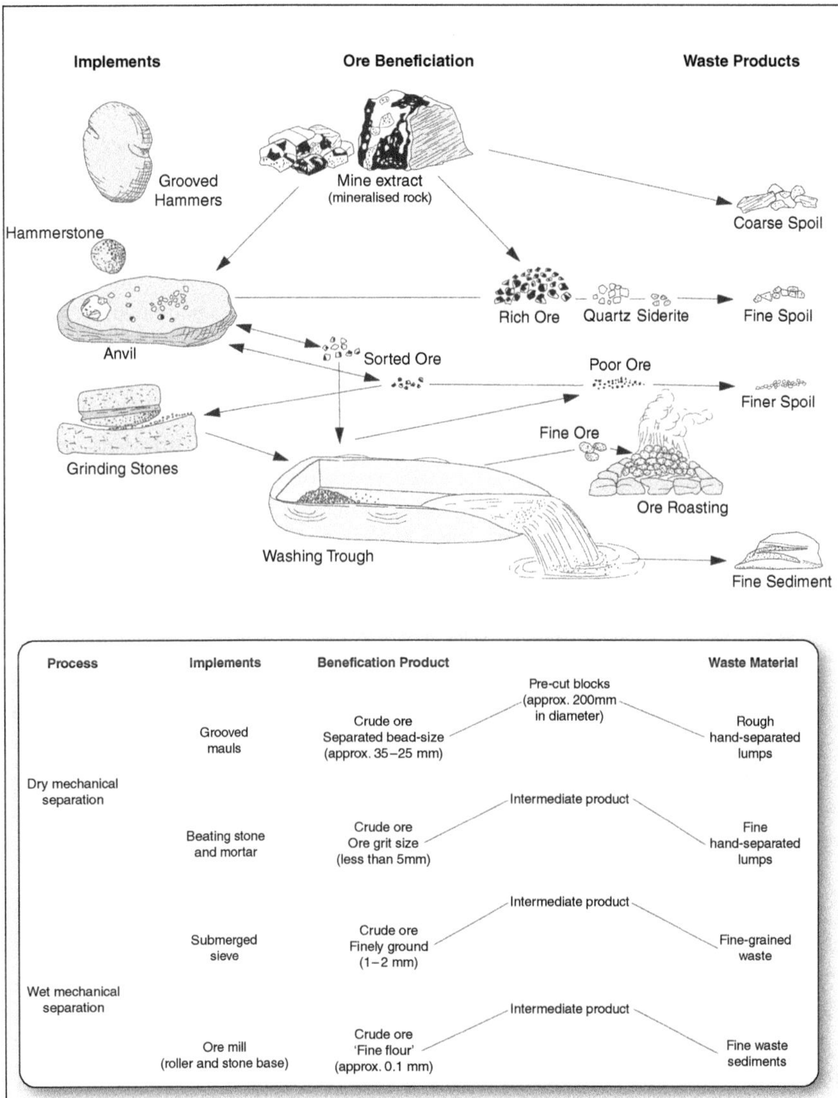

Fig. 8.21. Reconstruction of beneficiation process used to treat chalcopyrite ore in the Bronze Age mines of the Mitterberg, Austria

(*Source*: author, adapted from Eibner 1982).

mineralized fragments up to 200 mm in size taken to the surface for further treatment (Fig. 8.21). These were broken to 'bead' size (25–35 mm) using stone hammers (*rillenschlägel*), hand-sorted with the mineralized portion then reduced to grit size (< 5 mm) using smaller hammers (*klopfstein*) and mortars (*unterlagsplatte*). The ore was then ground using a milling stone (*bodenstein*) on a concave quern (*laufer*), from an initial 1–2 mm down to a 0.1 mm size fine 'flour'. This was then washed in wooden troughs to separate the heavier metal-bearing portion from barren rock minerals. This ore concentrate was then roasted in open fires to reduce the sulphur content and produce a friable material suitable for smelting.

A different type of copper-iron sulphide ore was processed at the Early Bronze Age copper mine of Saint-Véran in the French Alps (Fig. 5.9). The mineralization consists of very pure bornite (Cu_5FeS_4), with relatively low amounts of secondary mineralization. Details of the ore beneficiation were uncovered during excavation of a metallurgical workshop at the site of La Cabane des Clausis below the mine at Saint-Véran (Barge 1997; Ploquin et al. 1997). This process began with the crushing and continuous hand-sorting of mineralized rock from the mine. The ore was broken to a fine concentrate using stone hammers on anvil blocks of serpentine rock. Smaller anvil stones with surface depressions (*galets à cupules*) were used for this purpose, and also in the subsequent crushing of slag to extract copper prills.

What ore?

One of the major problems in considering any ancient metal mine is to identify exactly what parts of the ore deposit were exploited at different times. This is an issue even where the ore deposit comprised a single major constituent, as surface oxidation could create a range of secondary minerals. It is even more problematic in the case of ore deposits containing more than one metal ore constituent, particularly those with complex supergene mineralization.

The matter is further complicated because the rock spoil left behind is not fully representative of the ore removed by mining. While the mineralogy and chemistry of residual mineralization can be compared to the extracted ore, there is always a significant reduction in metal concentration (ore grade), as well as the selective dumping of undesired mineral associations. There is also the problem that the residual ore can be altered through oxidation processes in spoil contexts.

A good example is Cwmystwyth mine in mid-Wales, where excavation of the prehistoric spoil deposits uncovered significant amounts of lead ore (galena). That mineral was extracted in recent centuries, but was discarded by Bronze Age miners who were only interested in copper minerals. The initial interpretation of this mine argued for the extraction of chalcopyrite ore, small amounts of which were found in excavation (Timberlake 2002, 2003b). A further consideration of the evidence suggests that the miners were only interested in the oxidized portion of this orebody, extracting small amounts of malachite and other secondary copper minerals (Timberlake 2010).

Similar problems exist with other prehistoric copper mines, particularly where direct evidence of smelting of particular ore types is absent. The problem can be addressed where multi-context sampling from archaeological contexts is combined with geological sampling and mineralogical analysis to reconstruct the original mineral associations.

ORE TO METAL

The next stage of metal production involved the smelting of ore concentrate in a furnace operation at temperatures reaching 1,100° C to produce raw (primary) copper. The development of smelting processes in prehistoric Europe is a complex topic, outside the scope of this book (see Craddock 1995 and 1999 for review). It is, however, important to understand the organization of that activity, as smelting can be a proxy indicator of early copper mining in a locality.

The discovery of smelting sites depends on the nature of the process concerned and on whether visible residue in the form of slag was created, and how that material is exposed today. Slag production was determined by the type of copper ore used and the nature of the smelting process, while the amount created also reflects the duration of copper production at a location. The smelting of oxidized minerals or rich fahlore may produce little or no true slag, leaving only small amounts of heat-fused mineral behind. This was the case in the smelting of tennantite at Ross Island mine in Ireland (O'Brien 2004: 466–72). A reliance on oxidized ores with low iron content is one of the reasons why so few smelting sites of the Bronze Age have been discovered in Britain or Ireland. In contrast, an ability to process copper-iron sulphides, such as bornite or chalcopyrite, resulted in significant amounts of slag at smelting locations in Alpine mining areas during the Bronze Age.

The smelting of copper ore was mostly undertaken either within the mine site or in the general vicinity. Examples of the former are known from south-west Spain, at sites such as Chinflon and Cuchillares where deposits of smelting slag are close to the mine workings (Rothenberg and Blanco-Freijeiro 1981). This also occurred at Cabrières and Ross Island, where smelting was undertaken in work camp locations close to the mines (Ambert 1996b; O'Brien 2004).

Bronze Age smelting in Austria was often carried out a short distance downslope from the mines, often close to streams and woodland (Fig. 8.22). This occurred in the Mitterberg and is also the case with Bronze Age mines in North Tyrol and Styria. An example is the Mauken Valley at Brixlegg, where research has identified a chain of production that connected mine workings (Mauk B and E) and a mine camp (Mauk D) at higher altitudes with an ore processing site ('Mauk F') farther down the valley. There was also exceptions

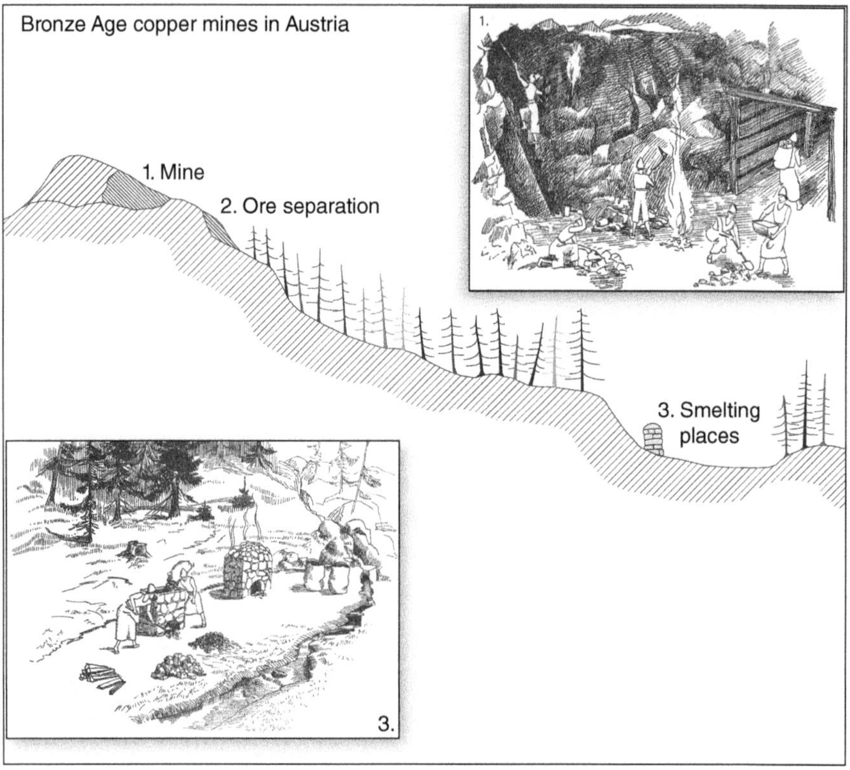

Fig. 8.22. Typical relationship of mining sites to smelting locations in the Austrian mines

(*Source*: author, after Pittioni 1951).

to this, as one of the smelting sites (Mauk A) in this valley is located on level ground close to a small stream, not far from the Mauk E mine (Goldenberg et al. 2011).

This separation of mining, ore processing, and smelting in Austria has been explained in terms of the logistics of fuel supply, haulage, and access to water. It is argued that much of the wood close to the mines was required for timbering and fire-setting, making it necessary to transport ore concentrates downslope to where other woodland was available to supply smelting fuel (Pittioni 1951). The association with streams in these locations may have been important in terms of metal recovery. An example is Mauk A where excavation revealed two stone-built furnaces and a roasting bed dated to the later twelfth and eleventh centuries BC (Figs 7.14 and 8.23). A large deposit of slag sand, up to 1.3 m in thickness, was uncovered, where almost 100 tonnes of crushed smelting slag was processed using water separation in wooden troughs to recover metal prills (Goldenberg et al. 2011).

Fig. 8.23. Remains of a copper smelting furnace, Mauk A, Brixlegg, Austria (*Source*: Gert Goldenberg).

Another example of downslope smelting is the site of Clausis at Saint-Véran (Fig. 5.9), located at the base of a steep hill where Early Bronze Age mining was undertaken close to the summit (Barge et al. 1998a; Barge 2003). This is a typical situation, but in many cases smelting sites in valley floor settings cannot be linked to mine workings on higher slopes with any certainty. This is true of the Eisenerz Valley in eastern Austria, where forest cover and landslides have made it difficult to identify copper mines connected to smelting sites on the valley floor (Fig. 7.14; see Klemm 2003, 2004). The reverse is true for British and Irish mines, where smelting locations cannot be identified for reasons outlined above. This is the case at the Mount Gabriel-type mines in south-west Ireland and those in the mid-Wales region and at locations such as Alderley Edge and Ecton. In other situations, opencast mining in the modern era has destroyed any evidence of early smelting. This is also true of Rio Tinto and other mining landscapes in the Iberian Pyrite Belt, for the Bronze Age mines on Parys Mountain, and for many opencasts in the Cypriot orefields.

Smelting was also undertaken at specialized settlements in the general vicinity of the mines. A good example is La Capitelle du Broum in the Cabrières district of southern France (Fig. 5.7). This permanent settlement of metalworkers was occupied *c.*3100–2400 BC, with various stages of copper production from the beneficiation of mined ore to smelting and metal casting. Smelting involved a one-stage reduction of copper minerals (tetrahedrite and some oxidized ore) in shallow pit furnaces, using ore from the Vallarade and Les Roussignole mines at Cabrières. This produced copper droplets, some of which had to be extracted from a crude slag consisting mostly of partially reduced ore. There is also evidence for the casting of metal objects in this settlement, and their finishing using processes of coldworking and annealing to make a range of objects (Ambert et al. 2009).

In some instances copper ore was taken a greater distance for smelting. This seems to have been the case for early metal production of the Vinča culture in the Balkans. Jovanović (1979) cited the discovery of finely ground malachite ore at the settlement of Fafos near Kosovska Mitrovica in support of this theory. Malachite ore is also recorded in the Vinča culture settlement at Selevac (Glumac and Tringham 1990), and also at Vinča-Belo Brdo itself. This seems to have been a wider practice as similar finds of copper ore are recorded from various settlements in eastern Bulgaria linked to the Ai Bunar mine (Chernykh 1978b).

In many cases mined copper ore was transported for smelting at nearby or distant settlements, which emerged as centres of craft specialization during the late sixth and fifth millennia BC. This is supported by recent investigation of an Early Vinča settlement at Belovode, some 50 km from Rudna Glava mine (Radivojević et al. 2010). A large amount of copper ore (malachite and azurite) was discovered, much of it heat-altered by smelting around 5400 BC.

Reference can also be made to the organization of copper smelting on Cyprus during the Late Bronze Age. In some instances this was undertaken close to the mines themselves, as in the miner's village excavated at Apliki-Karamallos close to the ore deposits at Apliki (Taylor 1952), or the settlement at Politiko-*Phorades* near the Kokkinorotsos orebody (Kassianidou 1999). These inland settlements close to the mines seem to have supplied raw copper to coastal urban centres where secondary processes such as refining and casting were carried out. The most important was Enkomi located on the eastern coast of Cyprus (Fig. 3.2), where there were large metallurgical workshops (Kassianidou 2012a). This emerged as a major centre for the export of Cypriot copper, a trade that involved cargoes such as those from the Uluburun and Cape Gelidonya wrecks.

The transport of copper ore over longer distances was exceptional, however, there were instances where it did occur. An example is the supply of ore from the Late Bronze Age mines at Kargaly in southern Russia across the wider settlement zone of the Srubnaya culture and affiliated groups. The settlements of these settled pastoralists extended across the steppe and forest steppe lands of eastern Europe, from the southern Urals to the Black Sea coast. While some smelting was undertaken at Kargaly, the limited reserves of local fuel meant that most of the copper ore was transported for smelting elsewhere over an area of up to 100,000 square kilometres (Chernykh 1998a: fig. 1).

ANCILLARY ACTIVITIES

The focus of activity in these prehistoric mines was the discovery, extraction, and processing of copper ore, leading to the smelting of metal within or outside of that location. Excavation at some mines has provided evidence for a range of support activities and facilities. These include the production and maintenance of mining equipment, the supply and processing of fuel for fire-setting, and other requirements to meet daily needs of food, hygiene, warmth, and shelter. Some of these activities may be inferred from material discovered inside the mine workings and in surface spoil. The most important information comes from habitation areas and work camps attached to these mines, where the final ore processing and smelting often took place. These locations can provide details on daily work routines and task specialization in the mine, as well as its connections to the external world.

The discovery of such ancillary areas continues to be a problem in the investigation of these mines. This is a problem of archaeological visibility, as the location of work camps and smelting locations is not always apparent on the ground. Key indicators include the discovery of metallurgical slag and surface artefacts, however, these are not always present or exposed in the environs of the mine workings.

An example of a work camp was discovered in the Beaker culture mine at Ross Island, southern Ireland (O'Brien 2004). As well as the beneficiation and smelting of copper ore, this central location within the mine provided logistical support to the mining operations through the provision of equipment and wood fuel. Excavation revealed a palimpsest of occupation surfaces relating to copper production and habitation over five centuries of mining (Figs 8.24 and 8.25). This location was used on a seasonal basis throughout that period, and so the layout of activity areas, structures, and facilities

Fig. 8.24. Miner's hut (Beaker culture), Ross Island, Ireland

(*Source*: author).

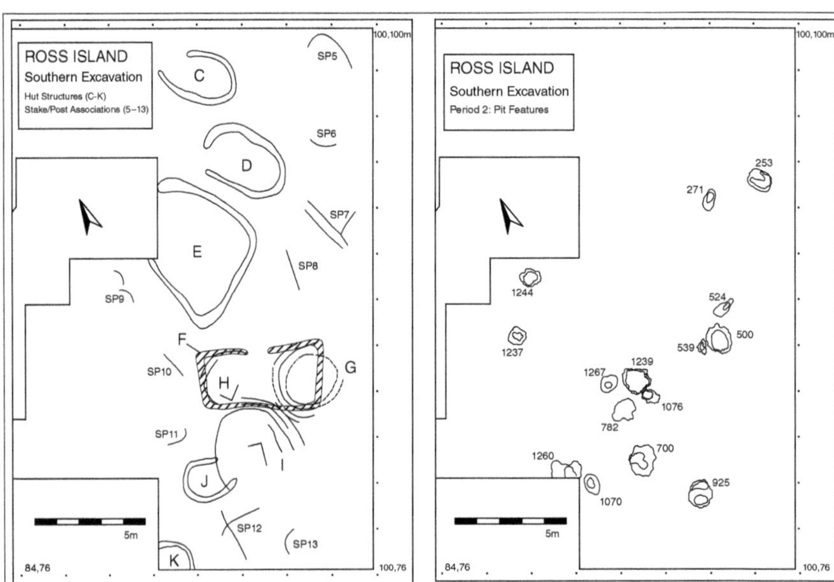

Fig. 8.25. Relationship of hut shelters to copper smelting furnaces in the miners' work camp, Ross Island, Ireland

(*Source*: author).

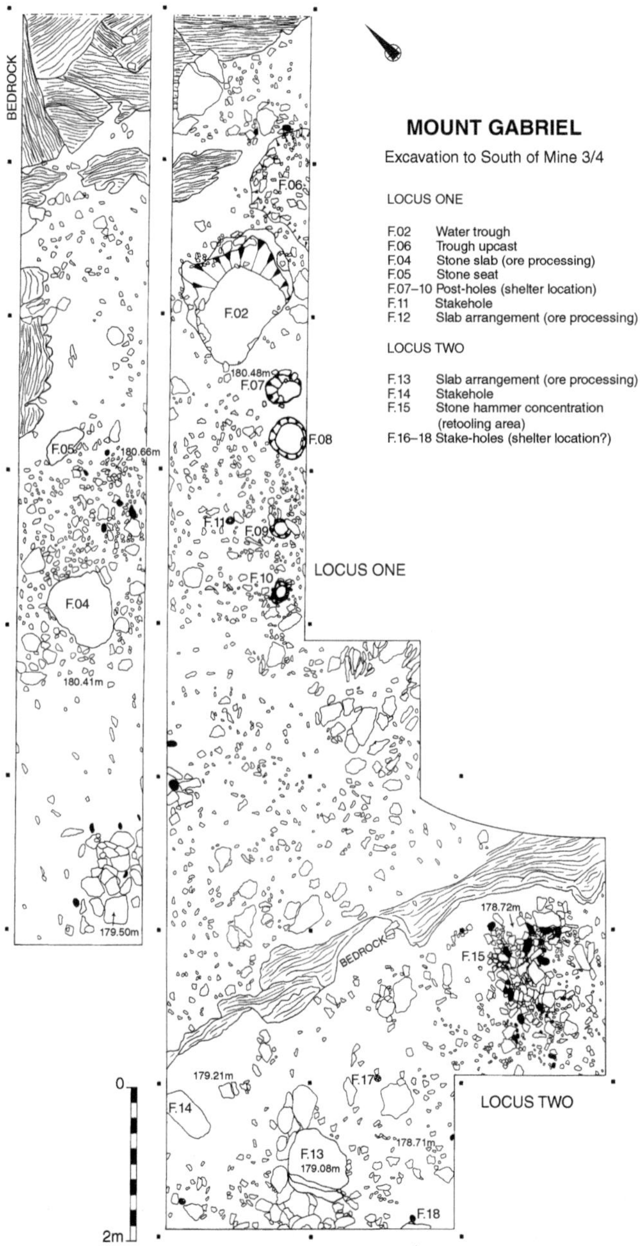

Fig. 8.26. Features discovered in mine work camp, Mount Gabriel, Ireland (*Source*: author).

changed many times. A different situation is recorded on Mount Gabriel, where a series of short-lived encampments followed the sequence of small-scale mining across the slopes of that mountain, again possibly organized on a seasonal basis (Fig. 8.26).

An investigation of these Irish mine camps has revealed different aspects of the daily work routine (Table 3). This includes the supply and repair of equipment used in mining. The high breakage rate of stone hammers meant that the sourcing, modification, and hafting of these cobble tools was an ongoing activity, the scale of which was dictated by the intensity of mining. This involved the regular transport of fresh cobbles from river, beach, or other sources within a 1–5 km distance of the mine, or occasionally farther away. These were extensively modified for hafting, which involved the preparation of handles, cordage, and withes. The discovery of thousands of broken examples reveals that the turn-over of these implements was considerable. Other tools prone to breakage (grinding stones, anvils, bone scoops, wooden implements) had to be produced from different sources on a regular basis. Some of this equipment may have been stock-piled prior to the commencement of mining, but supplies would have had to be maintained for the duration of the mining season.

The evidence from Ross Island and other early copper mines sheds light on other aspects of daily life in these mines. The supply of wood fuel was an important consideration, as fire-setting made considerable demands on a daily basis, as did domestic fires and charcoal required for smelting furnaces. Many operations had direct access to wood in the general vicinity of the mine, while others had to obtain supplies from longer distances. In certain environments these mines could be seriously constrained in terms of wood supply; for example, those at Kargaly and in arid parts of southern Spain.

The waterlogged wood and charcoal record points to deliberate selection of certain tree species for different tasks and equipment, suggesting a considerable knowledge of woodland resources. Young roundwood of mixed types was generally used for the fast-blazing fires required in mine fire-setting, and also domestic fires, whereas hardwood species were often used for smelting. The intermittent nature of many mines allowed for the regeneration of local trees to cope with these demands, with elements of woodland management in certain cases. The environmental impact of this mining will be considered further in Chapter 9.

Excavation provides evidence for various craft activities conducted within these mines and in external settlements. The production of wooden tools and equipment within or outside of the mine site is an obvious example. This included the production of timbers such as stemples used in mine support, as well as a range of equipment such as shovels and wedges, drainage launders, withes, lighting chips, and containers. This also extended to the domestic side of the mine, as seen in the discovery of wooden posts, roofing shingles,

Table 3. Range of work activities in a Beaker culture copper mine at Ross Island, and the Early/Middle Bronze Age mines on Mount Gabriel, south-west Ireland (after O'Brien 1994, 2004)

Work activities	Ross Island mine	Mount Gabriel mine
PROSPECTING	Discovery of mineralized outcrops (lakeshore exposures)	Discovery of mineralized outcrops (hillside exposures)
PREPARATION FOR MINING		
Organization of labour, equipment, and transport	No direct evidence; inferred from mining initiative and external sourcing of equipment	
MINING ACTIVITIES		
Fire-setting	Wallrock profiles; charcoal residues in surface spoil, Area 1–3	Wallrock profiles; fuel residues (waterlogged wood and charcoal inside mines, charcoal in surface spoil)
Mining tools	Stone hammers and bone scoops in spoil deposits, Area 1–3	Stone hammers and wooden shovels, picks, wedges and prise sticks
Access to underground mines; haulage of equipment, fuel, and rock extract	No direct evidence; inferred from physical requirements of underground mining	Footwall steps, wooden planks, and withies
Drainage, lighting, ventilation, roof support	No direct evidence; inferred from scale of mining, geological setting, proximity to lake	Pine lighting chips and oak planks possibly used as stemples
Mining rituals and superstitions	No direct evidence; inferred from anthropological sources	No direct evidence; inferred from anthropological data
ORE BENEFICIATION		
Preliminary sorting of rock extract	Surface spoil deposits, Area 1–3 mines	No evidence for underground sorting of rock extract
Coarse crushing and hand-sorting of mineralized rock	Discovery of spoil residues, stone hammers, and anvil stones in mine settlement	Spoil heaps directly outside mine entrances
Fine crushing, grinding, and hand-sorting of copper ores	Discovery of ground copper ore, stone hammers, and grinding stones in settlement	Crushing slabs on periphery of spoil heaps

(*continued*)

Table 3. Continued

Work activities	Ross Island mine	Mount Gabriel mine
METALLURGICAL TREATMENT OF ORES		
Roasting of sulphide ore	Possible use of Type 1 furnace pits; burnt ore residues; fuel ash residues	Not necessary due to use of oxidized ore
Smelting of ore	Use of furnace pits; fuel ash residues; sand flux deposits; metal droplet	Low-slagging processes and problem of blanket peat cover
Re-melting of metal droplets—ingot production	Copper slab find (Area 19A)	No evidence
CONSTRUCTION		
Building/repair of hut shelters	Hut foundations (Structures A–K) and stake-hole associations, mine settlement	Post-hole structure, Mine 3–4 work camp
Building of pit furnaces, hearths, and screens	Miscellaneous pit features, mine settlement	Water trough, Mine 3–4 work camp
FOOD CONSUMPTION		
On-site herding and butchery of farm animals	Animal bone record	No preservation of faunal remains
Hunting of wild game	Animal bone record	No preservation of faunal remains
Cooking and food consumption	Burnt bones; butchery marks; refuse disposal; possible hearth areas	No preservation of faunal remains
Use of ceramic vessels (off-site pottery production)	Beaker/ food vessel pottery, mine settlement	No evidence
ANCILLARY ACTIVITIES		
Modification and hafting of stone hammers	Scale of hammer use; pecking hammers	Re-tooling areas, Mine 3–4 work camp
Production of wooden mine tools	Wooden plank, Area 1/2; inferred from hafting of stone hammers	Waterlogged preservation of wooden shovels, picks, wedges, prise sticks, lighting chips, planks in Mine 3
Production of bone tools	Worked cattle bones, mine and adjacent settlement	No evidence
Flint-working	Worked flint pebbles; retouched flint tools, mine settlement	No evidence
Leather/skin working	On-site butchery of cattle; use of flint scrapers and knives	No evidence

Maintenance/repair of equipment	No direct evidence; inferred from mine activity	Sharpening stone for metal blades, Mine 3–4 work camp
Charcoal production	Inferred from selective use of fuel in metallurgical furnaces	Large amounts of charcoal produced by mine fire-setting
TRANSPORT		
Movement of workers, equipment, and food to mine	No direct evidence; movement by land and lake can be inferred	No direct evidence; movement by land from coastal lowlands can be inferred
Movement of copper ore and raw metal from mine	Local metalwork finds; ingot finds from mine camp and local settlement landscape; Ballydowny smelting site	No evidence
FUEL SUPPLY		
Collection and haulage of wood fuel for fire-setting, metallurgical treatment of ores, and domestic uses	Large amounts of wood indicated by charcoal/fuel ash residues in mine area	Large amounts of wood indicated by charcoal and waterlogged wood deposits in mine area
STORAGE		
Stock-piling of roundwood and charcoal fuel	No direct evidence; inferred from fuel demands	No direct evidence; inferred from fuel demands
Supply of stone cobbles	No direct evidence; inferred from external sourcing and transport of suitable cobbles	Haulage of stone cobbles from beach sources some 4 km from mines
Storage of ore concentrate and metal ingots	No direct evidence; inferred from entire mine operation	No direct evidence; inferred from entire mine operation

cooking spoons, pails and bowls, spindles, and dairying tools in the Bronze Age mines of the Kelchalpe near Kitzbühel, Austria (Pittioni 1951: plates 6–7).

As noted earlier, bone and antler tools were used in many prehistoric copper mines, particularly those in soft rock environments. In the case of Kargaly the enormous number of bone tools during the Late Bronze Age reflects the large-scale supply of cattle to the mine area during that period. Antler picks and hammers are recorded from many mines, including Rudna Glava and Ai Bunar in the Balkans, El Aramo and El Milagro in Spain, Cwmystwyth in Wales, Derrycarhoon in Ireland, among others. As with woodworking, the production of bone and antler tools must be inferred from finished objects and not from craft areas identified in these mine sites.

There is evidence of flint knapping in the Beaker mine camp at Ross Island, connected with the making of arrowheads for hunting, and knives and scrapers for leather working. This activity is not well documented in early copper mines where the use of flint tools seems to have been quite limited. In some instances it may be possible to infer the working of leather and repair of clothing, but the evidence is scant—the discovery of part of a leather jacket in the Kelchalpe mine is a rare find (Pittioni 1951: plate 7.2).

Pottery was used in some mines; however, there are no examples where it was actually produced at a mine or where mine sediment was used in that process. The discovery of slag-tempered pottery in Bronze Age settlements in Austria (e.g. Shennan 1995: 147) may in some instances provide a connection with local copper smelting, but not directly with the mines. Neither is there evidence for the casting or repair of metal implements used in these mines. This was probably undertaken in workshop settings, located in permanent settlements directly connected to the mine or farther down the line of metal exchange. An exception is the Kargaly complex where a limited amount of smelting was undertaken in the mine area. This was used in the production of mining tools, mainly picks and spades (Chernykh 2004a). The most common use of metal in most early copper mines involved axes and blades used in wood processing and domestic tasks. The use of these implements is seriously under-represented in the archaeological record. Whereas most mining tools were discarded once they broke, this did not occur with copper/bronze tools, which could be repaired or recycled and were rarely lost.

Food supply was an important consideration for workers engaged in the laborious, energy-sapping task of mining. Animal bones are the most important source of information on the food and equipment used in these copper mines. This relates both to food waste and the use of bones as mining tools. Bone preservation is an issue for mines located in upland acidic environments, with the best evidence coming from those in alkaline geology.

Detailed analysis of bone assemblages, such as those from Kargaly, the Great Orme, and Ross Island, suggest that live animals, most commonly cattle, sheep/ goat, and pig, were often brought to the work camp from farms in the vicinity, with immediate or delayed butchery in the mine area as required. A major consideration was the proximity of a mine to farming settlements in low-lying areas. This is examined further in Chapter 9 in relation to the agricultural background to many prehistoric copper mines.

Other aspects of daily life in these mines, such as washing, rest areas, and the maintenance of clothing, are not generally accessible through the archaeological record. Water supply may have been a concern for miners working in the Mediterranean zone, where abandoned workings may have later provided a source of usable water. This was not an issue in upland/montane environments of temperate Europe, or for those mines close to rivers and lakes.

Mine buildings

Depending on climate and the duration of the operations, most mines required some form of permanent or temporary structure to shelter their workers. The mine workings themselves may have provided some of these needs, through the use of abandoned tunnels and stopes. In the case of deeper workings there is evidence for the consumption of food underground, as seen, for example, in the Great Orme mine. In most cases there is no evidence that miners lived underground, voluntarily or not, for long periods. This is because many prehistoric workings had confined spaces or were subject to flooding, while continuous fire-setting required the miners to come to the surface on a regular basis.

The presence of roofed structures is likely in the vicinity of most early copper mines. This is borne out by the small number of work camps that have been excavated. In Ireland, excavations at Mount Gabriel uncovered part of a post-built structure used to shelter an ore processing area (Fig. 8.26). The Beaker miners at Ross Island built small structures, with circular, oval, or sub-rectangular ground plans, which provided adequate shelter for a few miners (Fig. 8.24). These huts did not have specialized functions, with a general occurrence of domestic refuse (animal bones, broken pottery, and flint debitage) in the vicinity. The amounts in question are small, suggesting some degree of site maintenance through organized refuse disposal. This is important in terms of archaeological site formation, as are the circumstances under which these mine camps were abandoned.

There is a distinction between the type of temporary structures recorded at Ross Island and the more permanent pit dwellings known from the Gorny settlement at Kargaly (Chernykh 2004a). Elsewhere, the evidence for buildings is limited, mainly due to a lack of excavation of possible work camp locations in the vicinity of these mines. Traces of post-built structures and hearths were identified in a settlement area adjacent to the Chinflon mine in Huelva, Spain (Pellicer and Hurtado 1980; Rothenberg and Blanco-Freijeiro 1981). Reference can also be made to the discovery of pottery and other domestic material in several Bronze Age copper mines in Austria. Wooden posts and roofing shingles are recorded from the Kelchalm mines (Pittioni 1951), however, evidence of mine buildings is mostly lacking.

Settlement sites with buildings have been identified close to a number of prehistoric copper mines in Europe. These include the Cabezo Juré evidence examined in Chapter 4, and the Middle Bronze Age settlement adjacent to the mine at La Loba, both in south-west Spain. Another example is the settlement at La Capitelle du Broum in the Cabrières mining complex, where hut structures have been identified (Fig. 5.7; Ambert et al. 2005).

CONCLUSIONS

While there are obvious differences in the scale and duration of prehistoric copper mines in Europe, the overall impression is an activity that was organized with considerable care and experience. There was an obvious ability to adapt to different orebodies and to experiment with different types of copper mineral. There was also a basic conservatism in the techniques employed, where 'Stone Age' methods were employed in the mining of metal ore. This changed in some regions as the Bronze Age progressed, with a distinction between an older tradition that applied basic methods to the mining of fahlores and oxidized copper ores, and a newer technology, first developed in Alpine Europe, designed for the mining of iron-rich sulphidic ores.

The archaeological record indicates a range of skill-sets and varying degrees of task specialization in these mines. An early interpretation of the Mitterberg operations proposed a division of labour where 'miners', 'timber-men', 'dressing personnel', and so on worked under the direction of a 'general manager', in a manner that 'closely followed the patterns of a modern industrial plant' (Pittioni 1951: 31). While modern principles of mine management are not appropriate to the understanding of an ancient mine, the nature of the activities involved makes some task specialization likely.

One obvious division may have been between those individuals involved in the extraction of mineralized rock in an underground mine, and those working on the surface on ore beneficiation. Records from most historic mines in Europe reveal a separation of surface and underground workers that, in some instances, includes age and sex distinctions. It is not certain to what extent this applied in prehistoric mines. One difference was the use of fire-setting, which created a work cycle where the miners had to vacate the mine for long periods, during which time they may have engaged in ore processing or rested.

Another separation of labour was between those involved in mining and beneficiation, and others who supported the operation through the supply of wood fuel and equipment, and the haulage of ore concentrate to smelting locations. The miners may have collected these materials themselves or used particular groups for this purpose during the course of a mine operation. Again, this need not be a separation of workforce, as fuel and equipment could be stock-piled before mining commenced, while smelting might be undertaken at the end of a mining season. In other situations, the haulage of service materials and ore concentrate involved a work routine where the miners moved between the mine and nearby settlements on a regular basis.

Whatever the specific details, it can be assumed that the deployment of labour in these pre-industrial mines was flexible around the cyclical nature of the operations. Some element of specialization is likely, but should not be over-stated. Tasks such as the stacking and lighting of fires, the hammering of rock faces, and the haulage of rock extract may have been generally shared among the miners with no particular division of labour. Other activities required greater experience, such as the sorting of mineralized rock and the smelting of copper.

While many workers engaged in laborious low-skill tasks, there must have been differences in experience and ability, with varying degrees of task specialization. Some individuals were more physically suited to the rigours of underground mining in confined spaces, while others were more adept at making equipment, woodworking, bone working, and so on. The number of workers underground was limited to some extent by the size of the workings. While the physical extent of a mine can be used to guesstimate the size of its workforce, this is problematic in the case of early mines worked over very long periods.

Work in these mines was undoubtedly hard and, in the case of some underground workings, fraught with dangers. While the archaeological record provides much detail on the technology and organization of these mines, this can lead to a somewhat mechanistic understanding of conditions in the underground workings. Bone waste found in these tunnels may represent food eaten underground, with some workers spending long periods, voluntarily or not, in very confined spaces. Reference can be made to a study of human palaeofaeces from the Iron Age salt mine at the Dürrnberg, close to the Mitterberg copper mining area, in Salzburg. This revealed that as much as 98 per cent of the workforce were infected with parasites (worms, lice, and liver fluke) due to the unhygienic conditions they worked in and the nature of the food supply (Stöllner 2003b: fig. 11). While there is no such information available for prehistoric copper mines, we can expect the working conditions in most instances to be just as difficult.

Many questions remain regarding the deployment of labour and the work environment in these mines. These issues, together with the distinctive nature of the communities involved and their role in a wider society, will be considered in the next chapter.

9

Mining, community, and environment

The previous chapters in this book have reviewed the extensive research carried out on prehistoric copper mines in Europe. Numerous site investigations provide detailed information on the geological setting of these mines and the types of ore minerals extracted. The same studies yield important insights into methods of rock extraction, ore beneficiation, and smelting, and the overall organization of those activities. Building on this solid base of information, it is possible to move beyond an understanding of mining as a technological process to an emphasis on the prehistoric communities involved. The technical details and logistics of the mining process continue to be important, but are now considered as part of a socially informed *chaîne opératoire* of early metal production.

Mining in the modern era is regarded as an economic activity; however, there is also a distinctive character to the individuals and communities who engage in that work. This is also true of prehistoric mining, which was undertaken within specific historical contexts that were structured by particular sets of cultural values. It was a highly social activity, involving closely-knit groups of individuals working together towards a common purpose, in situations where they depended on each other for their safety (Fig. 9.1). With no written records, the challenge for researchers is to understand these social dynamics using material correlates available in the archaeological record. As in other areas of prehistory, there are certain limits to the inferences that can be made using this type of evidence. That said, the significance of metal to later prehistoric societies provides a broad indication of the importance of mining as an activity.

A concern with the social background of these miners goes back to the earliest research in this field. The gradual move towards a 'social archaeology' of early copper mining reflects broader paradigm shifts in modern archaeology, away from culture historical explanations to more interpretative understandings of the material record. While recognizing the dangers of cross-cultural generalization, ethnohistoric studies serve to illustrate the vivid social history of mining communities. For prehistoric mining this helps to balance an emphasis on

Fig. 9.1. Reconstruction of Beaker culture copper mining at Ross Island, Ireland (*Source*: author and David Hill).

process and technology with a consideration of the social circumstances of the individuals and groups involved (Knapp 1998: 10).

Moving to a socially informed archaeology of mining, one that engages directly with the reality of the material evidence, is not easy. Outside of their working environment, miners can be largely invisible as a distinct social group. The absence of distinctive funerary rituals and settlement contexts makes it difficult to examine their standing in contemporary society. The starting point must be the mines themselves, their size and physical configuration, and the overall working environment. This, together with the organization of ore treatment and smelting, can allow some insight into the number of miners, how they were deployed, and the various skill-sets and experience required. The investigation of associated mining camps can provide insights into the type of social interactions arising from the provision of shelter, food consumption, and other domestic needs.

Empirical analysis of archaeological evidence reveals a range of activities and work routines in these mine operations. This allows some insight into the process of decision-making and the logistics of mine organization. The discovery of various symbolic deposits in mining contexts is significant in respect of cultural values and ideological belief. Further insights may be gained by considering local settlement landscapes that were contemporary with these mines. This can reveal aspects of the community background of those involved and the wider society that they were part of. Following the flow of metal from a mine can provide some understanding of trade and exchange networks, and the use and significance of metal in contemporary society.

Early research emphasized the role of migrant prospectors in the spread of copper mining across prehistoric Europe. Whether it was the arrival of 'colonists' of east Mediterranean origin in southern Iberia (Siret and Siret 1890; Blance 1961), or the movements of the Beaker prospectors in Atlantic Europe (Childe 1930), the search for copper has long been regarded as a major stimulus to trade and cultural exchange in prehistory. The emphasis today is towards an understanding of local and regional mining traditions operating within broader contexts of metal production and supply. While the separateness of mining communities is often stressed due to their remote location, distinctive work practices, and customs, such groups were not necessarily autonomous. In most cases their ethnicity lies in larger entities that controlled this activity over a wider social territory. These issues are at the core of what is now termed the 'social archaeology' of prehistoric copper mining.

MINING AS COMMUNITY

Historical and ethnographic sources confirm the distinctive character of mining communities the world over and the transient nature of their settlements. There is an important distinction between individuals from the same community who engaged in mining, and those with no specific social ties who came together for this purpose, to form 'a community of occupation, not a community of place' (Knapp 1998). There are many examples of the latter in mining history, groups that worked in extreme environments, motivated solely by profit, and who dispersed after each mining event. This image of the California prospector or Australian mine camp is certainly not appropriate to prehistoric copper mining. Most prehistoric miners probably worked within their own social territory, in close contact with home settlements and a wider social network that encouraged and supported their work in different ways.

The temporality of mining

The mining and production of copper in prehistory demanded considerable organization, both in terms of the activity itself and its place in a wider economy of metal circulation and value. The logistics of a mine operation could be considerable, involving the mobilization of a workforce, the preparation of equipment, and the supply of food and fuel. Much depended on the scale of operations, which ranged from part-time seasonal activity involving miner/ farmers to permanent mining communities on a full-time basis.

From a modern perspective there is a tendency to view many prehistoric copper mines as full-time operations employing professional miners. The archaeological evidence from the larger mines does certainly indicate a commitment to mining, which was partly determined by the richness of the ore deposit and the ease of metal extraction. More difficult to assess is the labour input and the intensity of mining over a defined period of time. Professional miners in the sense of permanent occupational groups may have been the exception, with most individuals and communities working on a part-time/ seasonal basis scheduled in accordance with the demands of the agricultural year. The intensity of mining could vary considerably depending on many factors, the most important being demand connected to the significance of copper in economic and/or social terms to that society.

A mine is not a fixed entity, but one that evolved over a period of time in terms of its physical space, extent, and internal environment. Unlike mining today, with its use of machinery and explosives, the extraction and processing of copper ore in prehistory was undertaken at a much slower pace, as rock was removed centimetre by centimetre, as mine tunnels were opened and abandoned, as ore was discovered and lost. Where this occurred at the same location over a long period, the routine of mining became part of a *tradition* of work at that place. This was personal to the individuals and groups concerned and would have shaped their identity to a considerable extent.

The size of a mine cannot be taken to indicate a large full-time operation, particularly if the physical remains accrued as a result of sporadic working over a long period. A good example is the Great Orme in Wales, often cited as one of the largest prehistoric copper mines in Europe. While this is certainly true in terms of its physical extent today, the visible workings are the product of a millennium or so of mining in a soft rock environment, where small groups of miners produced relatively small amounts of copper in each mining season. This contrasts with other situations where the scale and intensity of mining have been described as industrial, involving professional miners working on a full-time basis. The mines of the Mitterberg, Cyprus, and Kargaly are in this category, displaying levels of specialization, mining expertise, and output that were significantly greater than their seasonal counterparts. Stöllner (2003a) has observed how there are hardly any prehistoric mines whose technical installations (drainage, ventilation, and haulage) were so costly as to require year-round operation. One exception would be the Arthurstollen mine in the Mitterberg where work was conducted at depths of 200 m or more. He makes the point that such sophisticated deep mining could only develop within a stable and continuously settled landscape that supported a professional mining group.

Many of the same issues apply to the wider organization of metal production in a society. Prehistoric metalworking has long been considered as specialist in nature, with specific connections between miners, smelters, and smiths who operated a well-defined network, or else controlled by intermediaries trading in this valuable commodity (cf. Childe 1930). This interpretation was first challenged in a study of Middle Bronze Age metalworking in southern England (Rowlands 1976), which argued that such activity was often organized on a local and seasonal basis. Rowlands rejected the idea of specialized occupational castes in early metal production, citing many ethnographic examples where the smith is embedded in a particular social context and, even if to some extent itinerant, does not have to belong to a sub-group of different cultural identity (see also Rowlands 1971). An example is the Katanga tribe in the Congo, where the mining and smelting of copper was undertaken within a village-based system under the direction of a chief. This activity was confined to the three-month dry season at the end of the agrarian cycle when all people were free to engage in this work. This model of part-time mining connected to quiet periods in the agricultural year is also known from early historical contexts (e.g. Blanchard 1978).

It is also important to consider the *longue durée* of mining as a core economic activity in a given region. Stöllner (2003a) has proposed a model for the development of an early mining economy, beginning with an 'anterior phase' of part-time (seasonal) exploitation. This had to achieve a certain critical mass if mining was to become a full-time commitment, which was influenced by the dependencies that grew between the metal producers and consumers of metal, and the relationship between the success of mining and the wealth and priorities of the overall economy. This could lead to what Stöllner calls the 'initial' or 'establishing' phase, when new mining technologies and strategies were developed, leading to a 'consolidation phase' of industrial-scale production that dominated the economy of an entire region (Stöllner 2003a: 441). He listed a number of pre-conditions for such sustained mining, which included the quality, accessibility, and sustainability of an ore deposit, the technical skills to mine it, and the broader economic support for the operation.

The creation of a mining community

There is no doubt that participation in the practice of mining was a unique form of social reproduction that led to the emergence of distinctive communities. There is a growing research emphasis on understanding the social dynamics of Bronze Age mining (see Knapp 1998). This requires a consideration of how kinship and other social relations shaped the composition of a workforce, as well as leadership structures, the division of labour, and task specialization. Mining was not only a technological process, but also involved the making and reproduction of social relations, identities, and cultural ontologies. It was structured by particular societal values and circumstances, and woven into a complex web of social relations and work routines.

For many, the mining community can be defined as a community of labour, one that arises out of, and recognizes itself, through participation in the act of mining. This theme is explored by Wager (2002a) who points to the web of kinship and friendship ties and beliefs expressed and reproduced by people through relations engendered by their shared participation in mining. Mining was a forum for individuals or groups to come together in a cooperative venture that reinforced social bonds and helped create a sense of group identity and social structure. This would have been facilitated by kinship and community ties in small-scale mining involving collective effort by small acephalous groups. For such collaboration to succeed, the miners must have shared a similar world view and conformed to the same social norms. There are many examples in the ethnographic record of community-based mining undertaken on a collaborative basis (see Kienlin and Stöllner 2009: 73–5).

The fact that metal objects were commonly used as status signifiers during the Chalcolithic and Bronze Age in Europe means that the production of copper, including mining, would have been controlled in different ways. Multi-stage activities such as mining and metal production required planning and organization. The evidence for large-scale mining at centres such as the Mitterberg or Kargaly is consistent with focused leadership within those mining districts. This was probably ascribed on the basis of combined elements of experience, physical strength, and success. Ritual authority may have been significant where mining was undertaken in a charged atmosphere of superstition and religious belief (see Mining as Belief section). In many instances the control of a mine probably lay with an external authority who balanced the flow of metal from that source with demand in that society.

There is little reliable evidence on which to base estimates of the size or composition of a mine workforce and how this varied over time. The intensity of mining and the range of associated activities provide some indication of the number of workers involved, though any estimates are highly subjective. Depending on the scale of operation, as few as 10–20 individuals may have worked in a mine, with many times that number in very large operations, such as those on Cyprus or Kargaly. For example, Zschocke and Preuschen (1932) suggested that as many as 180 workers would have been involved in different capacities in the Mitterberg mines, with up to 500–600 required for the entire Salzburg mining region. Other estimates are higher, including a suggestion that annual production of 10 tonnes of copper from the Mitterberg required the involvement of 270–400 workers, not including those who supported the mining through farming (Shennan 1995: 300–2).

While the physical size of these early mine workings might dictate how many miners could work underground, it is more difficult to estimate the number engaged in surface activities. Where mine camps have been excavated, the evidence suggests small groups of fewer than 100 individuals. This evidence is also problematic, as no prehistoric mine settlement has been totally excavated anywhere in Europe.

For the most part there is little evidence on which to base accurate estimates of workforce in these mines. Most commentators take the approach of equating labour input with the physical extent of a mine. This is problematic as any assessment of labour input must consider the time-scale, intensity, and organization of the mining activity. The archaeological evidence tends to be equivocal on this matter. In the example of the Great Orme mentioned previously, there is evidence for a slow exploitation of an accessible ore resource over the course of a millennium, involving some 40 generations of mine workers whose communities were shaped by that activity. This commitment to long-term mining at one location may have involved 'professional' miners working on a permanent basis. Such a scenario is possible for large operations such as the Great Orme, however, it is just as likely that a particular mine resulted from short-term, seasonal mining over a long period.

There is little doubt that experienced and skilled individuals worked in these mines and in the associated smelting activity. The mining process did allow for less skilled labour where physical strength and endurance were required. Specialist work teams may have existed in some instances, to ensure a continuous supply of materials in support of the mining operation, such as wood fuel, hammerstones, antler picks, and food. The miners generally did not produce their own food, apart from occasional hunting forays. This is mainly because they were engaged in a labour-intensive activity that was supported by external settlements.

Any division of labour and skill-sets in these mines would have created an internal hierarchy in terms of authority, responsibilities, and roles. Underground mining and ore beneficiation required experienced individuals working alongside those with less knowledge. Kin relations were probably central to the passing down of skills from one generation to the next. Whether these miners always worked of their own free will should also be considered. There is no archaeological evidence for the use of slaves, and, though unlikely for small-scale seasonal mining, it cannot be excluded in the case of the larger intensive operations. The well-attested practice of gift giving during the Bronze Age would have created obligations between individuals and communities. Participation in the labour of mining may have been one way to discharge such debts, particularly if this involved the exchange of metal from the same mines.

While some tasks required particular skill-sets, there is no practical reason why there could not have been broad participation in the process of mining and its associated activities. Physical strength was not the only consideration, as tenacity and experience are essential in any mining context. These attributes, together with the inherent dangers, may have limited the involvement of children in underground work. That said, the use of child labour is well known in historic mining in different parts of the world and so cannot be excluded for the prehistoric period. An example is the Iron Age salt mines of the Dürrnberg in the Austrian Alps, where half of the leather shoes recovered belonged to children (Stöllner 2003b: fig. 12). Recent research on the Hallstatt cemetery itself suggests that children and juveniles took part in salt mining, working alongside adult men and women (Reschreiter et al. 2012). This may indicate an exploitative system, but could also reflect a learning environment where children worked with their experienced parents (Stöllner 2003a: 428). Another example is the involvement of children in the Bronze Age copper mine at the Great Orme, based on the discovery of underground tunnels considered to be too narrow for adult workers (Fig. 9.2). This is controversial, as many of those tunnels may be the product of karstic processes. While the number of burials is few, it should be noted that where human remains are recorded in prehistoric mines they do not include the remains of children (see Funerary Record section).

The aforementioned attributes required for mining do not exclude women from a direct involvement in this activity. Wager (2009) has challenged the prevailing androcentric consensus that prehistoric mining was primarily a male activity with women (and children) relegated to ancillary roles. The

Fig. 9.2. Narrow mine tunnels at Great Orme, Wales

(*Source*: Andy Lewis).

assumption that mining and metal production were highly gendered activities has its origin in historical mining traditions, supported by ethnographic sources from different parts of the world. The difficulty with such analogies is that they are the product of particular historical situations and socio-cultural contexts. They cannot be applied on a universal basis, and do not provide a direct insight into what happened in prehistory. Ethnographic sources are certainly useful in presenting a range of possibilities, such as how gender roles were influenced by magico-religious belief in activities such as underground mining or smelting. Other factors to be considered include social hierarchy in situations where mining (but not smelting or metalworking) may have been a low-status activity. Other work commitments must be taken into consideration, such as the labour demands of agriculture.

Wager (2009) argued for a move away from an emphasis on identifying the gender role of specific tasks to a consideration of how mining as a social phenomenon was caught up in sustaining and negotiating identities and understandings revolving around constructs of gender and age. It can no longer be assumed that men, women, and children laboured within separate task spheres in a mine, nor that this division exclusively revolved around conceptualizations of gender rather than the interplay between gender and other constructs such as age, skill, or kinship (Wager 2009: 110). She argues that writing a social narrative of early mining requires more than correcting implicit gender bias through careful use of language or by simply adding women (and children) to a visual portrayal of this activity. It requires an effort to actively investigate the ways that participation in mining was caught up in the making and reworking of gender and age identities within a specific socio-cultural context. This requires a shift from isolating the specific identities of people involved in different tasks to a concern with the dynamic social process involved in the construction and expression of such identities.

These ideas contribute to a richer understanding of how mining communities and their social relations were shaped by the distinctive nature of this work. The challenge is to apply such perspectives to a detailed understanding of particular mines operating in a specific historical and cultural context. Wager's analysis of the Bronze Age mining community at the Great Orme is a step in that direction, even if many of the underlying perspectives cannot be tested directly against the archaeological record.

MINING AS BELIEF

> Today, as of yore, the local Swazis fear the gods of the underworld, the greatest of them being Inkanyamba, the great horned snake. During mining, it was to this deity that they continually made offerings. His

> wrath could bring disaster to them all. According to the counsellors, the horned snake dwelt deep in the heart of the mountain. Consequently, mining on the surface was comparatively safe but even then they would not dare venture too deep. A rock-fall, particularly when it killed a miner, was attributed to the wrath of an offended spirit. In such a case the shaft was filled in. Included in this infill were ornaments and painted stones and a worked out mine was treated in the same manner . . .
>
> This mining was combined with a struggle to appease the gods. Even the ore 'food of the spirits' was carried miles away from the mine to be smelted. Another motive behind the filling in of mine workings was simply to hide their whereabouts from neighbouring tribes and foreigners . . . the Bantu, Bushmen and Hottentots regarded the earth as a living creature. Consequently, removal of parts of the earth had to be carried out with much ceremony and ritual.
>
> (Boshier 1965: 319)

Miners the world over are known to be highly superstitious, with many folk traditions and religious observances peculiar to their work. This was due to the dangerous nature of underground mining and the unpredictable manner in which the prized ore was found. Safety was always of concern to miners, expressed in formal religion by the cult of St Barbara in more recent mining in Europe. Across the world there are many traditions of underground creatures and spirits in mining folklore. An example would be the mythical creatures known as the Knockers in Cornish and Welsh mining. These tiny men wearing miner's garb were viewed as malevolent spirits or practical jokers, but also as protectors who alerted miners to impending cave-ins by knocking on the mine walls. Another example is that of the Kobolds in German mining folklore, tiny men who worked alongside the human miners and again were often blamed for accidents and for fooling the miners into taking worthless ore.

The mining of copper in prehistoric Europe was probably associated with similar elements of superstition and religious observance. These were designed to appease supernatural powers and to determine the most propitious time to commence mining. Such beliefs were intimately connected to the cosmology and world view of the groups in question. While this varied across different cultures, all miners would have had similar concerns over safety and would have developed particular understandings of their underground world and the spirit world therein. Such beliefs would have reflected the special relationship that miners have with the natural world, as they delved deep into the earth's surface to extract minerals in a sensory-deprived atmosphere of darkness and silence (see Blas Cortina 2010 for further discussion).

Special deposits

Ethnographic accounts, such as the above example from southern Africa, illustrate how religious beliefs can be manifest in the archaeological record, whether it was the backfilling of workings, the placing of offerings in expiation of thanksgiving, or the smelting of ore at a prescribed distance from the mine. The difficulty lies in recognizing material evidence connected to such beliefs, if indeed such remains were ever created. Most early copper mines have not yielded artefacts that can be directly connected to religious belief and superstition. Some finds continue to be enigmatic; for example, the discovery of notched sticks in the Kelchalm mines in Austria (Pittioni 1951: plate 7.5). These may have been tally counters, however a possible connection with divining has also been suggested (Pittioni 1951).

It is likely that their distinctive work practices, the mystery of their underground work and the magical aspect of metal production gave early miners and smelters a special standing in society (Budd and Taylor 1995). This is certainly substantiated by ethnographic evidence from Africa, where the smelting of ore was charged with sexual symbolism (Cline 1937; Herbert 1984). Fertility and sexuality metaphors may also have been applied to the act of mining, though evidence for such is generally lacking. Reference can be made to discoveries at the mining settlement of Gorny at Kargaly, southern Russia. The excavator stressed the importance of ritual in all aspects of daily life at Gorny. This included the discovery of a complex network of narrow and deep trenches in which '. . . different rituals performed by the miners took place in order to seek mercy and benevolence from the divine masters of the unpredictable Kargaly depths' (Chernykh 2004b: 236). The excavator goes on to propose male and female symbolism in the metallurgical hearths at Gorny, as part of fertility rituals in the production of copper. There is evidence for the mass slaughter of animals for sacrificial purposes. This is based on the discovery of pits full of animal bones that were carefully arranged with selected bones from which the meat had been carefully removed (Chernykh 2004a: 273).

The discovery of pottery and animal bone in most copper mines is consistent with domestic refuse rather than ritual deposition. This applies to pottery discovered in the Ross Island and Mitterberg mines, even if the actual use of these vessels is unclear. A different explanation may be applied to the discovery of a large pot in a small copper mine of Early Bronze Age date at La Vierge, Cabrières (Fig. 9.3; Ambert 1990: plates 4–5). This find was not associated with human remains and so may be a votive deposit. More problematic is the discovery of an articulated sheep skeleton in Bronze Age deposits in the Great Orme mine in Wales (Lewis 1996). Whether this was a sacrificial offering made as a closing deposit in the mine or has a more mundane explanation is uncertain.

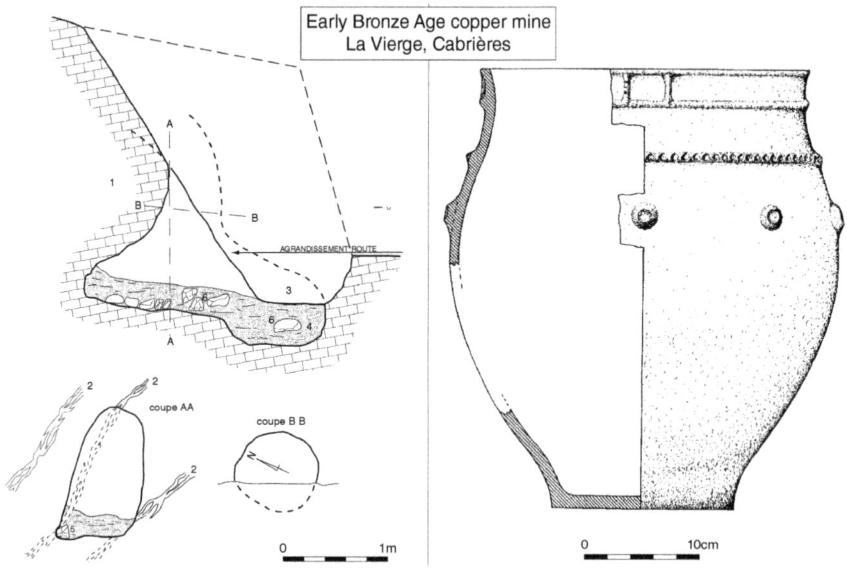

Fig. 9.3. Early Bronze Age pot discovered in a small copper mine at La Vierge, Cabrières (*Source*: author, adapted from Ambert 1990).

These possibilities are highlighted by a number of ceramic finds in the Rudna Glava mine in Serbia (Jovanović 1979). Excavation recovered three separate collections of pottery, as well as a number of individual vessels, inside the mine workings and on adjacent access platforms (Figs 2.4–5). Of particular interest are three ceramic altars of zoomorphic design found at the base of separate mine workings (Fig. 9.4). These were initially interpreted as representing deer or ibex, however Gimbutas (1983) argued that they represent the stylized head of a ram, possibly masked, with chevron and curvilinear line decoration. It is likely that these altars are representations of a mythical animal of religious significance. The find circumstances indicated that this pottery was placed in the mines as part of the closure process. The excavator suggested these ceramics were used in propitiatory offerings for the protection of the miners and in thanksgiving to an earth goddess for the ore. This may also explain why each working was backfilled on the completion of mining. A symbolic connection between the use of red deer antlers as mining tools and these zoomorphic altars has also been proposed (Borić 2009).

The deliberate backfilling of worked-out mines, and the placing of ritual offerings to placate the spirits, do not seem to have been widely practiced in early copper mines in Europe. Possible exceptions include the aforementioned mine deposits at Rudna Glava and possibly also the deliberate backfilling of trench workings at Ai Bunar in Bulgaria (Chernykh 1978b). In many instances

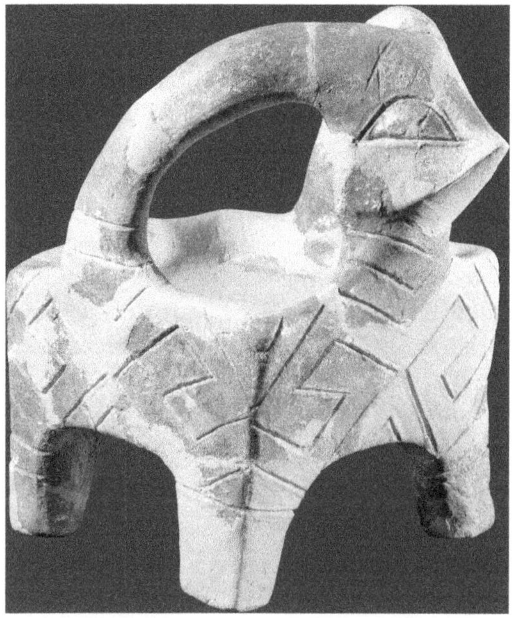

Fig. 9.4. Ceramic 'altar' found in the Rudna Glava copper mine

(*Source*: Borislav Jovanović).

the motivation behind such deliberate infilling is uncertain. A good example is the record from a Late Bronze Age working (Mine 4b) at Chinflon, south-west Spain (Fig. 4.5). This 15 m deep trench working was open for a short period before being deliberately backfilled with debris from an adjacent working (Andrews 1994: fig. 2). It may have been done for safety reasons or to dispose of surface spoil, though neither explanation is compelling for that particular situation. The fill contained a large number of broken stone hammers and some wooden objects, most of which can be interpreted as discarded equipment. The discovery of a human femur and a pottery bowl in the fill is less easy to explain and ritual deposition cannot be excluded.

There may also be a ritual aspect to the discovery of stone or copper/bronze axeheads in some prehistoric copper mines. During the Neolithic the stone axe was a potent symbol for agricultural communities and played an important role in various social transactions. It was often the case that only axes from particular sources were considered appropriate for such use, where the origin, individual history, and mythico-religious associations of the object were more important than functional considerations. In many societies, copper and bronze axeheads were invested with a similar significance, even if the metal in question was not as identifiable to a particular geological source as a stone equivalent.

Where metal axeheads are occasionally recorded in early copper mines, it is difficult to know whether they represent discarded tools or symbolic deposits. It is important not to over-interpret individual finds, but to instead look for consistencies in the archaeological record. This is illustrated by the discovery in 1854 of a hoard of twelve stone axeheads in a Bronze Age copper mine at Ballyrisode, south-west Ireland (O'Brien 2003). This find could be viewed as a ritual act, a closing deposit left by the miners in expiation or in thanksgiving for the removal of ore. Also interesting is the possibility that this mine provided a suitable setting for the substitution of stone axes for bronze equivalents. The difficulty is that no such finds were recorded in the excavation of twelve workings of similar age on nearby Mount Gabriel, suggesting that axe deposition was not a recurring feature of mine closure in that region. For that reason, the Ballyrisode find could as easily represent a forgotten tool cache as a ritual deposit.

Fig. 9.5. Votive metal deposit (decorated tin-bronze axehead and two copper ingots) placed outside the entrance to an Early Bronze Age megalithic tomb, Toormore, south-west Ireland

(*Source*: author).

This context is clearer in the case of a bronze axehead and two copper ingots found at the entrance to an Early Bronze Age megalithic tomb at Toormore, 2 km from the Ballyrisode mine (Fig. 9.5; O'Brien et al. 1989/90). The find context indicates a symbolic deposit made by local metalworkers at one of their sacred places. This is supported by the presence of decoration on the axehead, suggesting that it was invested with a symbolism connected to social display. The presence of ingots in this find points to a votive offering associated with metal production, and possibly with contemporary mining that has been identified in the surrounding hills.

Finally, mention must be made of the 'ingot god' found at Enkomi in eastern Cyprus. That settlement was a major production centre for copper during the period 1600–1100 BC, trading with settlements such as Ugarit on the Syrian coast. The object is a small bronze statue of a male warrior with helmet, greaves, spear, and shield, standing on what may be a miniature oxhide ingot (Fig. 9.6). Though not linked directly to mining, the object can

Fig. 9.6. The 'ingot god' from Enkomi, Cyprus
(*Source*: Department of Antiquities, Cyprus).

be associated with the importance of Enkomi as a centre of metal production during the Late Bronze Age (see Kassianidou 2012a).

Funerary record

The specialist nature of mining and metal production means that the individuals involved in those activities had a particular standing in society. This need not imply a position of high status, as there are few examples of ostentatious social display involving copper miners in funerary or other contexts. There are only a few instances of special mortuary treatment for miners in prehistoric Europe. This may be because the bodies of deceased miners were returned to their home settlements for burial, where their graves were not distinguished in any way. An example is the discovery of three Late Bronze Age burials at Volders in the lower Inn Valley of Austria, close to the copper mining area of Schwaz-Brixlegg. Each of these graves contained a single piece of copper ore, regarded by the excavator as a deliberate offering connected to their occupation (Kasseroler 1959). With so few examples, it is clear that miners were not regarded in the same way as smiths, whose visibility is much higher in the funerary record (cf. Jockenhövel 1982), and whose status as transformers of metal was mythologized in many early societies.

Human remains are recorded from a small number of prehistoric copper mines in Europe, mostly in situations indicating deliberate interment rather than casualties of mining accidents. Some finds are enigmatic, such as the discovery of a single leg bone from a refuse context at Ross Island (O'Brien 2004: plate 73). As cannibalism can be ruled out, this find relates either to a somewhat casual burial within the mine area or, less likely, the amputation of a miners leg following an accident. Similar explanations may apply to the human femur found in the fill of a trench mine at Chinflon (Andrews 1994) or the human skull fragment from Bouco-Payrol (Barge 1985).

Formal burials are recorded from a small number of sites. These include some examples found in abandoned workings at Ai Bunar in Bulgaria. The burial of an adult male was found in Mine 3 in a grave deliberately filled with fragments of copper ore. A double burial containing two adult inhumations was discovered in the upper fill of Mine 4b (Fig. 9.7; Chernykh 1978b: fig.6). This burial was discovered beneath the floor of a hut that was built in a partly-infilled trench mine, the significance of which is uncertain. A comparison may be made with two burials accompanied by Bronze Age pottery that were found in an infilled mine shaft at Tymnjanka, some 4 km west of Ai Bunar (Chernykh 1978b: 216).

The most important evidence for special mortuary treatment comes from the mine at El Aramo, Asturias, where up to 26 skeletons were discovered in the late nineteenth century (Blas Cortina 2005). Radiocarbon dating reveals

Fig. 9.7. Double burial in infill of Mine 4b, Ai Bunar, Bulgaria (*Source*: Chernykh 1978b).

these burials span a long period of mining, *c*.2500–1400 BC (Blas Cortina 2010, figs 5–9). They were found together with stone hammers in the deeper parts of the mine between the San Vicente and San Alejandro veins, and in the so-called 'Gallery of Skeletons' (Fig. 9.8). Blas Cortina argued that these burials are offerings made by the living to the underworld in compensation for the removal of its mineral wealth. Such beliefs may have been widespread in the Asturian mining region and might explain the discovery of human remains in the mine at El Milagro, where skulls from at least four burials were discovered in the nineteenth century (Blas Cortina 2010: figs 1–2).

In conclusion, while religious belief and industrial manufacture are very distinct spheres of activity today, this was not the case in prehistory when the making of metal was surrounded by much ritual and superstition. For prehistoric people, the act of mining and the subsequent smelting of copper would have been associated with magical powers, and so had to be undertaken in a suitably charged environment. Such beliefs are not always expressed in the archaeological record; however, there is enough evidence to indicate that this was an important concern for many early miners. The Toormore example is a reminder that rituals connected to the success of mining ventures and metal production may not always have been performed at the mine itself, but at other locations sacred to the mining community.

The ethnographic record emphasizes the importance of ritual control in relation to the care of special resources, involving individuals or groups who mediated with transcendental powers to gain access to the mine and to protect the miners in their dangerous work (Kienlin and Stöllner 2009). This ritual context may have been an essential precondition for mining, where younger members of a community had to prove their virtue and religious knowledge to be allowed to participate. The ethnographic sources reveal many complex

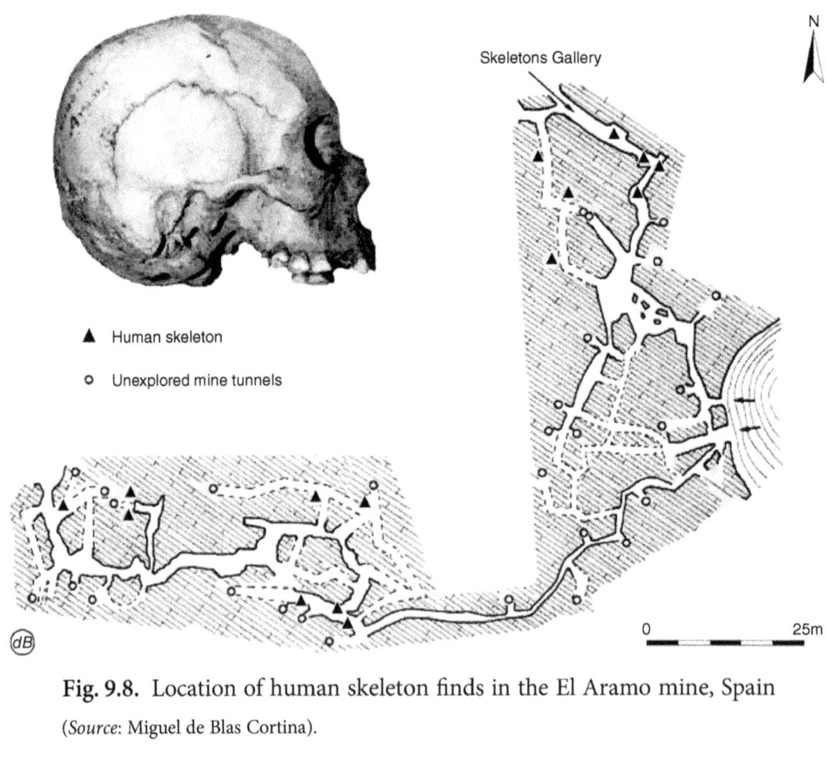

Fig. 9.8. Location of human skeleton finds in the El Aramo mine, Spain (*Source*: Miguel de Blas Cortina).

levels of social structure associated with periodic mining, where ritual control was interwoven with cultural categories such as age, gender, and kinship (Kienlin and Stöllner 2009: 75).

MINING AND SETTLEMENT

The social background of a prehistoric mining community is best explored through local settlement contexts. This is a neglected area of research as few settlements can be directly connected to early copper mines in Europe. Most early copper mines are devoid of artefacts, such as pottery, that could provide a cultural context for this activity. The proximity of mines to parent settlements can vary considerably depending on local environment and other factors. Whereas the former were located on a fixed mineral resource, the associated settlement(s) may have been close or some distance away.

The settlement potential of a landscape local to these mines may be investigated through resource catchment analysis. This requires certain assumptions to be made regarding the wider economy of the mining communities and the significance of monument and artefact distributions. Ritual monuments and artefacts finds provide some insight into the territorial organization of a landscape. Another approach is to use scientific characterization of metalwork to understand the spatial structure of a metal supply system involving mining, metalworking, and long-distance trade. This requires some understanding of the flow of metal from a mine source into wider contexts of fabrication, use, and disposal.

Miner-farmers

Whereas early studies tended to consider mining in isolation, the emphasis today is on how this activity related to other work routines in the landscape. Of particular concern is the manner in which local farming systems supported prehistoric copper mining and influenced its organization. Across Europe most mining ventures were undertaken in societies that had a strong commitment to agriculture. Both activities were labour intensive, and required careful organization and collaborative effort at a community level to be successful. These competing demands would have been significant in the conduct of part-time seasonal mining. They were also a factor in permanent mining, as the specialists involved had to be supported by an economy that produced a surplus of agricultural goods.

Knapp (1998: 4) observed how full-time, wage-earning, mining is mostly a phenomenon of the Industrial Age. Prior to this, informal mining as a part-

Fig. 9.9. St Crohane's cell, Iveragh, south-west Ireland

A fire-set copper mine of probable prehistoric date, supposedly used as a mountain hermitage by an early medieval saint. This religious association protected the mine from damage during subsequent mining at the same location in the nineteenth century.
(*Source*: author).

time activity was probably the norm. He cites ethnographic evidence from Africa that shows this as a predominantly seasonal activity, where entire families would move into bush camps to work at mining and smelting, following which they would return to their villages and traditional agricultural pursuits. This overlap between different modes of production can be explored through analysis of agricultural products found in mine work camps, and also the supply of pottery and other equipment from contemporary settlements.

The faunal record from excavated mines provides an insight into the supply of food from local agricultural systems. An example is the evidence from the Beaker mine camp at Ross Island, south-west Ireland. That bone assemblage consists almost entirely of domestic farm animals, dominated by cattle (67 per cent) and pig (25 per cent), with a low incidence of sheep/goat and dog (van Wijngaarden-Bakker 2004). The skeletal evidence points to the movement of live cattle and pigs, and occasionally sheep, from nearby farms to the mine where they were butchered as required. No evidence for the consumption of cereals or wild plant foods was found, nor did hunting contribute to food supply in any significant way. There is no evidence for consumption of fish, which is surprising given the lakeshore setting of this mine. This reflects the priorities of the miners, as a regular supply of meat from local farms allowed them to devote all of their energy to mining rather than having to engage in the equally time-consuming activities of hunting and fishing.

Similar conclusions have been drawn from a study of animal bones in the Mauken Valley near Radfeld/ Brixlegg in the lower Inn Valley of Austria. An analysis of animal bones from mine camp and smelting site locations dated 1200–800 BC revealed a reliance on the external supply of farm animals, principally pig and sheep/goat, with a lesser dependence on cattle (Schibler et al. 2011). The study concluded that the miners were supplied with live animals from farms located farther down the valley, which were butchered as required. Recent excavations conducted at the Late Bronze Age copper mine of Prigglitz-Gasteil in Lower Austria recovered an assemblage of 3,000 animal bones (Trebsche 2012). The remains of pig account for 64 per cent of identified specimens, consistent with many other prehistoric mining settlements in the Eastern Alps.

An analysis of animal bones from the Great Orme in Wales reveals a similar dependence on farm cattle, but with different management of food supply to the miners (James 2011). Excavation recovered some 30,000 fragments of animal bone, found in surface spoil and underground workings dated to the Bronze Age. These were well preserved in the alkaline geology; however, individual bones are very fragmented due to trampling and deliberate breakage. Many bear butchery marks and so can be regarded as food refuse, while a significant number were used as tools in the mining process.

There is a predominance of cattle bone in the assemblage, with lesser amounts of sheep/goat and pig bone (James 2011; see also Dutton and Fasham 1994: table 2). Red and roe deer are represented in small numbers, pointing to the occasional hunting of these wild animals. There is no evidence for the consumption of birds or fish, which is surprising given the proximity to the coast. The discovery of a small number of mussel and oyster shell in the mine suggests some shellfish collection, possibly during trips to collect beach cobbles for use as hammerstones. The assemblage also contains a few dog and horse bones, however those animals were not consumed as food.

An analysis of skeletal parts reveals an over-representation of meat-bearing elements, in particular long bones and ribs. The absence of skulls and hoof bones suggests that, unlike Ross Island, butchery did not occur in the immediate vicinity of the Great Orme mine (Fig. 9.10). The processed meat seems to have been consumed in small amounts while the miners worked underground, a high-protein diet that included the splitting of bones to extract nutritious marrow. James estimates that 5–10 per cent of the bones were utilized as tools in the mining process. These include long bones and ribs with rounded worn ends where they were used to pick out copper minerals from the rotten dolomite (James 2011: fig. 6.2). As was the case at Ross Island, shoulder-blade bones of cattle were used as shovels.

In general, the emphasis in many prehistoric copper mines seemed to be on the supply of meat from farmed animals rather than the consumption of other food products. An exception is the mining and metallurgical settlement at

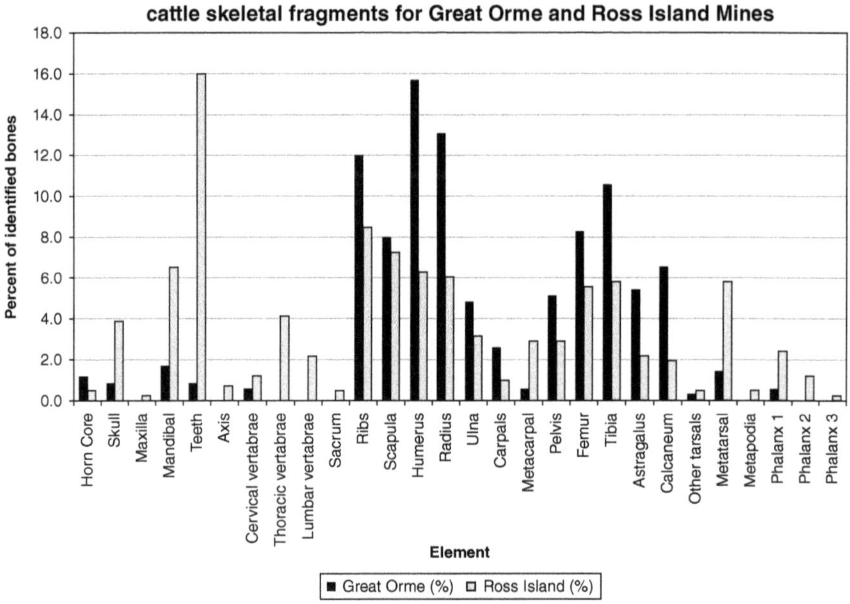

Fig. 9.10. Comparative graph representing cattle bone from the prehistoric copper mines of Ross Island, Ireland, and the Great Orme, Wales

(*Source*: James 2008).

Cabezo Juré in Huelva, south-west Spain. An analysis of animal bones from that site revealed that hunting of deer, boar, ibex, rabbits, and birds made up some 36 per cent of diet over a long occupation history in the third millennium BC (Nocete et al. 2005). It is the only example in Europe where a mining settlement was supplied with shellfish, brought from sources 30 km away in the Odiel estuary. This was connected with low levels of agricultural activity in the surrounding landscape during the mining period.

Trade may have been a significant factor in the supply of food to many mines. An example is provided by the enormous collections of animal bone from the mining settlement at Gorny, dated 1700–1400 BC, within the Kargaly mining district of the southern Urals. The bone assemblages were dominated by cattle (80 per cent), followed by sheep/goat (17 per cent), with the very occasional use of horse, pig, and wild animals (Chernykh 2004a: plate 7.1, table 7.2). The scale of meat consumption makes it likely that the miners obtained these animals through trade. During this Late Bronze Age period, the mining community was connected to the Srubnaya culture, a population of settled animal herders spread across the Great Eurasian Steppe. The discovery of many millions of animal bones in the Gorny settlement reveals an important link between these pastoralists and the supply of food to the miners.

The available evidence suggests that most prehistoric copper mining in Europe was a seasonal activity tied into the demands of the agricultural calendar. In some cases mining was a late-autumn/winter activity, which may have commenced each year following the harvest. At Ross Island mine the age profiles of pig and cattle provide slight evidence for winter activities (van Wijngaarden-Bakker 2004). Tree-ring analyses from the Mount Gabriel mines reveal that oak, alder, and ash were cut for fire-setting fuel in the winter months (O'Brien 1994). As with all seasonality evidence, this does not rule out mining at other times of the year. Each mine must be considered on its own evidence, which includes allowing for variation in the scheduling of ore extraction where work fluctuated between continuous and sporadic activity depending on the circumstances.

While the process of mining and its difficulties certainly distinguished life in those communities from agricultural settlements, the difference may be overstated. Farming and mining have much in common, not perhaps in terms of physical danger, but in the laborious and often unpredictable nature of the work and the transformative relationship with the natural world.

Mining and agriculture in upland environments

Some researchers have made a connection between the spread of copper mining and farming activity in upland environments. For example, Maggi and Pearce (2010) suggest that the emergence of copper mining in Liguria during the fourth millennium BC coincided with an increased investment in animal husbandry and the exploitation of new land by mobile pastoralism. They argue that a gradual opening up of the uplands for summer transhumance led to the discovery of copper deposits, at locations such as Monte Loreto and Libiola. They also believe that more intensive agricultural production based on secondary products (cf. Sherratt 1981) allowed Neolithic groups to move beyond a subsistence-based economy to one that prioritized the exploitation of copper and chert.

It has also been argued that a search for valuable resources such as copper and salt was central to the 'conquest of the Alps by Bronze Age people' (Wyss 1971 in Krause 2009). It is now recognized that a complex interplay of factors led to the progressive settlement of northern Alpine valleys by the Early Bronze Age. This was linked to their significance in trans-Alpine trade, and the growing importance of transhumance as farming settlement expanded out of agricultural lowlands. In some instances this was accompanied by the establishment of communities with a particular mining focus, following the discovery of promising mineral showings while herding animals.

A recent study by the author considered the *longue durée* of upland farming in south-west Ireland in relation to episodes of Bronze Age copper mining in

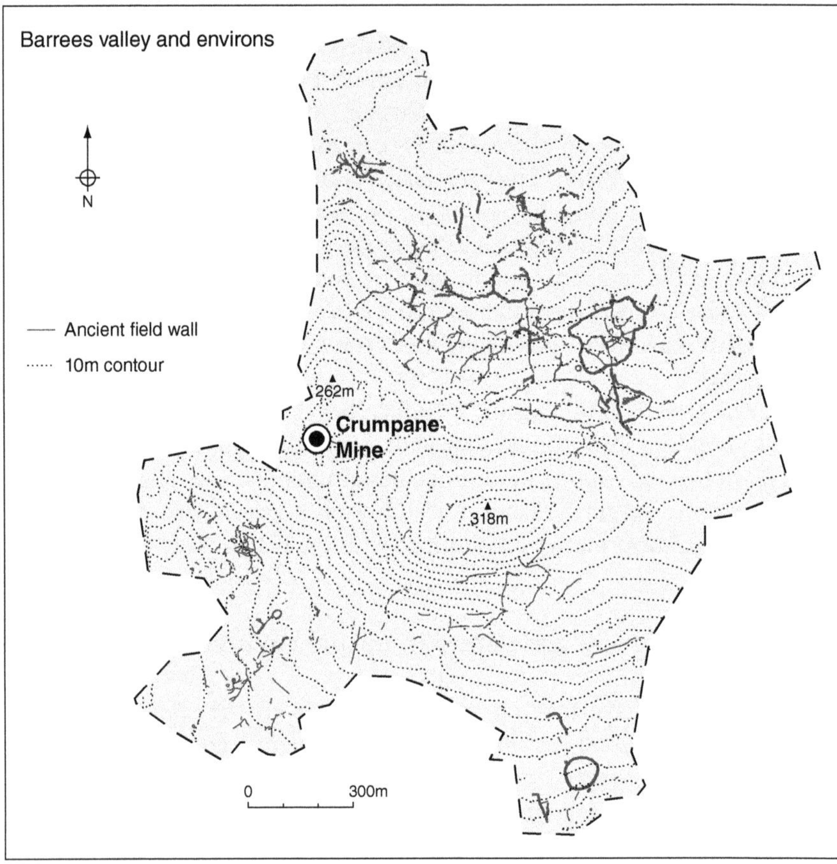

Fig. 9.11. Map showing the landscape relationship of a Bronze Age copper mine to prehistoric field walls, Eyeries area, Beara Peninsula, south-west Ireland

(*Source*: O'Brien 2009).

the same landscapes (O'Brien 2007, 2009). The investigation focused on the upland valleys of the Beara peninsula, where early field systems, habitation sites, and ritual monuments are preserved under blanket bog. A Mount Gabriel-type copper mine in one of these locations, the Barrees valley, has been dated 1700–1500 BC. There is evidence of contemporary settlement in the valley below this mine, represented by burnt mound cooking sites and a standing stone pair of ritual/funerary significance (Fig. 9.11). The pollen record reveals sustained human activity in the valley from the time of the copper mine onwards, mostly connected to pastoral farming with an arable component. While no direct link can be established, this farming seems to have supported the small copper mine on the adjacent ridge. This highlights the difficulty of linking small-scale mining events with a duration of a few decades to a continuum of human settlement and farming practice in the same landscape over thousands of years.

MINING AND THE ENVIRONMENT

Much of the public debate on modern mining tends to be on the negative consequences for the natural environment and human groups living in the vicinity. This is also relevant to the production of copper in prehistory, albeit on a much more limited scale than today. The type of environmental impact depended on the intensity and duration of mining and smelting at different locations and the processes involved. Of particular importance was the demand for wood/ charcoal fuel for use in mine fire-setting and smelting.

Scales of production

It is obvious from the examples presented in this book that prehistoric copper mining was undertaken at varying levels of intensity over different time periods. This ranged from one-off mining events of short duration producing small amounts of metal for local needs, to more sustained activity of greater output undertaken on a part-time or permanent basis, which supplied metal over a wide area as part of an organized trade. While sporadic mining may have been influenced by climate or other labour commitments, such as farming, it was also linked to the level of demand for metal in a society and the willingness or capability of different groups to engage in primary production. At the other end of the scale there were intensive operations, undertaken by full-time workers and specialists within permanent mining communities.

These different strategies and circumstances help to explain the variable output from prehistoric copper mines across Europe. An accurate assessment of production is essential to understanding the scale and organization of any mining venture. This should be feasible once the following variables are known:

1. The overall size of the mine workings as a measure of the total amount of rock extracted
2. The concentration of mineralized rock contained within this rock extract, focusing on those copper-bearing mineral associations (ore) that could be extracted by the prevailing technology
3. The percentage of this ore that could be concentrated through the beneficiation of rock extract

4. The amount of raw metal produced by the smelting of ore concentrate
5. The final amount of primary copper recovered in post-smelting treatment of slag and impure metal.

While these parameters provide a basis on which to estimate mine output, such calculations can be problematic for several reasons. First, the residual character of the archaeologically sampled mineralization makes it difficult to understand variability within the grade or type of ore extracted. While the total amount of extracted rock is indicated by the size of the mine workings (presuming that these are accessible), the mineralized portion is difficult to estimate when dealing with heterogeneous orebodies. Second, the mining environment would have affected the intensity of production, through variations in the structural setting, hardness, and mineralogy of the orebody, as well as operational difficulties caused by flooding, instability, and ventilation. Finally, the recovery of metal through various metallurgical processes is difficult to quantify, particularly in the absence of information about key elements such as the smelting process.

The amount of copper produced from any prehistoric mine must be viewed against the duration and intensity of that operation, which also requires a consideration of labour input and overall commitment. As already observed, while the physical size of a mine provides an indication of gross output, it does not reveal the temporal scale of production. It is not valid, therefore, to compare mines that operated on a small-scale over a many centuries with intensive operations worked with larger workforces over much shorter periods. A good example of the former is the Great Orme in Wales, which may have produced at least 175 tonnes of copper during the Bronze Age (Lewis 1996). This impressive output must be placed against the extended working of the mine over a millennium or so, with an annual output of possibly less than 200 kg of metal.

Many guesstimates of early copper production in Europe indicate that the amount of metal produced in different mines was far in excess of local needs. To justify continued mining, the surplus must have been traded over wider distances, in the form of ingots, finished artefacts, or, less commonly, ore concentrate. This was the case for many prehistoric mines in Europe, such as the Chalcolithic mines of Rudna Glava, Ai Bunar, Cabrières, and Ross Island, and the intensive Bronze Age operations at Kargaly, Cyprus, and the Mitterberg.

The scale of production in the larger mines is well known. Early estimates of copper production in the Austrian Alps suggested an output of approximately 18,000 tonnes of raw copper from the main Mitterberg deposit, with an estimate of 50,000 tonnes from the region as a whole (Zschocke and Preuschen 1932; Pittioni 1951). The total amount of copper contained in some 10,000 rib and ring ingots of Bronze Age date found within a distance of 700 km from these mines only accounts for a paltry two tonnes or 0.0005 per cent of this output (Harding 1983). The same disparity exists for early copper mines in other parts of Europe. For example Pearce (2009) has estimated a combined extraction of 4,500 tonnes of copper ore in the mines of Monte Loreto and Libiola in eastern Liguria, yielding as much as 788 tonnes of raw copper over a 1,500-year period of operation during the fourth and third millennia BC. The amount of recorded copper objects from the same period in that region represents only a fraction of this mine output. Another example is the early mine at Ai Bunar in Bulgaria, which is reckoned to have produced 2,000–3,000 tonnes of copper ore, yielding up to 500 tonnes of metal. Chernykh (1998a: 129) observed that the total amount of copper in south-east Europe that can be chemically linked to this mine hardly exceeds 5–6 kg.

This imbalance between the estimated output of early copper mines and the amount of contemporary metalwork in the archaeological record has been observed elsewhere in Europe and Eurasia. The most extreme example is the Kargaly mining district in the South Urals. Estimates of mining output suggest that 2–5 million tonnes of copper ore may have been mined there during the Bronze Age (*c*.3000–1400 BC). Smelting of this amount of ore would have yielded approximately 150,000 tonnes of metal, two-thirds of which was produced during the Late Bronze Age alone (Chernykh 1998a, 1998b). This copper was distributed over an area of one million square kilometres across the steppe and forest steppe of eastern Europe. Most of this has been lost, as the amount of Bronze Age metalwork recorded over that vast territory scarcely reaches one ten-thousandth of the total mine output (Chernykh 2004b).

Moving outside Europe, reference can be made to Late Bronze Age copper mines in the desert steppes of Kazakhstan. The mines at Kenkazghan produced an estimated 800,000 tonnes of copper ore during the second millennium BC, which may have yielded 25,000–30,000 tonnes of copper (Alekseev and Kuznetsova 1983). During the same period, as much as one million tonnes of copper ore yielding 30,000–33,000 tonnes of metal may have been mined at Jezkazgan (formerly Dzhezkazghan) (Satpaev 1941), with other Bronze Age mines also known in the region. The scale of production in Kazakhstan is extraordinary, particularly when as little as 60 kg of copper is recorded for that period in the entire region (Chernykh 1998b).

This great disparity between mine output and the amount of metal in the archaeological record must be explained (see Taylor 1999 for discussion). The nature of archaeological recovery is a factor, as is the characterization of metal from different sources. Other factors include the circulation of copper over great distances from the mine source, and various losses that accrued through the use of metal. The main explanation, however, must lie in the impact of recycling, which was a feature of metal supply from an early stage. Unlike wood or pottery, a finished metal product also represents a source of raw material in its own right, to be recycled over an indefinite period. This also means that a copper mine did not exert the same degree of control over raw material supply as a flint mine or rock quarry.

In some cases the scale of production may have been much lower than previously believed. An example is the Mount Gabriel-type mining in south-west Ireland. Early production estimates proposed a 'spectacular imbalance' between the copper output for these mines (estimated at 373 tonnes) and the calculated weight of contemporary metalwork (75 tonnes), leading to the conclusion that the region must have been a net exporter of metal (Jackson 1979, 1980). These estimates are not considered reliable, as they were based on limited survey data and over-optimistic estimations of ore grade. Depending on the parameters taken, a number of which are highly speculative, the total production of metal on Mount Gabriel itself in the period 1700–1400 BC is estimated at 1.47–29.43 tonnes, with likely output probably towards the lower end of this scale (O'Brien 1994: table 12). This was enough copper to produce 40–50 bronze axeheads a year, a scale of production adequate for local and regional needs, but not sufficient in terms of any wider export of copper. These revised estimates are important in adjusting perceptions of the importance of Irish copper sources during the Bronze Age in Europe (O'Brien 1999).

While estimates of mine output are often inflated, they provide a more plausible picture of metal use than the archaeological sample of finished objects. Taylor (1999) suggests the latter may represent a mere 0.01 per cent of the original production. This makes any calculation of the total amount of metal available to a society highly problematic.

Environmental impacts

Mining in the modern era is often controversial due to perceived or real consequences for the natural environment. This includes the physical alteration of landscape, any atmospheric and groundwater pollution, and resulting effects on vegetation and the food chain. The type of impact depended very much on the duration and intensity of the mining activity. A distinction may be made between mining and related processes that had a long-term environmental impact over a wide area, and those with a more localized impact of brief duration. Most prehistoric copper mines fall into the latter category; however, there are examples where the scale of operations had long-term consequences for the environment.

The issue can be examined in a number of ways, using palaeoecological data collected from the mine sites (waterlogged wood, charcoal, etc.) or from natural environments in the vicinity. The latter includes peat bogs and mires of various types, lakes, and glaciers. Peat bogs are important for temperate Europe, as they can provide a range of biological and geochemical proxies (pollen, heavy metals, and microscopic charcoal) connected to early mining and metallurgy. Pollen records are especially useful in providing an insight into interference with local woodland caused by mining, while a rise in heavy metal concentrations in bogs, lakes, and glaciers can be an indicator of mining and smelting. Some of the following examples demonstrate the value of a multi-proxy approach that combines palynological (pollen and micro charcoal) with geochemical data. This information can be used to examine the environmental context of early copper production, as well as its effect on the natural environment.

While there are many sources of environmental information, the interpretation of this data can be problematic. Even though farming creates particular signatures in the pollen record, it is generally difficult to distinguish evidence for a reduction in tree cover caused by mining from contemporary agricultural clearances. This is partly because many mining groups came from farming communities in the same general area. The interpretation of microscopic charcoal is similarly difficult, as is the source of heavy metal concentrations (see Mighall et al. 2012 for discussion). Much depends on the duration and intensity of the mining activity, and the quality of the environmental evidence.

In many cases the most severe impact on the environment was connected to the demand for wood as fuel in the mining and smelting operations. The intensive use of fire-setting in many mines consumed enormous amounts of fuel. Wood was also required for domestic use in the mine camps, to make equipment of various kinds, and, as seen in the Mitterberg mines, for roof support. The smelting of copper ore using charcoal was another significant demand. All of this would have impacted on tree growth in the immediate vicinity of the mine, and possibly farther away. Some mines were located in areas of naturally abundant woodland, while others, particularly those at higher altitudes, had only limited tree growth in the vicinity.

For those mines where fire-setting was used in an intensive manner, the supply of wood required careful management. This often involved thousands of fire-setting cycles leading to an enormous consumption of wood. An example is the small-scale mining conducted on Mount Gabriel, south-west Ireland, during the Early/Middle Bronze Age, where around 30 workings used an estimated 3,924–14,533 tonnes of roundwood fuel to extract an estimated 4,000 tonnes of rock (using wood–rock extract ratios of 1:12 and 1:0.27, respectively). Depending on the efficiency rate, a single 10 m deep mine working on this mountain required 100–200 fire-settings, which in turn consumed 138–509 tonnes of young roundwood fuel (O'Brien 1994: table 10).

Pollen data for Mount Gabriel reveals a pattern of sustained, but limited, woodland clearances during the earlier part of the Bronze Age (Mighall et al. 2000). The mining on this mountain seems to have had little impact on local woodland, with species such as oak and hazel actually increasing in abundance during the mining period. It can be explained by the small-scale and sporadic nature of this mining, and the selective manner in which fuel was gathered where wood species were used for different purposes. While the gross consumption of fuel was enormous, it must be borne in mind that this mining was conducted on a sporadic basis over three centuries or so. This reduced the immediate impact of wood collection on the local environment, allowing time for regeneration of local woodland. An analysis of waterlogged wood from the Mount Gabriel mines demonstrates that the miners avoided mature trunk wood as fire-setting fuel, opting for selective felling of roundwood branches from mature woodland, as well as the use of adventitious coppice from younger growth. There may have been occasional clear-felling of mature trees, but generally the miners extracted branch wood from trees that regenerated over a number of years. Tree-ring studies indicate that the same areas of woodland were cut over on several occasions, producing adventitious coppice rather than any formal woodland management (McKeown 1994).

Similar results were obtained for the Bronze Age mine at Copa Hill, Cwmystwyth, in mid-Wales (Mighall and Chambers 1993; Mighall et al. 2012). Analysis of peat cores sampled 500 m from the mine reveal a series of short-lived declines in tree pollen during the mining period, mainly oak and hazel, which correlate with higher copper concentrations in the bog. The overall evidence from those studies points to a limited impact of copper mining and metallurgy during the Bronze Age. The pollen records indicate small-scale, non-permanent phases of woodland disturbance confined to the general mine area. The Mount Gabriel and Copa Hill results indicate a series of small, short-lived declines in tree pollen probably connected to mining, where local woodland was not permanently affected by this activity.

There are similar findings from palynological analysis conducted near Bronze Age mining sites in Steiermark, eastern Austria (Marshall et al. 1999). More recent research in Austria, conducted at Kelchalm in Kitzbühel (Viehweider and Oeggl 2012) and in the Falkenstein mining district of north Tyrol (Breitenlechner et al. 2010), has identified changes in vegetation linked to Bronze Age mining. Again, this did not involve the wholesale destruction of tree cover, but rather local variations in woodland density and species representation. Similar results were obtained for pollen studies at the Iron Age copper mine of Campolungo, in the Lombard Alps of northern Italy (Mighall et al. 2003). None of these prehistoric mines can be linked to permanent reduction in tree cover or to any long-term impact on the environment.

The use of charcoal fuel for smelting was another significant demand on woodland resources, one that is well recorded from iron production in historic times. As with mining, this depended very much on the scale and duration of this activity. The smelting associated with small operations, such as Mount Gabriel in Ireland, had a limited impact on woodland resources compared to the larger Bronze Age mines in Europe. For example, it has been suggested that the forests of Cyprus had to be cut down 16 times to meet the fuel needs of Bronze Age copper production (Constantinou 1982). This indicates a considerable regenerative ability or else strong methods of woodland management to maintain the fuel resource.

Another example is the mining of several million tonnes of copper ore at Kargaly during the Bronze Age, leading to the production of an estimated 150,000 tonnes of metal. The smelting of this oxidized ore required large amounts of charcoal fuel, using methods not dissimilar to those employed by Russian industrialists in the eighteenth and nineteenth centuries (Chernykh 1994). Records from that period show that it was necessary to burn 300–500 cubic metres of pine or birch to produce enough charcoal to smelt a tonne of copper. Rovira (1999) estimates that 75 million tonnes of wood would have been consumed by the Bronze Age production. Chernykh (1998a) has calculated that one hectare of forest in the South Urals could provide 250–270 cubic metres of wood suitable for charcoal under conditions of continuous tree felling. This means that 1.5–2 hectares of good quality forest was required to smelt one tonne of copper metal. An estimated production of 150,000 tonnes of copper during the Bronze Age may have led to the deforestation of up to 3,000 square kilometres in the Kargaly region.

The enormous fuel demands made by these mines contrast with the limited amount of woodland available in the Kargaly region today. This is a steppe landscape with sporadic tree growth in deep ravines or watercourses. The nearest true forest is located 200–250 km to the north-east in the mountainous zone of the southern Urals. If the fuel estimates for the Bronze Age are accepted, this means that either the distribution of local forest has diminished considerably since that period or that the ore was transported to distant fuel sources, or the reverse.

Environmental studies suggest that tree growth in the Kargaly area during the Late Bronze Age mining period was almost identical to today (Vicent et al. 2000; Del Rio et al. 2006; Freire et al. 2012). There is evidence for intensive deforestation during the Early and Late Bronze Age phases of mining. While some woodland management may have been practiced, the available fuel resource at Kargaly would not have been sufficient to meet the demands of metal production. As a consequence, most of the copper ore was probably transported long distances to the forest zone of the South Urals for smelting (Chernykh 1998a; Rovira 1999). A similar strategy was employed during the eighteenth and nineteenth centuries when millions of tonnes of copper ore were carted from Kargaly to the Urals forest zone for smelting. In both instances this had the effect of spreading the ecological impact of copper production over a wide area.

Mining and smelting are processes that can result in atmospheric pollution, leading to elevated levels of heavy metals in soils and sediments. The interpretation of this geochemical data is complex and is best considered in combination with pollen and other evidence (see Mighall 2003 for discussion). The Cwmystwyth and Mount Gabriel studies identified some evidence of atmospheric pollution contemporary with early copper mining (Mighall 2003). Geochemical analysis of peat profiles at both locations identified enhanced metal concentrations that may have resulted from smelting emissions or dust blowing from exposed mine spoil of Bronze Age date. The spatial extent of that pollution is unclear, but was probably restricted to the vicinity of these mines.

Similar problems arise with the interpretation of microscopic charcoal in peat records. Higher levels of charcoal should in theory correlate with changes in arboreal pollen and pollution signals caused by early mining, particularly where fire-setting was employed. Smelting activity provides another source of wind-borne charcoal. This phenomenon has been studied at both Mount Gabriel and Cwmystwyth, where it has not proved to be a reliable indicator of mine fire-setting or smelting (Mighall et al. 2012: 127). The problem mostly lies in distinguishing charcoal produced by mining and metal production from contemporary agricultural and settlement activity.

Recent studies conducted at Lago Riane, in the Aveto Valley of north-west Italy, identified heavy metal concentrations believed to be airborne pollutants from early metal production (Braithwaite et al. 2012). The study suggests that this may have derived from the Copper Age mines at Monte Loreto and Libiola mines in this part of Liguria. There is similar evidence from the French Basque region, where palaeoenvironmental studies in the Nives des Aldudes indicate the presence of early metal production in this high altitude Pyrenean valley (Monna et al. 2004). This study revealed geochemical signals for air pollution possibly connected to copper mining and smelting at different times during the early third millennium BC, Middle Bronze Age (1500–1300 BC), and Late Bronze Age (900–800 BC).

High altitude lakes can also provide valuable pollution records, such as closed mountain glaciers where climate oscillation can be studied in relation to vegetation history. One such study connected to early mining has been undertaken at Lake Bramant in the Grandes Rousses massif of the western French Alps (Guyard et al. 2007; Carozza et al. 2010). Analysis of varve sediments reveals evidence of metal contamination in the period 2200–1700 BC (Guyard et al. 2007: fig. 14). This can be linked to archaeological evidence of Early Bronze Age copper mining in the same area (Bailly-Maitre and Gonon 2008). The environmental record indicates that this was a warm period of reduced glacial activity, which favoured the development of mining at altitudes of 2,300–2,600 m.

Metal pollution connected to prehistoric copper production can be recorded in other environments. A study of river sedimentation in the famous Rio Tinto mining district of Huelva, south-west Spain, revealed enhanced levels of copper dating to the later third millennium BC (Le Blanc 2005). The same sediments contained fragments of smelting slag and charcoal, indicating a direct connection with copper mining in the general area of Rio Tinto where most prehistoric mines were destroyed by later mining. This has also been assessed by borehole samples from the Guadiana River Estuary, which reveal significant enrichment of heavy metals over the last 4,500 years (Delgado et al. 2012: fig. 10). The higher levels of copper, lead, and zinc can be connected to intensive mining of sulphide deposits in the general area of the Iberian Pyrite Belt. Several phases of mining activity can be identified, commencing in the third millennium BC and continuing through the Bronze Age, followed by intensive mining of the Roman and early modern eras.

This pollution can be linked to discoveries at Cabezo Juré, a copper-producing settlement of the third millennium BC, adjacent to the Tharsis orebody in the Iberian Pyrite Belt (Nocete et al. 2005). The investigation of that site was part of a broader study of the environmental impact of early copper production in that part of south-west Spain. An analysis of pollen and charcoal data reveals a progressive decline in the local woodland from the earliest occupation at Cabezo Juré, which seems to be connected to mining and not agriculture. This correlates with marine molluscan data and sediment analysis from river estuaries, which indicates that heavy metal pollution associated with copper production commenced during the early third millennium BC. By 2500 BC the scale of this impact was considerable, with systematic deforestation, increased erosion, and heavy metal pollution in the estuaries of the Tinto and Odiel (Gulf of Cádiz) rivers. The same analyses reveal a gradual decrease in pollution towards the later third millennium BC coinciding with pollen evidence for woodland recovery. This was connected to a decline in copper production and a shift in economic strategy to agriculture (Nocete et al. 2005).

In conclusion, environmental records suggest several scenarios in relation to prehistoric copper mining, ranging from a sharp reduction of tree growth after mining commenced, to higher levels of tree growth suggesting some form of woodland management. In most cases the vegetation impacts are relatively localized, and cannot be easily distinguished from farming activity in those areas. The absence of evidence for substantial woodland clearance is surprising given the considerable demands of mine fire-setting. This may be explained either by an abundance of woodland close to some mines, or by careful management of limited resources to maintain regular fuel supply. There is much variation depending on the natural abundance of local woodland and the intensity and duration of the mining activity. In many instances the first significant impact on the natural vegetation of a mine location was due to agriculture. Where this coincided with the commencement of mining, as for example at Mount Gabriel and Ross Island in south-west Ireland, the pollen evidence indicates background agricultural settlement that sustained the mining activity.

At the other extreme are operations such as Kargaly and the mines on Cyprus, where copper production had a serious long-term impact on the environment. That said, these mines did not have the same environmental impact as mining operations in the Roman and historic period. Analysis of heavy metal contamination in Greenland ice reveals very low levels of atmospheric pollution connected to prehistoric metal production in the Northern Hemisphere. This pollution was significantly greater with the commencement of large-scale Roman mining in the Mediterranean basin from the second century BC onwards (Ferrari et al. 1999).

10

Mining, economy, and society

A MINING ECONOMY

The opening chapter of this book considered different factors that influenced the availability of copper resources in prehistory. While geological distribution and technological expertise were critical, consideration must also be given to the wider societal context of production. The operation of early mines must be explained in terms of access to ore deposits and the desire and ability of different population groups to become involved in primary metal production. The impact on local and regional economies is also relevant, in terms of wealth generation through trade and the repercussions for society as a whole.

Understanding the organization of this activity is a challenge. Key elements of the *chaîne opératoire* are often missing, such as the location of smelting sites or the workshops where objects were made. This makes it difficult to establish links between mines and the circulation of intermediate and final metal products in a wider settlement context. With stone tools it is possible to apply production indices to quantify the different stages involved in the use of a specific raw material, with a view to modelling a lithic production system in space (see Ericson 1984). This approach cannot be easily applied to metal objects, which generally have a more complex life cycle than stone tools (Fig. 10.1). This began with a fundamentally different use of a raw material to create a finished object, requiring chemical as well as physical transformation. For this reason, scientific analysis of prehistoric metalwork is problematic in terms of source provenancing to specific ore deposits and mines. There is the further complication of recycling, which in some instances involved the mixing of metal from different mine sources.

One approach has been to identify metal circulation zones where copper of a similar chemistry, lead isotope signature, and/ or alloy type was used (e.g. Northover 1982). Within these circulation zones various patterns of primary and secondary (recycled) metal use can be explored in the context of local workshop traditions. This provides a spatial and typochronological context in which to view the input of metal from particular mines. A good example is the identification of Ross Island copper mine as the principal source of arsenicated

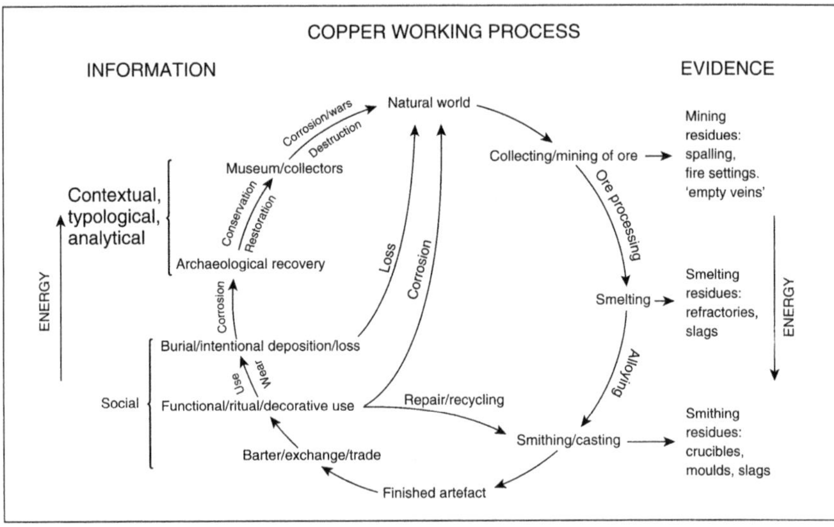

Fig. 10.1. Metallurgical cycle

(*Source*: Ottaway 2001).

copper used in the Irish-British Chalcolithic (Northover et al. 2001; see also Bray and Pollard 2012).

There has been a tendency to consider prehistoric metal production as a highly organized process, constituting what is often termed an industry (see Chapter 1, The Human Environment section). While modern terminology is problematic, it is also true that the scale and quality of metal production in many prehistoric societies reflects high levels of organization. Various models can be applied to an understanding of metal production and supply connected to prehistoric copper mines. The first is a mine-centred production system, where smelting and fabrication centres are located at or close to the ore source. The supply of metal from such a system may have involved down-the-line exchange, initially based on balanced reciprocity within socially connected groups, moving into trade-for-profit with greater social distance. Another model is where sequential production took place in the context of a regional distribution system. This is where production was taken to a particular stage in the mine and then resumed at distant settlement locations. It involved the transport of intermediate or final metal products (Fig. 10.2) to locations that acted as redistribution centres for finished metalwork.

Such models can be explored by considering the distribution of metal objects associated with a prehistoric mining tradition. In the case of mine-centred production there should be an exponential fall-off of metal products with distance from source. For regional distribution, the greatest concentration of metalwork may occur not concentric to the mine source, but around

Fig. 10.2. Copper ingot and tin-bronze axehead made of Ross Island copper, Knockasarnet, Killarney, south-west Ireland

(*Source*: author).

centres located a considerable distance away. In that situation the supply of metal was controlled by central place redistribution.

There are good examples of both supply systems in early lithic production (e.g. Cummins 1979: figs 6–7), but they are more difficult to identify in relation of metal supply. A good example is the problem of the Great Orme mine in Wales, a significant producer of copper over a millennium or so during the Bronze Age. Much of that metal was of a high chemical purity and so cannot be easily traced in the archaeological record. Other mines have distinctive chemical and isotopic signatures, such as the copper produced from fahlore sources at Cabrières and Ross Island. There are indications in the latter examples of strong regional circulation of metal. Some of this was mine-centred; however, in the case of Ross Island there is evidence for secondary distribution, operating at settlement foci such as Lough Gur and elsewhere in Ireland (O'Brien 2004: 567).

Procurement strategies

The organization of metal production in prehistory was influenced by the procurement strategies employed to source copper ore. These strategies, in turn, were closely linked to territorial awareness in a region and the perceived ownership of resources. It is likely that most prehistoric copper mining in Europe was undertaken in environments of relatively stable settlement within well-defined social territories. This landscape context implies a developed sense of resource ownership, where access to mineral deposits and the resulting production of metal was carefully controlled.

Some researchers question the idea that prehistoric groups could appropriate natural resources as their own property. For example, Stöllner (2003a: 439) argues that the legitimacy of exploiting a mineral deposit was partly based on magico-religious tradition. He cites numerous ethnographic sources where the ownership of ore deposits in traditional societies was considered to be the privilege of ancestors or deities. Such concerns may have influenced the mining process in different ways, such as determining the most propitious time to commence this work or limiting the amount of ore removed at any one time. On the latter point, Stöllner makes the interesting observation that this may have prevented the type of excessive mining of ore deposits that was taking place by the first millennium BC in the Mediterranean basin (Stöllner 2003a: 440).

Some metal-producing groups had free access to a particular ore deposit, while in other cases a particular group controlled the resource. Direct access procurement was a commonly used strategy in early lithic production, well suited to seasonal exploitation of hard rock sources on an ad hoc basis (see Torrence 1984). The production of copper is a more complex undertaking, involving a range of skills that required higher levels of specialization and greater time commitments for a community. The standardized output of many prehistoric metalworking traditions points to carefully organized production and more efficient sourcing of ore than uncontrolled direct access. This was also connected to the perceived value of metal, both as a trade commodity and as a signifier of status and wealth in early complex societies.

Copper mining in prehistoric Europe was mostly undertaken in societies with numerous expressions of territoriality and resource ownership. This applied to agricultural land, but also mineral resources. For that reason, it is likely that direct access procurement was the exception, with the mining of ore deposits carefully controlled by local population groups for whom this constituted a valuable resource. What is less clear is whether such miners were autonomous, or else operated under the influence of more powerful groups in agricultural hinterlands. It is possible the latter were able to mount expeditions to obtain ore and metal, which meant passing through different social territories. Such mining expeditions may have been the exception, with most distant consumers having to obtain metal through a network of trade partners.

Different models are possible, all influenced by the underlying social and economic structure of those societies. Some mines were connected to a coordinated system of copper supply, while others were part of a local pattern of consumption. In many instances there was no direct link between the producers of metal and distant consumers, imposing a strong spatial limit on effective demand with this type of one-to-one exchange. This may have been organized in the following manner (from Sherratt 1976):

Ore Source: Activities include the extraction and beneficiation of copper ore (occasionally smelting) in a mine site that is accessible to an area of agricultural production.

Production Zone: Local settlements involved in the active exploitation of the ore resource. Activities may include smelting, and also conversion of the primary metal into an exchangeable form such as ingots or finished metalwork.

Direct Contact Zone: Settlements linked directly to the production zone by face-to-face contact. Effective supply of metal as a result of close kinship links.

Indirect Supply Zone: Distant settlements without direct access to the production zone, receiving supplies of metal through intermediaries.

This basic structure of metal supply could be applied to many prehistoric copper mines in Europe. In some instances the scale of production is so limited that supply is entirely local, operating almost entirely within the production zone. The occurrence of mining on anything more than a very small scale implies production for the purpose of exchange into the Direct Contact Zone. For mines with greater output, there is the likelihood of trade for profit conducted in the Indirect Supply Zone. Such trade was generally conducted at a much greater social distance from the culture group that produced the metal, a good example being the commercial trade in oxhide ingots from Cyprus. There were, of course, instances where close cultural connections could exist over great distances, such as the place of Kargaly mine in the Srubnaya culture, or the significance of Rudna Glava for the Vinča culture.

The trade in copper

One of the distinctive features of the Bronze Age was the way that some regions lacking metal resources of their own were able to develop strong metalworking traditions, with Denmark an obvious example. This was possible through the development of trade networks in which the supply of metal played a significant part (Fig. 10.3; see Needham 1998 for discussion). The complexity of these exchanges lies in the movement of primary and recycled metal, in ingot, scrap, or finished object form, accompanied by the supply of tin ore or metal, and materials ranging from gold and amber to perishable commodities.

Metal could be exchanged in different contexts ranging from reciprocal gift exchange to organized trade-for-profit in market situations. Anthropological studies since Mauss (1923) have emphasized the importance of gift exchange

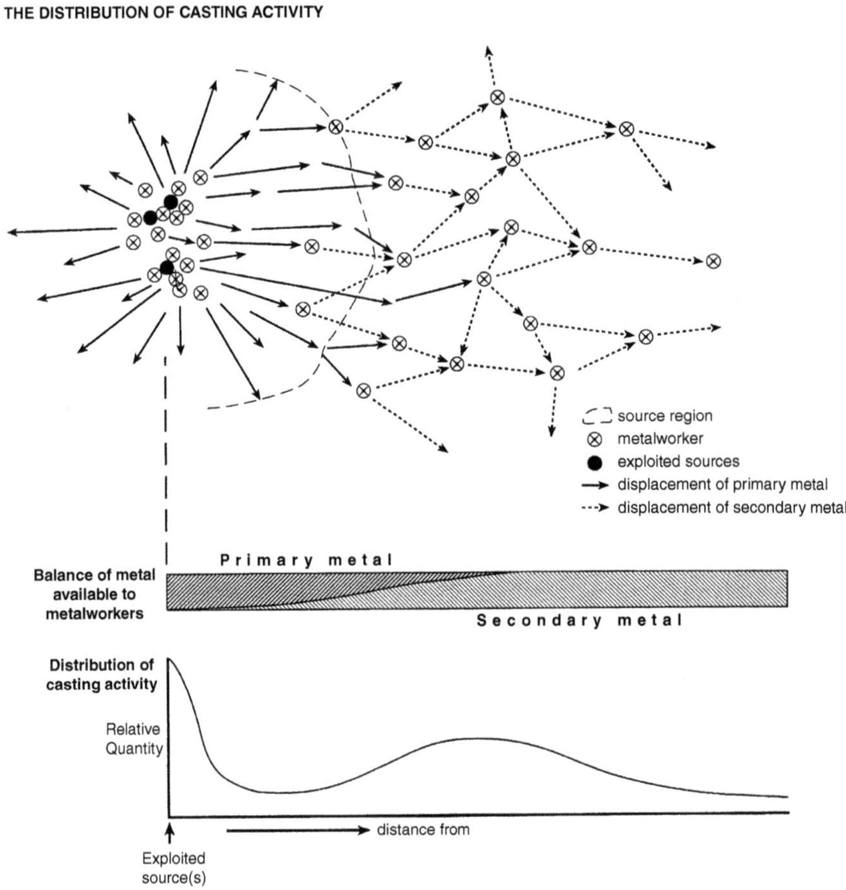

Fig. 10.3. Organization of metal production in relation to mine sources (*Source*: Needham 1998).

in early societies, while there has been much discussion on the role of prestige goods in the economies of the Chalcolithic and Bronze Age. There are also numerous examples of commercial trade involving freelance middlemen, vividly seen in the Uluburun and other Bronze Age shipwrecks (Bass 1967; Pulak 1998).

The question of demand is important when it comes to understanding regional economic systems specializing in a primary product such as copper. The motivation to engage in the laborious work of mining must be linked to perceived benefits for the society in question. This cannot be considered in isolation from other areas of the economy, such as agriculture, which are equally time-consuming. This question was considered by Shennan (1999, 2010) who stressed the relevance of cost benefit analysis in relation to the underlying motivation to engage in mining. Citing Ricardo's Law of Comparative Advantage (it is not worth producing commodity X if you are better off producing commodity Y and obtaining commodity X in exchange for it), he suggested that what mattered was the exchange value in relation to the costs of production and transportation.

Shennan argued that such thinking was central to the development of Early Bronze Age mining in the eastern Alps. Those mining communities lived in upland areas with limited potential for agricultural production beyond basic subsistence. They did, however, have a comparative advantage in respect of copper mining, based on productivity, technical expertise, and the necessary social arrangements to engage in that work. By contrast, communities in the Alpine foreland did not engage in copper production, not because they were unable to obtain supplies of ore for smelting, but because they could get higher returns on their labour by engaging in farming and obtaining copper by means of exchange.

Bartelheim (2009: 40) and others have made the point that metal production in excess of a community's own needs was only sustainable if a wider market demanded it. For that reason there is a strong connection between mine output and the operation of trade networks. Understanding that relationship is made difficult by the fact that mines are generally removed from the different contexts of metal circulation. This began in many instances with the physical separation of mining and smelting, and continued with the movement of primary metal to other centres for metalworking and redistribution.

The connection between mining, trading, and market-related activities played a pivotal role in the development of a mining district into an economic zone (Stöllner 2003a: 425). Mines were often located in geographically remote areas, a long way from potential markets and consumers. Stöllner notes that this did not favour the development of major settlements in mining regions, but rather intermediary trade and production centres. As with basic economic laws of supply and demand, trade was of great importance when it came to accelerating mining and metal production. In many cases, wealth was not generated by primary production, but by reprocessing and trade that involved proto-urban centres not linked directly to the mining areas (Stöllner 2003a: 427).

An example is the supply of copper during the Early Bronze Age in central Europe. The large number of rib and ring ingot hoards in the northern hills of the eastern Alps indicates significant levels of production of fahlore copper of both the *Singen* and *ösenring* varieties. The enormous amount of copper in the Alpine foreland (extending as far as the Danube) indicates an important transit area for copper coming from the Alpine mines (Moslein 1999; see also Winghart 2000). This involved the movement of raw copper from the mines to settlements at the exit of the Alpine valleys, for redistribution northwards in the form of ingots or finished metalwork. The evidence comes from the large number of ingot hoards discovered at the mouth of the copper-rich Inn and Salzach valleys (Menke 1978/9). Bartelheim (2009) has highlighted the large number of Early Bronze Age settlements in the Salzburg–Reichenhall region at the foot of the Alps. He suggests that metal from the Mitterberg mines was distributed through this pre-Alpine region for exchange to agrarian settlement centres across Bavaria and the lowlands of central Europe (Bartelheim 2009: 39).

The supply of primary copper from mine sources was always subject to interruption. In some instances the reasons were political, while in others new economic priorities and changes in trade relations were to the fore. These shifts in metal supply are best observed through changes in metal type, as determined by chemical and isotopic analyses. In many cases these cannot be directly linked to the opening or closure of specific mines; however, regional trends are generally apparent.

An example is a recent analysis of Early and Middle Bronze Age metal finds from the cemeteries in Lower Austria (Duberow et al. 2009). The lead isotope results indicate that Early Bronze Age objects from the Hainburg cemetery may derive from an ore source in the western Carpathian mountains, possibly in Slovakia. In contrast, the lead isotope analyses for Middle Bronze Age metalwork from the Mannersdorf cemetery reveal a shift in metal supply to the eastern Alps. This was part of a long-term trend in central Europe, connected to the rise of the Mitterberg in copper supply from the end of Early Bronze Age onwards.

Pearce (2000) examined this issue in the context of changes in metal supply to Bronze Age settlements in the western Po Plain of northern Italy. The initial supply of copper *c.*1700–1150 BC came from mines and smelters in the south Alpine Trentino area. During the final part of the Bronze Age (1150–950 BC) this source of copper was replaced by supply from as yet unidentified mines in Tuscany across the Apennines in central Italy. The reasons for this shift are unclear; however, Pearce has argued that the availability of tin in the latter region was a significant factor. Unlike their Trentino counterparts, the Tuscan copper producers could offer a supply of bronze and so may have been an attractive option for trade in spite of the greater distances involved.

This is one example of the impact that tin demand had on the supply of copper during the Bronze Age. That scarce raw material was critical to the growth of copper as a commodity in the Bronze Age and thus affected the importance of copper mining. The ability to integrate the supply of tin from a small number of sources with the output from copper mines located in distant regions stands as a particular achievement of Bronze Age trade. A good example is how the discovery of tin in south-west England, *c.*2200–2000 BC, stimulated the spread of copper mining in other parts of Britain. The development of tin-bronze in that period made the use of high purity copper from oxidized ores, such as that produced in the Great Orme, Alderley Edge, and other mines, both technically feasible and an attractive alternative to the older tradition of arsenical copper based on scarce fahlore deposits.

The trade network that developed across the east Mediterranean during the Late Bronze Age is another example of the dynamic and often unstable nature of metal supply. The Uluburun and Cape Gelidonya shipwrecks off the southern coast of Turkey were part of a highly organized trade in luxury goods in the east Mediterranean, *c.*1400–1200 BC. This required professional traders familiar with the risks of these sea crossings, and also with an understanding of the languages and value systems of different societies. These two shipwrecks illustrate the changing economic and political context of copper production on Cyprus in that period. During the late fourteenth century BC, ships such as Uluburun were part of a well-differentiated trade in prime or convertible materials (precious metals, bulk copper, and tin) that extended across the east Mediterranean. This trade was controlled centrally and operated through elite gift exchange and conspicuous consumption. A century or so later, these monopolies had broken down with the rise of free-trade entrepreneurialism centred on the urban coastal settlements of Cyprus. This shift is represented by the growing commercial trade in scrap metal as evidenced by the cargo of the Gelidonya ship (Sherratt 2000).

Maritime trade in metal and other valuable commodities was widespread in Europe during the Bronze Age. Examples include several Bronze Age metal cargoes lost off the southern coast of England (Muckelroy 1981). These include the 1974 Langdon Bay find, a collection of scrap metal that was part of a cross-Channel trade during the Middle Bronze Age. There are two such finds from Salcombe off the Devon coast, the most recent in 2009 when a Late Bronze Age cargo containing 259 copper ingots and 27 tin ingots was discovered. Another example is Scandinavia where maritime connections involving the importation of metal may be emphasized by the importance of ships on rock art and on bronze items (Ling et al. 2013). In most cases these trade networks were not connected directly with mining areas, but instead operated through intermediary traders and secondary redistribution.

Mining and the hoarding of metal

In terms of dryland equivalents, reference should be made to the widespread practice of metal hoarding during the Chalcolithic and Bronze Age (see Harding 2000: 352–68). Many explanations for the phenomenon are possible, depending on the type of hoard, the circumstances of deposition, and the wider historical context. The practice has been linked to the supply of metal, in the form of merchants' hoards connected to trade in ingots or finished metalwork, or scrap hoards where items of broken and worn metalwork were gathered together for recycling. Hoards may represent a surplus of metal at times of strong production or can be linked to a scarcity of metal at other times. More controversial is the suggestion that hoarding was a mechanism to control the amount of metal in circulation, thus maintaining its value in the face of fluctuating demand. This is a modern concept relevant to market economies, which is unlikely to apply to the redistribution systems of the prehistoric period.

Other interpretations of hoarding can be considered in relation to the circulation of primary metal, whether as security deposits hidden in times of turmoil or as ritual offerings. In some parts of Europe, such as Britain and Ireland, hoarding was most prevalent during the Late Bronze Age, a period when warfare and security considerations are manifest in the archaeological record. This was also a time when the distribution of copper mining across Europe was reduced to a small number of very large producers, notably those in the eastern Alps. In circumstances of restricted supply, the practice may have been closely connected with trade networks and recycling systems.

The ritual dimensions of metal hoarding are complex (see Bradley 1990), but include the notion of returning metal to the earth as an act of expiation for the removal of ore minerals. While such beliefs may have existed, there is no indication that this was ever done at the mines themselves where metal hoards are generally not found.

Copper and stone

In many parts of prehistoric Europe the mining of copper had a significant impact on established systems of lithic production. There is evidence from many regions for surface quarrying and underground mining of flint and other hard rocks during the Neolithic. This ranged from small, short-lived, operations serving local needs, to large-scale production of longer duration tied into regional exchange. It is possible that many of these networks became unstable with the circulation of metal, as competition increased between groups who controlled the movement and consumption of different classes of material. This helps to explain the sudden shifts in stone axe production in the Late Neolithic and the increasing demand for material of distant origin. The circulation for metal added to this instability, considering its scarcity and high value in social exchanges, all of which impacted on the perceived value of stone axes.

Taking Britain and Ireland as an example, many stone axe factories had ceased operation prior to the introduction of metal in the mid-third millennium BC, or shortly afterwards as the demand for stone axes declined. Production did expand at a number of sources during the Chalcolithic (e.g. Grimes Graves flint mine), however, none of the major stone axe producers survived the widespread adoption of tin-bronze after *c.*2000 BC. The response of established stone-working traditions cannot be explained solely in terms of a popular demand for a technologically superior material. Though the practical advantages of metal would become important with the development of tin-bronze, the explanation for the earlier use of copper (and gold) was linked to new regimes of value established during the Chalcolithic. The use of copper and gold was socially determined to an even greater extent than stone, as the scarcity of these objects, combined with their 'magical' production and aesthetic appeal, served to limit access to important tokens of value. This reflects new social values and ideological change, connected to the spread of material values of the Beaker culture.

The older networks of lithic production and exchange must have influenced the circulation of copper during the Chalcolithic, though to what extent is unclear. Geological factors limited the ability of most groups to engage in primary metal production. While most parts of Britain and Ireland had local options in respect of lithic raw material, there were now clear distinctions between areas with direct access to primary metal and those forced to engage in regional exchange. Even where ores were locally available, many regions never developed a mining tradition, opting to use other resources to source metal through regional exchange. This includes the copper (and tin) rich region of Cornwall where stone axe production was well established during the Neolithic.

The demand for metal created new possibilities in terms of control of resources and the creation of economic power and status. This was clearly important during the Chalcolithic when there was a restricted supply of primary copper due to the dominant position of a small number of regional sources, such as Rudna Glava in Serbia, Cabrières in France, or Ross Island in Ireland. As the main producer of primary copper within Ireland and Britain *c.*2400–2100 BC, the latter was more influential than even the largest stone axe producers. The implications were significant, both for the Beaker groups who controlled the Ross Island resource and for the stone axe producers whose products were now increasingly devalued in social exchanges. While many quarries and flint mines ceased operation, some areas with a tradition of hard rock extraction would, in time, adapt to the changing circumstances by developing their own metal resources. This partly explains the development of Bronze Age copper mines in parts of north Wales with established quarrying traditions; for example, the Great Orme mine near the Penmaenmawr/Graig Lwyd stone axe source. This did not occur in Cornwall where the exploitation of tin resources offered a better economic strategy in the long term.

In conclusion, the adoption of metal had a significant impact on lithic production and exchange systems during the Chalcolithic in Europe. This impact varied both temporally and geographically as metal use gradually expanded, depending on whether a particular region was metal producing or receiving and the status of existing lithic traditions. The reason stone axe networks began to contract during the Chalcolithic was not connected to a scarcity of raw material or perceptions of efficiency, but rather to reduced demand for such artefacts in the social exchanges that distinguished that period. The move to greater metal dependency was linked to the development of bronze metallurgy around 2000 BC, when copper mining began to expand in many parts of Europe to meet the growing demand for metal.

COPPER RESOURCES AND SOCIAL POWER

> It is only because miners sweat their guts out that superior persons can remain superior.
>
> George Orwell (*The Road to Wigan Pier*)

Metal is widely regarded as having played an important role in the development of later prehistoric societies in Europe. This is particularly true of those groups who had direct or indirect control over the supply of copper. Many of these emerged as primary producers of copper due to a combination of favourable geological factors and technological expertise, and a willingness to invest the time and resources required. In some instances this was a part-time activity involving groups who had a greater commitment to agriculture. For others, it represented a long-term commitment to mining and copper production that shaped their own development as a community and their relations with neighbouring groups.

The Chalcolithic and Bronze Age in Europe are generally associated with hierarchical social structures, where powerful individuals or families held sway over the mass of a peasant farming population. Control over the supply of metal is widely viewed as critical to the exercise of social power in that period, contributing to the aggrandizement of certain individuals and groups. One widely held belief is that these elites created and maintained positions of social dominance by control over the production and supply of prestige goods made of metal. This was expressed in lavish funerary display, gift exchanges, votive offerings, or metal hoarding. The importance of copper and bronze objects lay not only in their practical value or aesthetic qualities, but also in their cosmological references and connection to distant sources.

The involvement of controlling elites in early copper mining and metal production derives from a belief that these were complex processes that required careful organization (e.g. Strahm 1994). It is based on the specialist nature of copper production and the division of labour required by the different stages of production and supply, as well as the mobilization of a workforce and other resources. This required strong leadership and organizational ability, which could be used to explain the rise to prominence of certain individuals. The position of these local elites is reflected in the use of prestige metalwork, often objects of cosmological significance and exotic origins. The difficulty is that these status signifiers cannot be directly linked to mining, nor is there any reason to believe that the individuals or groups concerned were directly involved in the mining operations. The material benefits of this activity did not necessarily pass down to the miners themselves, whose work in many instances was controlled and exploited by other groups. The degree of autonomy enjoyed by mining communities was influenced by the need to participate in wider networks of exchange to obtain a tangible material return for their work.

The cause-effect relationship between the adoption of metallurgy and growing social complexity has long been debated. While early research favoured a close connection between the two phenomena, recent studies emphasize that access to metal was one of many factors that shaped the development of a prehistoric society. The following examples serve to illustrate this point.

Cabrières and early copper mining in southern France

An example of the cause/effect approach is Strahm's (1994) analysis of the wider economic and social impact of metal in relation to central Europe (see also Strahm and Hauptmann 2009). He proposed two systems: a Chalcolithic model where production was undertaken by small domestic units, and an Early Bronze Age model, termed Metallikum, where metal supply was controlled by a centralized political power (Strahm 2005: fig. 4). In this analysis the evolution from the Chalcolithic to the Metallikum stage signified the emergence of chiefdom-type societies.

Strahm (2005) subsequently applied this model of growing social complexity linked with copper production to the evolution of early metal-using societies in France. Mille and Carozza (2009) have questioned the approach, arguing that the adoption of copper did not have major economic and social consequence for most Late Neolithic farming communities in France. They point to the use of metal during the third millennium BC as largely symbolic, having little real impact in economic terms. They do acknowledge that the new technology had a significant impact in those regions that emerged as primary production centres. This includes Cabrières, where the output of copper by the mid-third millennium BC can be linked to changes in settlement organization across the central valley of the Hérault. These centred on the emergence of a new social hierarchy in which fortified settlements had a central role in terms of resource control and economic competition (Mille and Carozza 2009: figs 19–20). The integrated nature of that production, and the systematic exploitation of the resource over a long period, confirms the high social value of copper in that period.

The economic and social impact of the Cabrières production was largely contained within a specific regional setting. Mille and Carozza point to a strong correlation between the distribution of Chalcolithic metalwork in the south of France and the occurrence of copper resources (Mille and Carozza 2009: fig. 3). They believe that the range of different metal compositions indicates a proliferation of local centres, where metal production was aimed at local supply and not long-distance exchange. Cabrières presents a model of centralized production connected to regional supply, while the evidence from the Al Claus settlement (Tarn-et-Garonne) indicates metallurgy on a smaller domestic scale, undertaken some distance from likely mine sources.

Mille and Carozza observed that the social impact of metal production linked to long-distance exchange would have been greater than that associated with local use. The ability of any mining group to establish a network of long-distance exchange would have depended on the amount of copper that they could produce. This presents a further challenge to Strahm's evolutionary model, as does the evidence for a decline in copper production at Cabrières from *c.*2200 BC, at the beginning of the Bronze Age (Mille and Carozza 2009: fig. 21). That coincided with an increase in copper mining in the western Alps, at centres such as St Véran. The decline of copper production in the Languedoc suggests that metal did not hold the same significance for Early Bronze Age groups there as it did for their Chalcolithic predecessors. That earlier history of copper production did not lead to the emergence of hierarchical societies controlled by a central political power in the Early Bronze Age.

Bronze Age mining elites and strongholds in the eastern Alps

Chapter 7 examined how copper mining was underway at a number of centres in the eastern Alps during the early part of the Bronze Age (2200–1800 BC). The general picture is one of small-scale mining, undertaken on a seasonal basis by small expeditions sent from the agricultural hinterland. Krause (2009) argued that this supply chain was directed by individuals or groups who acquired a higher status by virtue of this activity. He attached much significance to the discovery of halberds and solid-hilted daggers in a number of mining districts as indicators of powerful individuals and groups in the mining district and Alpine foreland (Fig. 10.4).

These changes in society have been linked to the fortification of hill-top settlements in several mining districts towards the end of the Early Bronze Age. Examples in the general area of the Mitterberg include the Götschenberg near Bischofshofen (Lippert 1992), the Höchbauer at the Einödberg, St Johann and the Klinglberg near St Veit (Shennan 1995). Other examples include the

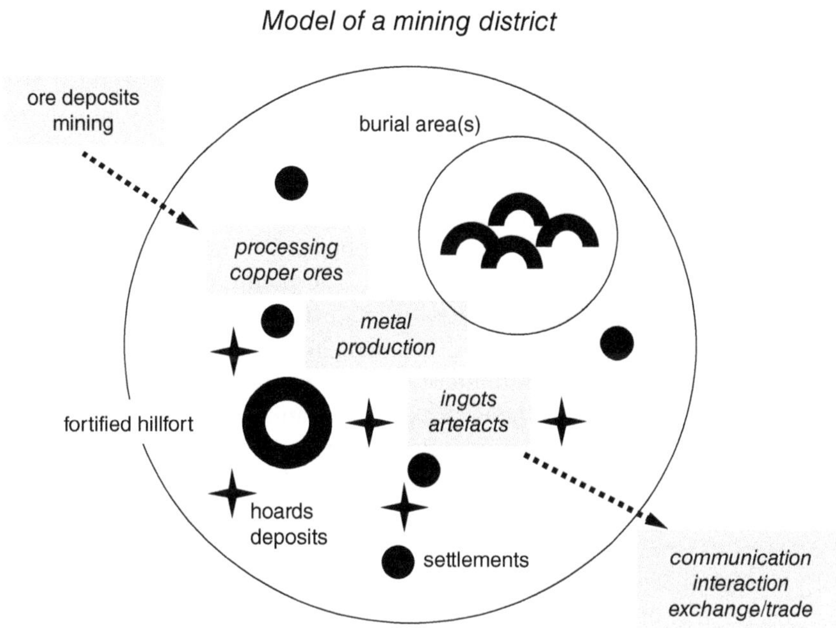

Fig. 10.4. Model of a Middle/ Late Bronze Age mining district in Austria (*Source*: Krause 2009).

Buchberg near Wiesing in the lower Inn Valley of North Tyrol (Martinek 1995; Martinek and Sydow 2004) and the recent excavations conducted at Bartholomäberg in the Montafon region (Krause 2007).

The presence of these hierarchically structured settlements in the period 1800–1600 BC coincided with intensified copper mining in the eastern Alps. Krause argued that these developments were closely connected, in that the Alpine mining communities were controlled by elites based in the fortified settlements. These individuals and groups had a major interest in the production and supply of copper ingots into the north Alpine foreland. A control of supply at a time of growing demand for metal would have had economic and social repercussions, leading to wealth accumulation and status differences among the settlements of the mining districts.

This view of copper mining as the main driver of social change in the eastern Alps during the Early Bronze Age has been questioned. Several researchers have noted the absence of visible wealth or status differentiation within the mining districts, with no evidence of elite burials or conspicuous consumption. Kienlin and Stöllner (2009) cite many examples in the ethnographic record where small-scale tribal societies were able to engage in seasonal mining without any coercive force being applied (e.g. Petrequin and Petrequin 1993).

The hypothesis that fortified settlements in Alpine valleys were connected to elites with a controlling interest in metal production was tested by excavations at the Klinglberg near St Veit (Shennan 1995). That particular settlement was fortified at the beginning of the Middle Bronze Age, at a time when copper mining expanded in the Salzach valley. Shennan interpreted the Klinglberg as a mining settlement, based on proximity to the mines, the discovery of ore and raw copper, and the presence of slag-tempered pottery. He argued that this group was essentially autonomous, but was connected to external groups through the supply of copper. This explains the discovery of imported cereals in that settlement obtained from agrarian communities in the Alpine forelands to the north. The nature of those exchanges remains in question; however, there is no evidence that this trade was under the coercion of ruling elites. Shennan interpreted the Klinglberg as the settlement of a small community engaged in copper production, pursuing their own interests without any coercion, and displaying few obvious signs of wealth or social differentiation (Shennan 1995).

To explore this further, Kienlin and Stöllner (2009) considered the long-term development of Alpine mining communities. They identified a gradual movement of small farming communities into the inner Alpine valleys in the early second millennium BC. This included the Saalach and Salzach valleys in the general area of the Mitterberg, as well as the lower Inn Valley in North Tyrol close to the ore resources of Schwaz-Brixlegg. The earliest mining was in the form of small seasonal expeditions from the north Alpine foreland, with some of those groups settling in the inner Alpine valleys by 1800 BC. Kienlin and Stöllner acknowledge that the choice of hilltops for important settlements in these valleys may reflect an element of competition, but do not believe this was connected to the control of ore resources. In the early stages the main economic priority was agriculture, with only seasonal interest in small-scale copper production. Some of that production was considerable, as attested to by the large number of rib and ring ingot finds in central Europe.

Over time, some of the Alpine mining groups were in a position to engage in more intensive production, such as that recorded for the Mitterberg *c.*1600–1300 BC. Even then, there is no strong evidence for controlling elites associated with those mines. Stöllner et al. (2011) observed that the proliferation of smelting sites across that area does not support centralized organization. This is difficult to confirm when dealing with an archaeological landscape formed by a millennium or so of continuous mining. It must also be acknowledged that high levels of organizational ability are evident in the deep underground workings of the Mitterberg veins.

While copper production and the metal trade were significant elements of Bronze Age economy, most researchers would agree that agriculture was the basis of settlement in that period. Bartelheim (2009) has argued convincingly that it was the agriculture-based prosperity of Early Bronze Age populations in central Europe that stimulated the development of metal production. He makes the point that the settlement centres of dominant social groups tend to be located close to the most important economic resources. The major concentrations of settlement and population in that period are found in areas of fertile soil with good agricultural potential (Bartelheim 2009: fig. 1), and not close to ore deposits or mining centres. Bronze Age settlements in the Alpine mining districts do not have significant indications of wealth compared to their counterparts in the agrarian lowlands. It was agricultural surplus generated by the latter that provided the means to acquire metal and so acted as the main driving force behind this mining 'industry'.

Mount Gabriel and wedge tomb society

A final example relates to a much lower scale of copper mining undertaken during the Early/Middle Bronze Age in south-west Ireland. The Mount Gabriel-type mines are found in a landscape where contemporary settlements have not been identified. Understanding the social context of these mines has centred on the analysis of a type of megalithic monument termed the wedge tomb (O'Brien 2000). The territorial significance of wedge tomb distributions as settlement proxies has been considered, as have the implications of resource control for social relations in what are believed to have been small-scale segmentary societies.

The underlying assumption is that the distribution of wedge tombs reflects kin-based residence patterns, where each wedge tomb or tomb cluster represents a unilineal descent group, bound to neighbouring groups by common descent or marriage relationships. The distribution of wedge tombs in the Cork/Kerry region represents the geographical spread of unilocal descent groups as part of an expanding tribal/lineage network. There is a spacing of similarly-sized tombs or tomb clusters, with no indication of hierarchical ordering of monuments on the basis of size, architecture, or location. This tendency to dispersal and uniformity reflects a territorial constraint on the growth of this society, in a cellular pattern repeated across those parts of the landscape with agricultural potential (Fig. 10.5).

It is argued that the social territories suggested by this monument distribution correspond to the home areas of population groups, some of whom were involved in copper mining. Central to this is the idea that these monuments were part of a system of socially recognized claims to valuable resources (Renfrew 1976; Chapman 1981, 1995).

Each wedge tomb-associated community was a corporate group in the sense that it had proprietary rights to land and other resources, like mineral ores, within its own territory. These rights were primarily established by residence and lineal descent from ancestors whose remains may have been enshrined in the community tomb.

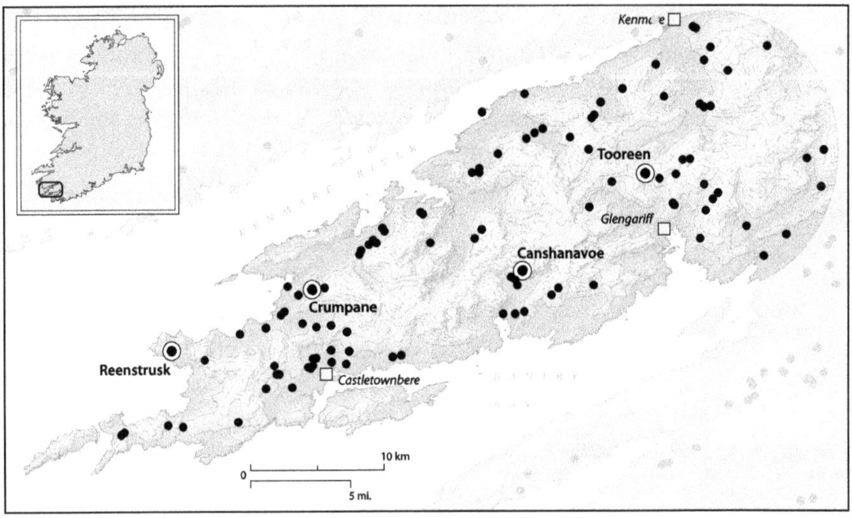

Fig. 10.5. Distribution of Bronze Age copper mines (Tooreen, Canshanavoe, Crumpane, and Reentrusk) and megalithic monuments (wedge tombs, stone circles, stone rows and pairs, boulder-burials) in the Beara Peninsula, south-west Ireland

(*Source*: O'Brien 2009).

The Mount Gabriel-type mines represent a particular approach to the sourcing of metal in the Bronze Age, one that reflects the particular social context in which this activity was undertaken. There was a proliferation of small surface workings, each representing a limited investment of time and material resources by groups of 10–20 individuals, possibly working on a seasonal or sporadic basis. Copper mining in these wedge tomb territories was characterized by the efforts of small local communities, consisting of extended or multi-family groups (primary tribal segments), with minimal inter-group collaboration. The supply of metal was structured by the segmentary nature of society, with metal objects exchanged across a cellular tribal network through kin relations and other social ties. Instead of centralized control, metal production may have been in the hands of local descent groups (primary tribal segments) who occasionally collaborated in mining ventures and exchanged metal products and expertise, possibly within wider clan-like affiliations. The circulation of metal was determined in the first instance by social distance and subsequently by a desire to obtain other scarce materials such as gold and tin through trade. Tin supply is likely to be of particular significance as the first bronze-making groups in south-west Ireland had to trade with Cornwall to obtain this raw material.

Most Bronze Age copper mines in Britain and Ireland are located on rich orebodies where there was a considerable commitment to mining over a long period. The Mount Gabriel-type mining presents a different approach where metal needs were met by short-term commitments to mining across an entire landscape. This approach does not reflect geological factors as there are many rich veins of copper ore in the same area as Mount Gabriel, but was instead rooted in the social context of this mining, the low level of demand for metal and a limited ability to mobilize resources to obtain same. While individual workings only returned small amounts of copper mineral, taken together they represent a significant mining initiative during the Bronze Age.

Staying in southern Ireland, Ross Island presents a different social setting from the Mount Gabriel mines, where the scale and duration of mining indicates higher levels of organization. A Beaker culture group, whose origins lay in Atlantic France or possibly northern Spain, settled in that area and worked that mine. The dominant position of Ross Island in the supply of copper across Chalcolithic Ireland gave an important impetus to social differentiation at a local level. The discovery of two gold lunulae in the Killarney area points to the emergence of individuals and groups with a controlling interest in this metal resource (O'Brien 2004: 570).

In conclusion, the intensification of copper production during the Early Bronze Age in Europe has often been linked to the growing importance of metalwork in contemporary society. What is less clear is the extent to which control of metal supply was a prime mover towards increased social complexity in that period. For many, the question is whether copper constituted a critical resource in prehistory. This seems to have been the case for Cyprus, an island economy dominated by copper production during the Late Bronze Age. That society, however, was part of a different 'world system' than the rest of prehistoric Europe where most mines operated on a lower scale.

Many researchers would now agree with Gilman's (1981) assertion that the development of metallurgy must be considered against a background of wider resource intensification in the Bronze Age. The expansion of agriculture was central to this development, and in many cases gave rise to 'protector elites' who exploited the security needs of those groups. In some instances these elites had a controlling role in metal production, to the point that they were the primary beneficiaries in terms of wealth accumulation. As they consolidated their position, the display of power created a demand for prestige metalwork and other items of social display. This process began in Europe as early as the fifth millennium BC, with cemeteries such as Varna anticipating the important role of metal in the chiefdom societies that emerged during the Bronze Age.

An ability to obtain metal through established trade meant that the mining of ore deposits was never an essential activity, but one to be pursued and exploited as a strategic choice where circumstances allowed. The development of this specialized technology can be seen as reflecting rather than causing social differentiation in this period. Metal provided an important medium for the accumulation of wealth in those societies and had a central role in the emergence of trade networks that connected the different regions of late prehistoric Europe. Most people during the Chalcolithic and Bronze Age lived their entire lives in local communities that were territorially defined and largely self-sufficient. Yet, these same groups would have encountered materials from distant sources, objects of metal, amber, jet, and other exotic substances, which gave a strong sense of connection to distant regions. The exchange of these materials was an important vehicle for cultural exchange, with the supply of copper being particularly significant. For this reason, it is reasonable to suggest that copper mines played a significant role in the shaping of late prehistoric societies.

BOOM TO BUST

This book has reviewed some 5,000 years of copper mining in prehistoric Europe, using the evidence of the mines themselves, the metal they supplied, and how it was regarded in contemporary societies. The evidence suggests that mining was conducted at different scales, ranging from local production on a seasonal basis to larger permanent operations that became important regional centres of copper production, with trade connections extending across Europe. Those regions had their own history of mining, which varied in duration and intensity for different reasons. In some instances there was a slow growth of indigenous mining expertise, such as occurred in the Balkans between the sixth and fourth millennia BC, or in Iberia from the fourth to the second millennia BC. There is also evidence for a sudden transmission of mining and metallurgical expertise from one region to another, as occurred in the twenty-fifth century BC in southern Ireland or around 2100 BC in Wales.

While individual mines could be of short duration, the practice of copper mining generally led to an enduring strong mining tradition in the regions concerned. This meant more or less continuous mining over long periods, with notable examples in Cyprus, the Austrian Alps, and the Iberian Pyrite Belt. The duration of mining in other regions is similarly impressive, lasting for a millennium or more at Kargaly in Russia, the Languedoc, south-west Ireland, and Wales. In some instances this centred on a single mine, such as the Great Orme or El Aramo, which provided a steady source of copper over a long period. In other cases, it is possible to identify a mining district, where there was a progressive expansion of activity within a geologically defined area.

However successful, all of these mines eventually declined, to be replaced in some instances by others in the same region. This happened in the case of Ross Island and Mount Gabriel in southern Ireland, and with the overlapping production of Cabrières and St Véran in southern France (Fig. 10.6). In other situations, several mines commenced at more or less the same time in

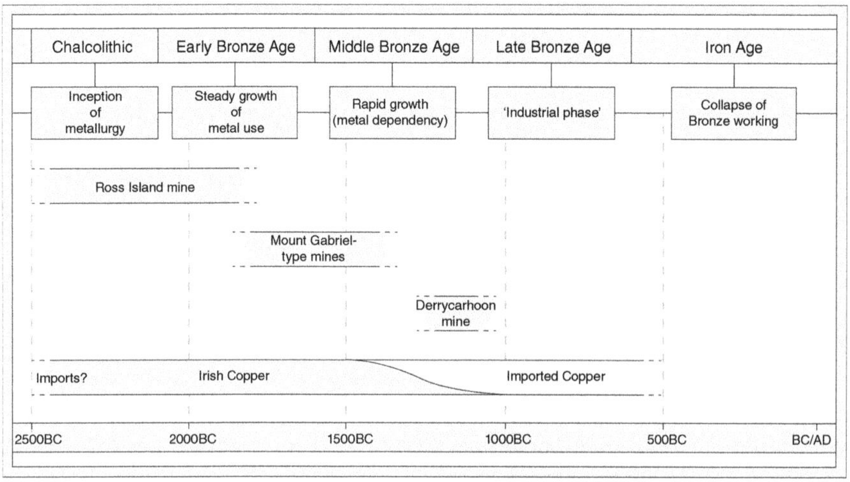

Fig. 10.6. Chronology of prehistoric copper mining in south-west Ireland (*Source*: author).

a region, as happened in Wales in the Early Bronze Age and in Austria during the Late Bronze Age. Some of these constitute what might be termed 'mining districts', where copper production was practised over many centuries by distinctive communities. The relationships that existed between individual mines that were contemporaneously worked in these districts are unknown. The extent to which mines cooperated or competed with each other depended on local factors particular to those societies.

While the end of prehistoric copper mining in Europe should not be considered as the collapse of an industry, this essentially did occur at different times across Europe. An example is Ireland where copper mining contracted significantly after 1400 BC, and had more or less ceased by the end of that millennium (Fig. 10.6). This included those parts of south-west Ireland, mid and north Wales, and the English midlands where copper mining was well established earlier in the Bronze Age. Metal analyses indicate a growing reliance on imported Continental metal during the Middle Bronze Age, with declining supply from British and Irish sources (Northover 1982: fig.7). Imported metal from the Continent, such as that represented in the Langdon Bay cargo from Dover, Kent, and the Salcombe find from Devon, grew in importance in the Penard phase, to eventually dominate copper supply during the Wilburton and Ewart Park phases of the Late Bronze Age (Northover 1982: figs.11 and 13).

This decline of copper mining was a wider phenomenon across Atlantic Europe, which is all the more interesting as it coincided with a growing demand for metal in societies with extensive trade connections. Several theories can be advanced to explain the changes in copper supply that occurred during the later Bronze Age. Some mines are known from this period, for example Chinflon in south-west Spain, and the Great Orme in Wales, while other examples may not have not survived or cannot be recognized today. The destruction of evidence as a result of later mining is a factor, but unlikely to be the entire explanation.

Archaeological visibility may be a problem, particularly if there were significant changes in mining technology. The replacement of fire-setting would remove distinctive traces of this technique, such as the characteristic wallrock surfaces and the availability of wood/charcoal for the purposes of radiocarbon dating. The occurrence of stone hammers in large numbers in earlier copper mines is relevant, as their replacement by specialized bronze (or iron) tools would remove an important surface indicator of this activity. However, with no evidence that the latter were used to any extent during the Late Bronze Age, depending on the geological setting there may have been few options but to continue with the older technology.

It does seem that copper mining did actually decline during the Middle Bronze Age in both Britain and Ireland, before finally ending around 1000 BC. This is the picture presented by radiocarbon data from the archaeologically sampled copper mines. The rise and fall of different mining regions could be understood in terms of boom/bust cycles that would be familiar from historic times. The reasons behind the closure of any prehistoric mine were connected to the geological circumstances of that copper source, various technological constraints, and the wider socioeconomic and cultural context. The earliest copper mining was based on the extraction of oxidized and fahlore mineralization; the ability to exploit more abundant types of copper, in particular chalcopyrite, was developed in central Europe by the Late Bronze Age. While the Mitterberg-type smelting process did spread across the Alpine zone during the second millennium BC, it did not enter other copper mining regions possibly because the secrets of the technology were carefully guarded. The exhaustion of oxidized ores, combined with an inability to process primary sulphidic ore, was a factor in the cessation of copper mining in Britain and Ireland by 1000 BC. This, of course, is not the entire explanation, as many sources of surface mineralization were still untouched by that time.

As the Bronze Age progressed, new economic priorities emerged in some of the major copper mining centres in Europe. The most significant of these was the adoption of iron technology, which happened at different times across the Continent *c.*1500–500 BC. The transition to iron use was gradual and in the early stages did not impact significantly on the main centres of copper mining activity. A growing dependency on iron placed a new value on copper and, by extension, on the significance of copper mining. In most parts of Europe the use of copper and bronze by the developed Iron Age was confined to the production of ornaments and prestige objects. The amount of bronze in circulation was mostly adequate to be recycled for such specialized purposes.

This created a significantly reduced demand for copper in comparison to earlier periods, and so there was less need to engage in the laborious process of mining. The greater availability of iron at a local level also impacted on the long-distance supply networks established during the Late Bronze Age. These same networks were vulnerable to political tensions and conflicts of different kinds, which were commonplace during the Late Bronze Age and Early Iron Age.

Mining did continue in some regions during the pre-Roman Iron Age, notably in the Iberian Pyrite Belt and in Cyprus. Few copper mines of that period have been archaeologically investigated, and so it is difficult to make direct comparisons with Bronze Age operations. Where Iron Age mines have been identified, as at Campolungo in the Lombardy Alps, a connection with earlier operations in the same region is uncertain. There are also hints that some older copper mines were re-worked at different times during the later Bronze Age. These later phases of mining tend not to be well understood, possibly because of changed technology and other factors mentioned above.

It is interesting to consider the parallel development of copper and salt mining during the Late Bronze Age in Austria. By the twelfth century BC the mining of copper had intensified in Salzburg province and in North Tyrol, driven by new developments in the smelting of chalcopyrite and fahlore. This period also witnessed the first extraction of rock salt in the so-called Salzkammergut region of Upper Austria. This commenced as early as 1000 BC in the northern group of the Hallstatt mines (Barth 1992). By the seventh century BC this had emerged as a major producer of salt, overtaken in the fifth century by the rise of the Dürrnberg mines in the same area. The latter became a major economic centre with a considerable population, where the production and trade of salt was controlled by a wealthy elite (Stöllner 2003b, 2010).

It was previously thought that the rise of the Alpine salt industry coincided with, and possibly contributed to, the decline in copper mining at the end of the Bronze Age. This was also explained by the adoption of iron technology; however, the latter did not become widespread until the Late Iron Age (La Tène culture). While there is relatively little evidence for copper mining and smelting during the Iron Age, new dating evidence suggests that some established mines in Austria continued to produce copper in this period (Stöllner 2010: fig 2). This is supported by analytical evidence, such as a recent study of an Early Iron Age hoard of 385 bronzes from Fliess near Landeck in the Tyrol (Lutz and Schwab 2012). Assuming that recycled metal was not used, the results reveal the continued use of both chalcopyrite and fahlore, with lead isotope results pointing to the lower Inn Valley as a major source of copper. The same study suggests that the major break in production may have occurred, instead, during the Late Iron Age, when new copper sources were used by the La Tène culture. At that stage, salt production was a major economic priority, generating sufficient wealth to obtain the relatively limited amounts of copper required for specialist use by trade alone.

In conclusion, this book has hopefully demonstrated the importance of copper mining in the development of prehistoric societies in Europe. This is true not only of those groups directly involved in mining but also for wider populations, because of the central place this activity had in the supply of metal and the establishment of cultural connections. While these mines should be celebrated for their technological achievement, they were at the centre of a new type of commodity-based economy that emerged in the Bronze Age. The trade routes that resulted served to connect different regions across Europe, bound together by a common desire for the golden metal that is copper.

References

Alekseev, V. and Kuznetsova, E. 1983. 'Kenkazghan—drevnij mednyj rudnik v tsentralnom Kazakhstane', *Sovetskaja Archaeology* 1983, 2, 203–12.

Allard, P., Andre, M., Chambon, P., Lafage, F., and Praud, I. 1998. 'La sépulture collective de Vignely, la Porte aux Bergers (Seine-et-Marne)', in Gutherz, X. and Joussaume, R. (eds) *Le Néolithique du Centre-ouest de la France.* Acte du XXleme colloque inter-régional sur le Néolithique. Association des Publications Chauvignoises, Poitiers, 395–401.

Almagro Gorgea, M. and Collado Villalba, O. 1981. 'La Loma de la Tejeria. Un asentamiento Campaniforme en Teruel (Albarracín)', *Teruel* 66, 13–25.

Ambert. P. 1990. 'Evidence, age, and technical level of prehistoric cupriferous exploitations in Cabrières-Hèrault (south of France)', in Ambert, P. (ed.) *Cabrières-Herault le plus vieux centre Minier Métallurgique de France.* Archéologie en Languedoc, Colloque International, livret-guide de l'excursion du 24 Septembre 1990. Federation Archeologique de l'Herault, 9–12.

Ambert, P, 1995. 'Les mines préhistoriques de Cabrières (Herault): quinze ans de recherches. Etat de la question', *Bulletin de la Société Préhistorique Française* 92, 4, 499–508.

Ambert, P. 1996a. 'The copper mines of Cabrières (France) and results of the experimental metallurgy', *Archéologie en Languedoc* 20, 13–19.

Ambert, P. 1996b. 'Cabrières (France), mines et métallurgie au III millénaire BC: apports de la métallurgie expérimentale', *Archéologie en Languedoc* 20, 21–6.

Ambert, P. 1998. 'Importance de la métallurgie campaniforme en France. In L'Enigmatique Civilisation Campaniforme', *Archéologia* 9, 36–41.

Ambert, P. 1999. 'Les minerais de cuivre et les objets metalliques en cuivre à antimoine-arget su sud de la France', in Hauptmann, A. et al. (eds) *The Beginnings of Metallurgy.* Proceedings of the 1995 International Symposium. Der Anschnitt beiheft 9, Bochum, 193–210.

Ambert, P. 2001. 'La place de la métallurgie campaniforme dans la première métallurgie française', in Harrison, R. et al. (eds) *Bell Beakers Today*, Proceedings of the Riva del Garda conference, May 1998, Trento, 577–88.

Ambert, P. 2002. 'Utilisation préhistorique de la technique minière d'abattage au feu dans le district cuprifère de Cabrières (Hérault)', *Comptes Rendus Palevol* 1, 711–16.

Ambert, P. and Barge, H. 1990. 'Les mines préhistoriques de Cabrières (Hérault): leur importance pour la métallurgie chalcolithique languedocienne', in Ambert, P. (ed.) *Cabrières-Herault le plus Vieux Centre Minier Métallurgique de France.* Archéologie en Languedoc, Colloque International, livret-guide de l'excursion du 24 Septembre 1990. Federation Archeologique de l'Herault, 21–6.

Ambert, P., Bourhis, J. R., and Houles, N. 1990. 'Ube pointe de Palmela près des mines de Cabrières (Hérault)', in Ambert, P. (ed.) *Cabrières-Herault le plus vieux centre Minier Métallurgique de France.* Archéologie en Languedoc, Colloque International, livret-guide de l'excursion du 24 Septembre 1990. Federation Archeologique de l'Herault, 17–18.

Ambert, P., Barge, H., Bourhis, J., and Esperou, J. 1984. 'Mines de cuivre préhistorique de Cabrières (Hérault): premiers resultants', *Bulletin de la Société Préhistorique Française* 81, 3, 83–9.

Ambert, P., Guendon, J., Laroche, M., and Mischka, D. 2002. 'Données nouvelles concernant le plus vieil etablissement métallurgique de France. La Capitelle du Broum (district minier de Cabrières-Péret, Herault)', *Archéologie en Languedoc* 26, 45–53.

Ambert, P., Figueroa-Larre, V., Guendon, J. L., Klemm, V., Laroche, M., Rovira, S., and Strahm, C. 2009. 'The copper mines of Cabrières (Hérault) in southern France and the Chalcolithic metallurgy', in Kienlin, T. and Roberts, B. (eds) *Metals and Societies*. Verlag Dr Rudolf Habelt GMBH, Bonn, 285–95.

Ambert, P., Bouquet, L., Guendon, J., and Mischka, D. 2005. 'La Capitelle du Broum (district minier de Cabrières-Peret, Hérault): établissement industriel de l'aurore de la métallurgie française, 3100–2400 BC', in Ambert, P. and Vaquer, J. (eds) *La Première Métallurgie en France et dans les Pays Limitrophes*. Mémoire 37 de la Société Préhistorique Française, Carcassonne, 83–96.

Ancel, B., Cottet, M., Kammenthaler, E., Morin, D., Oppizzi, P., and Tizzoni, M. 1998. 'La mine de cuivre de Campolungo (Bienno-Brescia): une exploitation révolutionnaire de l'Age du Fer', *Notizie Archeologiche Bergomensi* 6, 163–83.

Andrew, C. 2003. 'Europe's base metal mining industry: a status report', in Kelly, J., Andrew, C., Ashton, J., Bland, M., Earls, G., Fusciardi, L., and Stanley, G. (eds) *Europe's Major Base Metal Deposits*. Irish Association for Economic Geology, Dublin, 31–47.

Andrews, P. 1994. 'Excavating mines (Copa Hill, Chinflon, Kestel)', *Mining before Powder: Peak District Mines Historical Society Bulletin* 12, 3, 13–21.

Anreiter, P., Goldenberg, G., Hanke, K., Krause, R., Leitner, W., Mathis, F., Nicolussi, K., Oeggl, K., Pernicka, E., Prast, M., Schibler, J., Schneider, I., Stadler, H., Stöllner, T., Tomedi, G., and Tropper, P. (eds) 2010. *Mining in European History and its Impact on Environment and Human Societies*. Proceedings of the 2009 SFB-HIMAT conference, University of Innsbruck. Innsbruck.

Bailly-Maitre, M-C. and Gonon, T. 2008. 'L'exploitation de la chalcopyrite à l'Âge du Bronze dans le massif des Rousses en Oisans (Isere)', in De Hervé, R. and Garcia, D. (eds) *Le Peuplement de l'arc Alpin*, Congrès National des Sociétés Historiques et Scientifiques, Grenoble, 207–23.

Bar-Adon, P. 1980. *The Cave of the Treasure*. The Israel Exploration Society, Jerusalem.

Barge, H. 1985. 'Les indices de minerai de cuivre et les traces des anciennes exploitations minieres de la bordure meridionale du Massif Central', *Bulletin de la Société Méridionale de Spéléologie et de Préhistoire* 25, 61–7.

Barge, H. 1997. 'L'installation métallurgique préhistorique de la cabane des Clausis à Saint-Véran (Hautes-Alpes)', in Ambert, P. (ed.) *Actes du Colloque 'Mines et Métallurgie de la Préhistoire au Moyen-Age en Languedoc-Roussillon et régions périphériques'. Archéologie en Languedoc* 21, 99–110.

Barge, H. 2003. *Saint-Véran, la Montagne, le Cuivre et l'Homme.* Actilia Multimedia, Theix.

Barge, H., Ancel, B., Rostan, P., and Guendon, J. L. 1998a. 'La mine des Clausis a Saint-Véran (Hautes-Alpes) exploitation et aire de reduction du minerai de cuivre d'époque prehistorique', in Mordant, C., Pernot, M., and Rychner, V. (eds) *L'Atelier du Bronzier en Europe du XX au VIII siècle avant notre ère* (vol. II). Actes du Colloque International Bronze '96, Neuchâtel and Dijon, Paris, 71–82.

Barge, H., Bourhis, J-R., and Rostan, P. 1998b. 'Métallurgie préhistorique et gîtes cuprifères dans le sud-est de la France. Premiers résultats', *Rencontres Méridionales de Préhistoire Récente,* Editions APDCA, Antibes, 65–79.

Barge, H. and Talon, B. 2012. 'Attaque au feu au Bronze ancien sur le gîte de cuivre de Saint-Véran (Hautes-Alpes)', *Bulletin de la Societe Prehistorique Française* 109, 1, 145–54.

Barnatt, J. and Thomas, G. 1998. 'Prehistoric mining at Ecton, Staffordshire: a dated antler tool and its context', *Mining History: Bulletin of the Peak District Mines Historical Society* 13, 5, 72–8.

Bartelheim, M. 2009. 'Elites and metals in the central European Early Bronze Age', in Kienlin, T. and Roberts, B. (eds) *Metals and Societies.* Verlag Dr Rudolf Habelt GMBH, Bonn, 34–46.

Bartelheim, M., Eckstein, K., Huijsmans, M., Krauss, R., and Pernicka, E. 2002. 'Kupferzeitliche metallgewinnung in Brixlegg, Österreich', in Bartelheim, M., Krause, R., and Pernicka, E. (eds) *Die Anfänge der Metallurgie in der Alten Welt.* Freiberger Forschungen zur Archäometrie und Kulturgesch 1, Rahden, 33–82.

Bartelheim, M., Eckstein, K., Huijsmans, M., Krauss, R., and Pernicka, E. 2003. 'Chalcolithic metal extraction in Brixlegg, Austria', in *Archaeometallurgy in Europe,* Milan conference proceedings volume 1. Milan: Associazione Italiana di Metallurgia, Milan, 441–7.

Bartelheim, M. and Montero Ruiz, I. 2009. 'Many ores and little water—the use of resources during the Bronze Age on the Iberian Peninsula', in Bartelheim, M. and Stäuble, H. (eds) *Die wirtschaftlichen Grundlagen der Bronzezeit Europas.* Forschungen zur Archäometrie und Altertumnwissenschaft 4, Verlag Marie Leidorf GmbH, Leidorf, 5–21.

Bartelheim, M., Niederschlag, E., and Rehren, T. 1998. 'Research into prehistoric metallurgy in the Bohemian/Saxon Erzgebirge', in Hänsel, B. (ed.) *Man and Environment in Bronze Age Europe.* Oetker-Voges-Verlag, Kiel, 225–30.

Barth, F. 1992. 'Zu den Tragsäcken aus dem Salzbergwerk Hallstatt', *Archaeologica Austriaca* 76, 121–7.

Bass, G. 1967. *Cape Gelidonya: a Bronze Age Shipwreck.* Transactions of the American Philosophical Society 57, 8, Philadelphia.

Bass, G. 1986. 'A Bronze Age shipwreck at Ulu Burun (Kas), 1984 campaign', *American Journal of Archaeology* 96, 269–96.

Bassiakos, Y. and Philaniotou, O. 2007. 'Early copper production on Kythnos: archaeological evidence and analytical approaches to the reconstruction of metallurgical process', in Day, P. and Doonan, R. (eds) *Metallurgy in the Early Bronze Age Aegean.* Oxbow Books, Oxford, 207–22.

Bassiakos, Y. and Tselios, T. 2012. 'On the cessation of local copper production in the Aegean in the 2nd millennium BC', in Kassianidou, V. and Papasavvas, G. (eds) *Eastern Mediterranean Metallurgy and Metalwork in the Second Millennium BC.* Oxbow Books, Oxford, 151–61.

Baumgarten, B., Folie, K., and Stedingt, K. 1998. *Auf den Spuren der Knappen: Bergbau und Mineralien aus Südtirol.* Tappeiner, Lana.

Begemann, F., Schmitt-Strecker, S., Pernicka, E., and Lo Schiavo, F. 2001. 'Chemical composition and lead isotopy of copper and bronze from Nuragic Sardinia', *European Journal of Archaeology* 4, 1, 43–85.

Bengtsson, L. 1986. 'Försök med kopparframställning', *Forntida Teknik* 12, 5–16.

Betancourt, P. 2006. *The Chrysokamino Metallurgy Workshop and its Territory.* American School of Classical Studies at Athens, Princeton.

Beyrie, A. 2005. 'L'industrie minière antique dans les Pyrénées occidentales', *L'Archéologue* 78, 19–21.

Beyrie, A. and Kammenthaler, E. 2008. 'Aux origines de l'activité minière dans les Pyrénées occidentales'. On line publication: <http://inrap.fr/userdata/c_bloc_file/6/6961/6961_fichier_dossier22-beyrie.pdf>

Billard, C. 1991. 'L'habitat des Florentins à Val-de-Reuil (Eure)', *Gallia Prehistoire* 33, 140–71.

Bird, J. 1975. *The 'Copper Man': A Prehistoric Miner and his Tools from Northern Chile.* Dumbarton Oaks Conference on Pre-Columbian Metallurgy of South America. Washington. Harvard, 105–32.

Blance, B. 1961. 'Early Bronze Age colonists in Iberia', *Antiquity* 35, 192–202.

Blanchard, I. 1978. 'Labour, productivity and work psychology in the English mining industry, 1400–1600', *Economic History Review* 31, 1–24.

Blas Cortina, M.A. De 1992. 'Minas prehistóricas del Aramo (Riosa): campaña arqueologica de 1987', *Excavaciones Arqueológicas en Asturias* 1987–90, 59–68.

Blas Cortina, M.A. De 1998. 'Producción e intercambio de metal: la singularidad de las minas de cobre prehistóricas del Aramo y El Milagro (Asturias)', *Minerales y Metales en la Prehistoria Reciente.* Studia Archaeologica 88, Universidad de Valladolid y Fundación Duques de Soria, 71–103.

Blas Cortina, M.A. De 1999. 'Asturias y Cantabria', in Delibes de Castro, G. and Montero Ruiz, I. (eds) *Las Primeras Etapas Metalúrgicas en la Península Ibérica: II Estudios Regionales.* Instituto Universitario Ortega y Gasset y Ministerio de Educación y Cultura, Madrid, 41–62.

Blas Cortina, M.A. De 2003. 'La mina como ámbito infraterreno y el cadaver como ofrenda ritual. A propósito de los esqueletos humanos hallados en las explotaciones cupríferas del Aramo', in Manzano, J. and Martínez, J. (eds) *Mineros y Fundidores en el Inicio de la Edad de los Metales.* Madrid, 32–48.

Blas Cortina, M.A. De 2005. 'Un témoignage probant de l'exploitation préhistorique du cuivre dans le nord de la Péninsule Iberique: le complexe minier d'El Aramo (Asturias)', in Ambert, P. and Vaquer, J. (eds) *La Première Métallurgie en France et dans les Pays Limitrophes.* Mémoire 37 de la Société Préhistorique Française, Carcassonne, 195–206.

Blas Cortina, M.A. De 2007–8. 'Minería prehistórica del cobre en el reborde septentrional de los picos de Europa: las olvidadas labores de El Milagro (Onís, Asturias)', *Veleia* 24/25, 723–53.

Blas Cortina, M.A. De 2009. 'Utillaje faunístico inédito de las labores de cobre prehistóricas de La Profunda (León) y su datación C14 (AMS)', *Zephyrus, Revista de Prehistoria y Arqueologia* 64, 5–18.

Blas Cortina, M.A. De 2010. 'El expolio del subsuelo y las prácticas rituales en la minería prehistórica: a proposito del hallazgo de esqueletos humanos en las explotaciones de cobre en Asturias', in Blas Cortina, M.A. De, Delibes de Castro, G., Villa Valdes, A., and Suárez Fernández, M. (eds) *Cobre y Oro. Minería y Metalurgia en la Asturias Prehistórica y Antigua.* Real Instituto de Estudios Asturianos, Oviedo, 127–69.

Blas Cortina, M.A. De 2013. 'De las labores subterráneas a las actividades metalúrgicas en el exterior: investigaciones 2007–2012 en las minas de cobre prehistóricas de la Sierra del Aramo', *Excavaciones Arqueológicas en Asturias 2007–2012.* Consejería de Cultura del Principado de Asturias, Oviedo, 169–87.

Blas Cortina, M.A. De and Suárez Fernández, M. 2009. 'Utillaje faunístico inédito de las labores de cobre prehistóricas de La Profunda (León)', *Zephyrus* 64 (July–December 2009), 5–18.

Blas Cortina, M.A. De and Suárez Fernández, M. 2010. 'La minería subterránea del cobre en Asturias: un capítulo esencial en la prehistoria reciente del norte de Espana', in Blas Cortina, M.A. De, Delibes de Castro, G., Villa Valdes, A., and Suárez Fernández, M. (eds). *Cobre y Oro. Minería y Metalurgia en la Asturias Prehistórica y Antigua.* Real Instituto de Estudios Asturianos, Oviedo, 43–82.

Blazquez, J. 1988. 'La Loba. Mina y almacenes de finales de la República Romana (120–80 BC) en Fuenteovejuna (Córdoba)', *Historia* 16, 118–28.

Bogosavljević, V. 1995. 'Mining hammerstones of Prljusa—Mali Sturac site', in Petrović, P.and Djurdjekanović, S. (eds) *Ancient Mining and Metallurgy in Southeast Europe.* Archaeological Institute, Museum of Mining and Metallurgy, Bor, 37–44.

Borić, D. 2009. 'Absolute dating of metallurgical innovations in the Vinča culture of the Balkans', in Kienlin, T. and Roberts, B. (eds) *Metals and Societies.* Verlag Dr Rudolf Habelt GMBH, Bonn, 191–245.

Bosch, J. 2005. 'Les techniques d'exploitation des plus anciennes mines d'Europe méditerranéenne: l'exemple de Gavá, Barcelone', in Ambert, P. and Vaquer, J. (eds) *La Première Métallurgie en France et dans les Pays Limitrophes.* Mémoire 37 de La Société Préhistorique Française, Carcassonne, 207–10.

Bosch Gimpera, P. and Luxan, F. 1935. 'Explotación de yacimientos argentíferous el el eneolítico de Almizaraque, provincia de Almería', *Investigación y Progresso* 9, 112–17.

Boshier, A. 1965. 'Ancient mining of Bomvu ridge', *Scientific South Africa* 2, 317–20.

Bouquet, L., Figueroa-Larre, V., Laroche, M., Guendon, J., and Ambert, P. 2006. 'Les Neuf-Bouches (district minier de Cabrières-Péret), la plus ancienne exploitation minière de cuivre de France: travaux récents, consequences', *Bulletin de la Sociéte Préhistorique Française* 103, 1, 143–59.

Bourgarit, D. and Mille, B. 2005. 'Les nouvelles données de l'atelier metallurgique chalcolithique de La Capitelle du Broum dans le district de Cabrières (Hérault)', in Ambert, P. and Vaquer, J. (eds) *La Première Métallurgie en France et dans les Pays*

Limitrophes. Mémoire 37 de La Société Préhistorique Française, Carcassonne, 97–108.
Bourgarit, D., Rostan, P., Burger, E., Carozza, L., Mille, B., and Artiolo, G. 2008. 'The beginnings of copper mass production in the western Alps: the Saint-Véran mining area reconsidered', *Historical Metallurgy* 42, 1, 1–11.
Boyd Dawkins, W. 1875. 'On the stone mining tools from Alderley Edge', *Proceedings of the Manchester Literary and Philosophical Society* 14, 74–9.
Bradley, R. 1990. *The Passage of Arms*. Cambridge University Press, Cambridge.
Braithwaite, R., Black, S., Branch, N., and Maggi, R. 2012. 'Evaluating the environmental impact of metallurgical activities during the Copper Age and Bronze Age (*c*.5800–2900 cal yrs BP) in the Ligurian Apennines, north-west Italy: a pilot study', in Anreiter, P. et al. (eds) *2nd Mining in European History Conference*. Abstract, conference guide, Innsbruck University Press, Innsbruck, 115.
Brandherm, D. and Maass, A. 2010. 'Copper mining, settlement, and society in the Earlier Bronze Age of Southeast Spain: prospects for new research in the Lower Segura Valley', in Anreiter, P. et al. (eds) *Mining in European History and its Impact on Environment and Human Societies*. Proceedings of the 2009 SFB-HiMAT conference, University of Innsbruck. Innsbruck, 17–22.
Bray, P. and Pollard, A.M. 2012. 'A new interpretative approach to the chemistry of copper alloy objects: source, recycling and technology', *Antiquity* 86, 853–67.
Breitenlechner, E., Hilber, M., Lutz, J., Kathrein, Y., Unterkircher, A., and Oeggl, K. 2010. 'The impact of mining activities on the environment reflected in pollen, charcoal, and geochemical analyses', *Journal of Archaeological Science* 37, 1458–67.
Briard, J. and Roussot-Larroque, J. 2002. 'Les debuts de la metallurgie dans la France Atlantique', in Bartelheim, M., Pernicka, E., and Krause, R. (eds) *The Beginnings of Metallurgy in the Old World*. Verlag Marie Leidorf GmbH , Rahden, 135–60.
Briggs, C.S. 1976. 'Notes on the distribution of some raw materials in later prehistoric Britain', in Burgess, C. and Miket, R. (eds) *Settlement and Economy in the Third and Second Millennium* BC. British Archaeological Reports (British series) 33, Oxford, 267–82.
Briggs, C.S. 1983. 'Copper mining at Mount Gabriel, Co. Cork: Bronze Age bonanza or post-Famine fiasco?', *Proceedings of the Prehistoric Society* 49, 317–35.
Briggs, C.S. 1984. 'The discovery and description of trench mines at Derricarhoon townland, Co. Cork, in 1846', *The Journal of Irish Archaeology* 2, 33–9.
Briggs, C.S. 1991. 'Early mines in Wales: the date of Copa Hill', *Archaeology in Wales* 31, 5–7.
Brill, R. and Wampler, J. 1965. 'Isotope studies of ancient lead', *American Journal of Archaeology* 69, 165–6.
Brill, R. and Wampler, J. 1967. 'Isotope studies of ancient lead', *American Journal of Archaeology* 71, 63–77.
Bruce, J., Marglis, C., and Creveling, D. 1937. 'Antiquities in the mines of Cyprus', Appendix V in *The Swedish Cyprus Expedition* (Volume 3). Stockholm, 637–71.
Budd, P. and Gale, D. (eds) 1997. *Prehistoric Extractive Metallurgy in Cornwall*. Camborne School of Mines, Redruth, Cornwall.

Budd, P. and Taylor, T. 1995. 'The faerie smith meets the bronze industry: magic versus science in the interpretation of prehistoric metal-making', *World Archaeology* 27, 1, 133–43.

Budd, P., Pollard, A.M., Scaife, B., and Thomas, R.G. 1995. 'Ox-hide ingots, recycling and the Mediterranean metals trade', *Journal of Mediterranean Archaeology* 8, 1–32.

Budd, P., Gale, D., Pollard, A.M., Thomas, R.G., and Williams, P.A. 1992. 'The early development of metallurgy in the British Isles', *Antiquity* 66, 677–86.

Budd, P., Gale, D., Pollard, A.M., Thomas, R.G., and Williams, P.A. 1993. 'Evaluating lead isotope data: further observations', *Archaeometry* 35, 241–63.

Budd, P., Haggerty, R., Pollard, A.M., Scaife, B., and Thomas, R.G. 1996. 'Rethinking the quest for provenance', *Antiquity* 70, 168–74.

Budd, P., Haggerty, R., Ixer, R., Scaife, B. and Thomas, R. 1999. 'Copper deposits in south-west England identified as a source of Copper Age metalwork'. Available online: <http://www.archaeotrace.co.uk/provenance.html>.

Campana, N. and Franceschi, E. 1997. 'Analyses of the metal objects', in Maggi, R. (ed.) *Arene Candide: A Functional and Environmental Assessment of the Holocene Sequence.* Memorie dell'Istituto Italiano di Paleontologia Umana, nuova serie 5, Rome. 611–21.

Campana, N., Maggi, R., Stos Gale, Z., and Houghton, J. 1996. 'Miniere e metallurgia in Liguria fra IV millenio e IV secolo BC', in *La Miniera l'Uomo e l'Ambiente.* Convegno di Studi, Cassino 1994, All'insegna del Giglio, Firenze, 15–52.

Campos, A., Herrero, E., Rodríguez, R., Prieto, N., Ortíz, L., and Fernández, F. 2007. 'La más antigua minería metálica en tierras de León', in Sanchez, J. C., De Castro, G. D., Manzano, J. F., and Lobo, L. G. (eds) *El Hallazgo Leonés de Valdevimbre y los Depósitos del Bronze Final Atlántico en la Península Ibérica.* Estudios y Catálogos 17, Museo de León, 221–36.

Carozza, L., Bourgarit, D., and Mille, B. 1997. 'L'habitat et l'atelier de métallurgiste chalcolithique d'Al Claus: analyse et interprétation des témoins d'activité métallurgique', *Archéologie en Languedoc* 21, 147–60.

Carozza, L., Chapron, E., Simonneau, A., Mille, B., Guyard, H., St-Onge, G., Rostan, P., Bourgarit, D., and Burens, A. 2010. 'Glacial fluctuations and exploitation of copper resources in high mountain during the Late Neolithic and Bronze Age in the French Alps (2500–1500 BC)', in Anreiter, P. et al. (eds) *Mining in European History and its Impact on Environment and Human Societies.* Proceedings of the 2009 SFB-HiMAT conference, University of Innsbruck,. Innsbruck, 81–90.

Case, H. 1966. 'Were Beaker-people the first metallurgists in Ireland?', *Palaeohistoria* 12, 141–77.

Castaing, J., Mille, B., Zink, A., Bourgarit, D., and Ambert, P. 2005. 'L'abattage préhistorique au feu dans le district minier de Cabrieres (Hérault): évidences par thermoluminescence (TL)', in Ambert, P. and Vaquer, J. (eds) *La Première Métallurgie en France et dans les Pays Limitrophes.* Mémoire 37 de La Société Préhistorique Française, Carcassonne, 53–61.

Catapotis, M. 2007. 'On the spatial organisation of copper smelting activities in the southern Aegean during the Early Bronze Age', in Day, P. and Doonan, R. (eds) *Metallurgy in the Early Bronze Age Aegean.* Oxbow Books, Oxford, 207–22.

Cert, C. 2005. 'Les outils de métallurgiste du site du Néolithique Final de La Capitelle du Broum (Péret, Hérault)', in Ambert, P. and Vaquer, J. (eds) *La Première Métallurgie en France et dans les Pays Limitrophes*. Mémoire 37 de La Société Préhistorique Française, Carcassonne, 109–15.

Chapman, D, 1997. 'Great Orme smelting site, Llandudno', *Archaeology in Wales* 37, 56–7.

Chapman, R. 1981. 'The emergence of formal disposal areas and the 'problem' of megalithic tombs in prehistoric Europe', in Chapman, R., Kinnes, I., and Randsborg, K. (eds) *The Archaeology of Death*. Cambridge University Press, Cambridge, 71–82.

Chapman, R. 1995. 'Ten years after: megaliths, mortuary practices and the territorial model', in Beck, L. A. (ed.) *Regional Approaches to Mortuary Analysis*. Plenum Press, London, 29–51.

Charles, J. A. 1979. 'From copper to iron: the origin of metallic materials', *Journal of Metals* 30, 8–13.

Charles, J. A. 1985 (reprinted 1992). 'Determinative mineralogy and the origins of metallurgy', in Craddock, P. and Hughes, M. 1992. *Furnaces and Smelting Technology in Antiquity*. British Museum Occasional Papers 48. British Museum, London, 21–8.

Chernykh, E. 1978a. *Gornoe Delo I Metallurgia v Drevneishei Boulgarii*. Nauka, Sofia.

Chernykh, E. 1978b. 'Ai Bunar, a Balkan copper mine of the fourth millennium BC', *Proceedings of the Prehistoric Society* 44, 203–17.

Chernykh, E. 1982. 'Die ältesten Bergleute und Metallurgen Europas', *Das Altertum* 28, 1, 5–15.

Chernykh, E. 1992. *Ancient Metallurgy in the USSR: The Early Metal Age*. Cambridge University Press, Cambridge.

Chernykh, E. 1994. 'L'Ancienne production minière et métallurgique et les catastrophes écologiques anthropogènes: introduction au problème (1)', *Trabajos de Prehistoria* 51, 2, 55–68.

Chernykh, E. 1998a. 'Ancient mining and metallurgy in Eastern Europe: ecological problems', in Hänsel, B. (ed.) *Mensch und Umwelt in der Bronzezeit Europas*. Oetker-Voges Verlag, Kiel, 129–34.

Chernykh, E. 1998b. 'Kargaly: le plus grand ancien complexe minier et de métallurgie à la frontière de l'Europe et de l'Asie', in Frère-Sautot, M-C (ed.) *Paléomètallurgie des Cuivres*. Éditions Monique Mergoil, Montagnac, 71–6.

Chernykh, E. (ed.) 2002 *Kargaly, Volume I: Geological and Geographical Characteristics; History of Discoveries, Exploitation and Investigations; Archaeological Sites*. Languages of Slavonic Culture, Moscow.

Chernykh, E. 2003. 'Die vorgeschischtlichen Montanreviere an der Grenze von Europa und Asien: das Produkzionszentrum Kargaly', in Stöllner, T., Körlin, G., Steffens, G., and Cierny, J. (eds) *Man and Mining—Mensch und Bergbau*. Der Anschnitt. Beiheft 16. Bochum: Deutsches Bergbau-Museum, 79–92.

Chernykh, E. (ed.) 2004a. *Kargaly, Volume III: archaeological materials; mining and metallurgy technology; archaeobiological studies*. Languages of Slavonic Culture, Moscow.

Chernykh, E. 2004b. 'Kargaly: the largest and most ancient metallurgical complex on the border of Europe and Asia', in Linduff, K. (ed.) *Metallurgy in Ancient Eastern Eurasia from the Urals to the Yellow River*. The Edwin Mellen Press, Lampeter, 223–49.

Childe, V.G. 1930. *The Bronze Age*. Cambridge University Press, Cambridge.

Cierny, J. 2008. *Prähistorische Kupferproduktion in den Südlichen Alpen*. Der Anschnitt Beiheft 22, Bochum: Deutsches Bergbau-Museum.

Cierny, J., Marzatico, F., Perini, R., and Weisgerber, G. 1998. 'Prehistoric copper metallurgy in the southern Alpine region', in Mordant, C., Pernot, M., and Rychner, V. (eds) *L'Atelier du Bronzier en Europe du XX au VIII siècle avant notre ère*. Actes du Colloque International Bronze' 96 (vol. II). Neuchâtel and Dijon. Paris, 25–34.

Cline, W. 1937. *Mining and Metallurgy in Negro Africa*. George Banta Publishing, Wisconsin.

Coghlan, H., Butler, J. R., and Parker, G. 1963. *Ores and Metals. A Report of the Ancient Mining and Metallurgy Committee*, Royal Anthropological Institute. London.

Coghlan, H. H. and Case, H. J. 1957. 'Early metallurgy of copper in Britain and Ireland', *Proceedings of the Prehistoric Society* 23, 91–123.

Constantinou, G. 1982. 'Geological features and ancient exploitation of the cupriferous sulphide orebodies of Cyprus', in Muhly, J., Maddin, R., and Karageorghis, V. (eds) *Early Metallurgy in Cyprus 4000–500 BC*. Pierides Foundation, Nicosia, 13–24.

Constantinou, G. 2012. 'Late Bronze Age copper production on Cyprus from a mining geologist's perspective', in Kassianidou, V. and Papasavvas, G. (eds) *Eastern Mediterranean Metallurgy and Metalwork in the Second Millennium BC*. Oxbow Books, Oxford, 4–13.

Cortesogno, L. and Gaggera, L. 2002. 'Valle Lagorara: analisi delle conoscenze litologiche nell'Età del Rame', in Campana, N. and Maggi, R. (eds) *Archeologia in Valle Lagorara: Diecimila Anni di Storia Intorno a una Cava do Diaspro*. Istituto Italiano di Pristoria e Protostoria, Florence, 31–52.

Craddock, P. T. 1995. *Early Mining and Metal Production*. Edinburgh University Press, Edinburgh.

Craddock, P. T. 1999. 'Paradigms of metallurgical innovation in prehistoric Europe', in Hauptmann, A. et al. (eds) *The Beginnings of Metallurgy*. Der Anschnitt, Beihaft 9, 175–92.

Crew, P. 1994. 'Review of 'Mount Gabriel: Bronze Age Copper Mining in Ireland', *Historical Metallurgy* 28, 2.

Crew, P. and Crew, S. (eds) 1990. *Early Mining in the British Isles*. Plas Tan y Bwlch Occasional Paper 1, Snowdonia Press, Porthmadog.

Cucini Tizzoni, C., Ferrario, A., Morin, D., Ixer, R., Ruffa, M., Tizzoni, M., and Zahova, A. 2001. 'Campolungo (Bienno-Collio, Brescia). Campagna di scavi anno 2000', *Notizie Archeologiche Bergomensi* 9, 113–50.

Cummins, W. A. 1979. 'Neolithic stone axes: distribution and trade in England and Wales', in Clough, T. M. and Cummins, W. (eds) *Stone Axe Studies*. Research Report 23, Council for British Archaeology, London, 5–12.

Davies, O. 1928–30. 'The copper mines of Cyprus', *The Annual of the British School at Athens* 30, 74–85.

Davies, O. 1932. 'Bronze Age mining around the Aegean', *Nature* 130 (3296), 985–7.

Davies, O. 1935. *Roman Mines in Europe*. Oxford University Press, Oxford.

Davies, O. 1937. *Mining Sites in Wales*. Report of the British Association for the Advancement of Science, Section H, 1–3, London.

Davies, O. 1938a. 'Ancient mining in the central Balkans', *Revue Internationale des Études Balkaniques* 6, 405–18.

Davies, O. 1938b. 'Ancient mines in Montgomeryshire', *Montgomeryshire Collections* 45, 55–60.

Davies, O. 1939. 'Excavations on Parys Mountain', *Transactions of the Anglesey Antiquarian Society and Field Club*, 40–2.

Davies, O. 1947. 'Cwmystwyth mines', *Archaeologia Cambrensis* 99, 57–63.

Davies, O. 1948. 'The copper mines on Great Ormes Head, Caerns', *Archaeologia Cambrensis* 100, 61–6.

De Jesus, P. 1980. *The Development of Prehistoric Mining and Metallurgy in Anatolia* (two vols). British Archaeological Reports (international series) 74, Archaeopress, Oxford.

De Melo, A., Álves, H., and Araúfo, F. 2002. 'The bronze palstave from the Quarta Feira copper mine, central Portugal', in Ottaway, B. and Wager, E. (eds) *Metals and Society*. 'British Archaeological Reports (international series) S1061, Archaeopress, Oxford, 109–15.

Del Río, P., García, O., López, J., Navarrete, I., Rovira-Llorens, S., García, J., and Morencos, I. 2006. 'Understanding the productive economy during the Bronze Age through archaeometallurgical and palaeo-environmental research at Kargaly (Southern Urals, Orenburg, Russia)', in Peterson, D., Popova, L., and Smith, A. (eds) *Beyond the Steppe and the Sown: Proceedings of the 2002 University of Chicago Conference on Eurasian Archaeology*. Colloquia Pontica 13, Brill, Boston, 343–57.

Delgado, J., Boski, T., Nieto, J., Pereira, L., Moura, D., Gomes, A., Sousa, C., and García-Tenorio, R. 2012. 'Sea-level rise and anthropogenic activities recorded in the late Pleistocene/Holocene sedimentary infill of the Guadiana Estuary (SW Iberia)', *Quaternary Science Reviews* 33, 121–41.

Delibes de Castro, G., Fernandez Manzano, J., and Herran Martinez, J. 2005. 'La métallurgie du debut de l'Age du cuivre dans le basin du Duero (Espagne): les défis de l'approvisionnement', in Ambert, P. and Vaquer, J. (eds) *La Première Métallurgie en France et dans les Pays Limitrophes*. Mémoire 37 de La Société Préhistorique Française, Carcassonne, 225–32.

Delibes de Castro, G. and Montero Ruiz, I. (eds) 1999. *Las Primeras Etapas Metalúrgicas en la Península Ibérica: II Estudios Regionales*. Instituto Universitario Ortega y Gasset y Ministerio de Educación y Cultura, Madrid.

Derikonjić, S., Radivojević, M., Pernicka, E., and Rehren, Th. 2011. 'The Vinča culture mining complex in Jarmovac, southwest Serbia', in Hauptmann, A., Modarressi-Tehrani, D., and Prange, M. (eds) *Archaeometallurgy in Europe III*. Deutsches Bergbau Museum, Bochum, 39.

Dolfina, A. 2010. 'The origins of metallurgy in central Italy: new radiometric evidence', *Antiquity* 84, 707–23.

Dolfina, A. 2013. 'The emergence of metallurgy in the central Mediterranean region: a new model', *European Journal of Archaeology* 16, 1, 21–62.

Domergue, C. 1987. *Catalogue des Mines et des Fonderies Antiques de la Península Ibérique*. Publications de la Casa de Velazquez, Madrid.

Doonan, R. and Eley, T. 2000. 'The Langness ancient mining survey', in Darvill, T. (ed.) *Billown Neolithic Landscape Project, Isle of Man, 5th Report*. Bournemouth University School of Conservation Sciences Research Report 7, Bournemouth and Douglas, 45–53.

Doonan, R., Klemm, S., Ottaway, B., Sperl, G., and Weinek, H. 1996. 'The East Alpine Bronze Age copper smelting process: evidence from the Ramsau Valley, Eisenerz, Austria', in Demirci, S., Özer, A., and Sumers, G. (eds) *Archaeometry 94: Proceedings of the 29th International Symposium on Archaeometry*. Tübitak, Ankara, 17–22.

Dory, A. 1893. 'Las minas antiguas de cobre y cobalto del Aramo descubiertas por el ingeniero Sr Van Straalen', *Revista Minera y de Ingeniería* 44, 1463, 333–62; 1466, 861–6. Madrid.

Duberow, E., Pernicka, E., and Krenn-Leeb, A. 2009. 'Eastern Alps or Western Carpathians: Early Bronze Age metal within the Wieselburg culture', in Kienlin, T. and Roberts, B. (eds) *Metals and Societies*. Verlag Dr Rudolf Habelt GMBH, Bonn, 336–49.

Dubois, C. and Guilbaut, J. 1980. 'Antiques Mines de Cuivre du Seronais (Pyrenees Ariegeoises)', *Mines et Fonderies Antique de la Gaule*. Centre Regional de Publications de Toulouse, 95–122.

Duffy, T. 1932. *Copper Deposits in South-west Ireland*. Unpublished report, Geological Survey of Ireland, Dublin.

Dutton, A. and Fasham, P. J. 1994. 'Prehistoric copper mining on the Great Orme, Llandudno, Gwynedd', *Proceedings of the Prehistoric Society* 60, 245–86.

Eibner, C. 1972. 'Mitterberg-Grabung 1971', *Der Anschnitt* 24, 2, 3–15.

Eibner, C. 1974. 'Mitterberg-Grabung 1972', *Der Anschnitt* 26, 2, 14–22.

Eibner, C. 1979. 'Zum Stammbaum einer urzeitlichen Kupfererzaufbereitung', *Berg-Hüttenmännische Monatsh.* Wien 124, 157–61.

Eibner, C. 1982. 'Kupfererzbergbau in Österreichs Alpen', in Hansel, B. (ed.) *Südosteuropa Zwischen 1600 und 1000 v. Chr.* Prahistorische Archäologie in Südosteuropa 1, Berlin, 399–408.

Eibner, C. 1992. 'Der kupferbergbau in den österreichischen Alpen in der Urzeit', *Archaeologia Österreichs* 3/1, 12–16.

Eibner Persy, A. and Eibner, C. 1970. 'Erste Grossgrabung auf dem bronzezeitlichen Bergbaugelände von Mitterberg', *Der Anschnitt* 22, 5, 12.

Ericson, J. 1984. 'Towards the analysis of lithic production systems', in Ericson, J. and Purdy, B. (eds) *Prehistoric Quarries and Lithic Production*. Cambridge University Press, Cambridge, 1–9.

Espérou, J. 1981. *La Structure Minière de Roque-Fenestre (Cabrières-Hérault)*. Rapport de D.E.A., Universite de Provence.

Espérou, J. 1990. 'Un outil de métallurgiste aux fonctions imprécises', in Ambert, P. (ed.) *Cabrières-Herault le plus vieux centre Minier Métallurgique de France*. Archéologie en Languedoc, Colloque International, livret-guide de l'excursion du 24 Septembre 1990. Federation Archeologique de l'Herault, 19–20.

Espérou, J-L., Ambert, P., Bourhis, J., Roques, P., Gilot, E., and Chabal, L. 1994. 'La fosse chalcolithique Pioch-Farrus 448', *Bulletin du Musée d'Anthropologie Préhistorique de Monaco* 37, 53–62.

Evans, A. 1980. *An Introduction to Ore Geology*. Blackwell Scientific Publications, Oxford.

Fasnacht, W. 1999. *Excavations at Ayia Varvara Almyras: A Review of 12 years of Research*. Report of the Department of Antiquities, Cyprus, 179–84.

Fasnacht, W. 2004. 'Prähistorischer Kupferbergbau in den Schweizer Alpen I', *Der Anschnitt* 17, 107–11.

Fasnacht, W. and Kassianidou, V. 1992. 'Copper at Almyras: a mining and smelting site on Cyprus', in Marangou, A. and Psillides, K. (eds) *Cyprus, Copper and the Sea*. 'The Government of Cyprus, Nicosia, 77–90.

Ferrari, C., Hong, S., and Boutron, C. 1999. 'Ice archives of atmospheric pollution from mining and smelting activities during antiquity', in Young, S. et al. (eds) *Metals in Antiquity*, Proceedings of the 1995 Harvard Conference. 'British Archaeological Reports (international series) 792, 211–17.

Ferrario, A. and Garuti, G. 1980. 'Copper deposits in the basal breccias and volcano-sedimentary sequences of the eastern Ligurian ophiolotes (Italy)', *Mineralium Deposita* 15, 291–303.

Freire, C., González, A., Garcia, J., and Navarrete, I. 2012. 'Bronze Age economies and landscape resources in the Kargaly steppe (Orenburg, Russia). Remote sensing and palynological data for ancient landscape resources modeling', *EARSel eProceedings* 11, 1/2012, 87–97.

Gago Rabanal, E. 1902. *Estudios de Arqueología Protohistorica y Etnografia de los Astures Lancienses (hoy Leoneses).* León.

Gale, D. 1990. 'Prehistoric stone mining tools from Alderley Edge', in Crew, P. and Crew, S. (eds) *Early Mining in the British Isles*. Plas Tan y Bwlch Occasional Paper 1, Snowdonia Press, Porthmadog, 47–48.

Gale, D. 1995. 'A comparative study of the earliest European copper mining tools', in Petrović, P.and Djurdjekanović, S. (eds) *Ancient Mining and Metallurgy in South-east Europe*. Archaeological Institute, Museum of Mining and Metallurgy, Bor, 47–53.

Gale, N. 1990. 'The provenance of metals for Early Bronze Age Crete: local or Cycladic?', *Pepragmena* I, 299–316.

Gale, N. 1999. 'Lead isotope characterization of the ore deposits of Cyprus and Sardinia and its application to the discovery of the sources of copper for Late Bronze Age oxhide ingots', in Young, S., Pollard, M., Budd, P., and Ixer, R. (eds) *Metals in Antiquity*, Proceedings of the 1995 Harvard Conference. British Archaeological Reports (international series) 792, 110–21.

Gale, N. 2009. 'A response to the paper of A. M. Pollard: What a long, strange trip it's been: lead isotopes and archaeology', in Shortland, A. J., Freestone, I. C., and Pollards, A. M. (eds) *From Mine to Microscope: Advances in the Study of Ancient Technology*. Oxbow Books, London, 191–6.

Gale, N. and Stos-Gale, S. 1982. 'Bronze Age copper sources in the Mediterranean a new approach', *Science* 216, 11–19.

Gale, N. and Stos-Gale, S. 1986. 'Oxhide copper ingots in Crete and Cyprus and the Bronze Age metals trade', *The Annual of the British School at Athens* 81, 81–100.

Gale, N. and Stos-Gale, S. 2000. 'Lead isotope analyses applied to provenance studies', in Ciliberto, E. and Spoto, G. (eds) *Modern Analytical Methods in Art and Archaeology*. Chemical Analyses Series 155, New York, 503–84.

Gale, N. and Stos-Gale, S. 2002. 'Archaeometallurgical research in the Aegean', in Bartelheim, M., Pernicka, E., and Krause, R. (eds) *The Beginnings of Metallurgy in the Old World*. Verlag Marie Leidorf GmbH , Rahden, 277–302.

Gale, N. and Stos-Gale, S. 2012. 'The role of the Apliki mine region in the post *c.*1400 BC copper production and trade networks in Cyprus and the wider Mediterranean', in Kassianidou, V. and Papasavvas, G. (eds) *Eastern Mediterranean Metallurgy and Metalwork in the Second Millennium BC*. Oxbow Books, Oxford, 70–82.

Gale, N., Stos-Gale, S., Lilov, P., Dimitrov, M., and Todorov, T. 1991. 'Recent studies of Eneolithic copper ores and artifacts in Bulgaria', in Mohen, J. P. and Eluère, C. (eds) *Découverte du Métal.* Amis du Musee des Antiquites Nationales, Picard, 49–76.

Gale, N., Stos-Gale, S., Raduncheva, A., Panayotov, I., Ivanov, I., Lilov, P., and Todorov, T. 2003. 'Early metallurgy in Bulgaria', in Craddock, P. and Lang, J. (eds) *Mining and Metal Production through the Ages*. The British Museum Press, London, 122–73.

Garavelli, A., Larocca, F., Levato, C., and Pinto, D. 2012. 'The prehistoric mines of the upper Esara Valley (Calabria, Italy)', in Anreiter, P. et al. (eds) *Second Mining in European History Conference*. Conference abstracts, University of Innsbruck, Innsbruck, 93.

Geiger, T. 1984. 'Zusammensetzung und mikrogefüge von schlackenfunden aus dem Oberhalbstein', *Bergknappe* 28, 2, 2–11.

Georgakopoulou, M., Bassiakos, Y., and Philaniotou, O. 2011. 'Seriphos surfaces: a study of copper slag heaps and copper sources in the context of Early Bronze Age Aegean metal production', *Archaeometry* 53, 1, 123–45.

Giardino, C. and Steiniger, D. 2011. 'Evidenze di miniere preistoriche nell'Etrutia meridionale', in Giardino, C. (ed.) *Archeometallurgia: dalla Conoscenza all Fruizione*. Edipuglia, Bari, 289–92.

Giles, D. and Kuijpers, E. 1974. 'Stratiform copper deposit, Northern Anatolia, Turkey: evidence for Early Bronze Age I (2800 BC) mining activity', *Science* 186, 823–5.

Gilman, A. 1981. 'The development of social stratification in Bronze Age Europe', *Current Anthropology* 22, 1, 70–85.

Gimbutas, M. 1983. 'Copper mining in Old Europe, 5000–4000 BC', *Quarterly Review of Archaeology* (March 1983), 2–3.

Glumac, P. and Tringham, R. 1990. 'The exploitation of copper minerals', in Tringham, R. and Krstić, D. (eds) *Selevac: A Neolithic Village in Yugoslavia*', institute of Archaeology, University of California, Los Angeles, 549–65.

Glumac, P. and Todd, J. 1991. 'Early metallurgy in south-east Europe: the evidence of production', in Glumac, P. (ed.) *Recent Trends in Archaeometallurgical Research.* MASCA Research Papers in Science and Archaeology 8. University Museum, Philadelphia, 8–19.

Golden, J., Levy, T., and Hauptmann, A. 2001. 'Recent discoveries concerning Chalcolithic metallurgy at Shiqmim, Israel', *Journal of Archaeological Science* 28, 951–63.

Goldenberg, G. 1998. 'L'exploitation du cuivre dans les Alpes Autrichiennes à l'âge du Bronze', in Mordant, C., Pernot, M., and Rychner, V. (eds) *L'Atelier du Bronzier en Europe du XX au VIII siècle avant notre ère.* Actes du colloque international Bronze '96 (vol. 2). Neuchâtel and Dijon, Paris, 9–23.

Goldenberg, G. and Hanning, E. in press. 'Mining archaeological prospection for prehistoric copper ore mining activities in Portugal (Alentejo and Algarve, 2004–7)', *Madrider Mitteilungen* publication of Alqueva 2005 conference.

Goldenberg, G. and Rieser, B. 2004. 'Die fahlerzlagerstätten von Schwaz/Brixlegg (Nordtirol). Ein weiteres Zentrum urgeschichtlicher Kupferproduktion in den österreichischen Alpen', in Weisgerber, G. and Goldenberg, G. (eds) *Alpenkupfer—Rame delle Alpi*. Der Anschnitt 17, 37–52.

Goldenberg, G., Breitenlechner, E., Deschler-Erb, S., Hanke, K., Hiebel, G., Hüster-Plogmann, H., Hye, S., Klaunzer, M., Kovács, K., Krismar, M., Lutz, J., Maass, A., Moser, M., Nicolussi, K., Oeggl, K., Pernicka, E., Pichler, T., Pöllath, N., Schibler, J., Staudt, M., Stopp, B., Thurner, A., Töchterle, U., Tomedi, G., Tropper, P., Vavtar, F., and Weinold, T. 2011. 'Prähistorischer kupfererzbergbau im Maukental bei Radfeld/Brixlegg', in Goldenberg, G., Töchterle, U., Oeggl, K., and Krenn-Leeb, A. (eds) *Forschungsprogramm HiMAT: Neues zur Bergbaugeschichte der Ostalpen*. Archäologie Österreichs Spezial 4, Vienna, 61–110.

Gómez Ramos, P. 1999. *Obtención de Metales en la Prehistoria de la Península Ibérica*, British Archaeological Reports (international series) 753, Oxford.

Griffith, R. 1828. *Report to the Royal Dublin Society on the Metallic Mines of Leinster*. Royal Dublin Society, Dublin.

Groer, C. 2008. *Früher Kupferbergbau in Westeuropa*. Band 157, Aus der Abteil für Ur- und Frühgeschichtliche Archäologie der Universität Münster. Bonn: Verlag Dr Rudolf Habelt GmbH.

Gruber, F. 1991. 'Der bergbau im Gemeindegebiet St Veit', in Lindenthaler, K. (ed.) *Heimatbuch St Veit*, St Veit, Austria, 43–52.

Gstrein, P. 1988. 'Neuaufnahme eines vorgeschichtlichen Abbaues im Arthurstollen (Bergbau Mitterberg)', *Mitteilungen der Gesellschaft für Salzburger Landeskunde* 128, 425–38.

Guilaine, J. 1991. 'Roquemengarde et les débuts de la metallurgie en France méditeranéene', in Mohen, J. P. and Eluère, C. (eds) *Découverte du Métal*. Amis du Musee des Antiquites Nationales, Picard, 279–94.

Guyard, H., Chapron, E., St-Onge, G., Anselmetti, F., Arnaud, F., Magand, O., Francus, P., and Mélières, M. 2007. 'High-altitude varve records of abrupt environmental changes and mining activity over the past 4000 years in the Western French Alps (Lake Bramant, Grandes Rousses Massif)', *Quaternary Science Reviews* 26, 2644–60.

Hadjianastasiou, O. and MacGillivray, J. 1988. 'An Early Bronze Age copper smelting site on the Aegean island of Kythnos II: archaeological evidence', in Ellis-Jones, J. (ed.) *Aspects of Ancient Mining and Metallurgy*. British School at Athens Centenary Conference, University of North Wales, Bangor, 31–4.

Halén, O. 1994. 'Sedentariness during the Stone Age of Northern Sweden in light of the Alträsket site, *c.*5000 BC and the Comb Ware site Lillberget. *c.*3900 BC'. Source Critical Problems of Representativity in Archaeology. *Acta Archaeologica Lundensia* 4, 20, Stockholm.

Hampl, F. 1976. 'Die bronzezeitliche kupfergewinnung in Niederösterreich. Forschungsstand Ende 1974', *Archaeologia Austriaca* 14, 58–67.

Hampl, F. and Mayerhofer, R. 1958. 'Die ur- und frühgeschichtliche Bergbauforschung in Niederösterreich', in *Studia Palaeometallurgica in honorem Ernesti Preuschen*. Archaeologia Austriaca beiheft 3, Wien, 46–56.

Hampl, F. and Mayerhofer, R. 1963. 'Urnenfelderzeitlicher Kupferbergbau und mittelalterlicher Eisenbergbau in Niederösterreich', *Archaeologia Austriaca* 33, 50–106.

Hanning, E., Stöllner, T., Hornschuch, A., Sikorski, B., and Röttger, K. 2012. 'Quantifying Bronze Age smelting sites in the Mitterberg mining district', in Anreiter, P. et al. (eds) *2nd Mining in European History Conference*. Abstract, conference guide, Innsbruck University Press, Innsbruck, 43.

Harding, A. 1983. 'The Bronze Age in central and eastern Europe: advances and prospects', *Advances in World Archaeology* 2, 1–50.

Harding, A. 2000. *European Societies in the Bronze Age*. Cambridge University Press, Cambridge.

Hauptmann, A. 2003. 'Developments in copper metallurgy during the fourth and third millennia BC at Feinan, Jordan', in Craddock, P. and Lang, J. (eds) *Mining and Metal Production through the Ages*. The British Museum Press, London, 90–100.

Hauptmann, A. 2009. 'Lead isotope analysis and the origin of Sardinian metal objects', in Lo Schiavo et al. (eds) *Oxhide Ingots in the Central Mediterranean*. A. G. Leventis Foundation, Cyrprus, and the Instituto di Studi Sulle Civilta Dell'Egeo e del Vicino Oriente, Rome.

Herbert, E. 1984. *Red Gold of Africa: Copper in Precolonial History and Culture*. University of Wisconsin Press. Madison.

Herdits, H. 2003. 'Bronze Age smelting site in the Mitterberg mining area in Austria', in Craddock, P. and Ling, J. (eds) *Mining and Metal Production through the Ages*. The British Museum Press, London, 69–75.

Herdits, H. and Löcker, K. 2004. 'Eine bronzezeitliche Kupferhutte im Mitterberger Kupferkies-Revier (Salzburg)', in Weisgerber, G. and Goldenberg, G. (eds) *Rame delle Alpi: Alpenkupfer*, Der Anschnitt, Beiheft 16. Deutsches Bergbau-Museum, Bochum, 177–88.

Holzer, H. and Momenzadeh, M. 1971. 'Ancient copper mines in the Veshnoveh area, Kuhestan-e-Qom, west-central Iran', *Archaeologia Austriaca* 49, 1–47.

Höppner, B., Bartelheim, M., Husijmans, M., Krause, R., Martinek, K., Pernicka, E., and Schwab, R. 2005. 'Prehistoric copper production in the Inn Valley, (Austria), and the earliest copper in Central Europe', *Archaeometry* 47, 2, 293–315.

Huelga-Suarez, G., Moldovan, M., Suárez Fernández, M., Blas Cortina, M. De, Vanhaecke, F., and García Alonso, J. 2012a. 'Lead isotopic analysis of copper ores from the Sierra El Aramo (Asturias, Spain)', *Archaeometry* 54, 4, 685–97.

Huelga-Suarez, G., Moldovan, M., Suárez Fernández, M., Blas Cortina, M. De, and García Alonso, J. 2012b. 'Defining the lead isotopic fingerprint of copper ores from north-west Spain: the El Milagro mine (Asturias,)', *Archaeometry* 56, 1, 88–101.

Huelga-Suarez, G., Moldovan, M., Suárez Fernández, M., Blas Cortina and García Alonso, J. 2013. 'Isotopic composition of lead in copper ores and a copper artefact from the La Profunda mine (Leon, Spain)', *Archaeometry* May. Available at: <http://onlinelibrary.wiley.com/doi/10.1111/arcm.12040/full>.

Hunt Ortiz, M. 2003. *Prehistoric Mining and Metallurgy in the South-west Iberian Peninsula*. British Archaeological Reports (international series) 1188. Archaeopress, Oxford.

Hunt Ortiz, M. and Hurtado Pérez, V. 1999. 'Suroeste', in Delibes de Castro, G. and Montero Ruiz, I. (eds) *Las Primeras Etapas Metalúrgicas en la Península Ibérica: II Estudios Regionales*', instituto Universitario Ortega y Gasset y Ministerio de Educación y Cultura, Madrid, 275–331.

Huurre, M. 1981. *Suomussalmen Varhaista Metallikautta*. Suomen Museo, Finland.

Issel. A. 1879. 'Sulle trace di antichissima lavorazione osservate in alcune miniere della Liguria', *Rassegna Settimanale* 3, 70, 348–9.

Issel, A. 1892. *Liguria Geologica e Preistorica* (two vols). Donath, Genoa.

Ivanov, I. and Avramova, M. 2000. *Varna Necropolis: the Dawn of European Civization*. Agató, Sofia.

Ixer, R. A. 1994. 'A study of Mine 3/4: ore mineralogy, Mount Gabriel', in O'Brien, W. *Mount Gabriel, Bronze Age Mining in Ireland*. Galway University Press, Galway, 253–63.

Ixer, R. A. 1999. 'The role of ore geology and ores in the archaeological provenancing of metals', in Young, S. M., Pollard, M., Budd, P., and Ixer, R. A. (eds) *Metals in Antiquity*. British Archaeological Reports (international series) 792, Archaeopress, Oxford, 43–52.

Ixer, R. A. and Budd, P. 1998. 'The mineralogy of Bronze Age copper ores from the British Isles: implications for the composition of early metalwork', *Oxford Journal of Archaeology* 17, 15–41.

Ixer, R. A. and Pattrick, R. A. D. 2003. 'Copper-arsenic ores and Bronze Age mining and metallurgy with special reference to the British Isles', in Craddock, P. T. and Lang, J. (eds) *Mining and Metal Production Through the Ages*. British Museum Press, London, 9–21.

Jackson, J. 1968. 'Bronze Age copper mines on Mount Gabriel, west county Cork', *Archaeologica Austriaca* 43, 92–114.

Jackson, J. 1979. 'Metallic ores in Irish prehistory: copper and tin', in Ryan, M. (ed.) *The Origins of Metallurgy in Atlantic Europe*. Proceedings of the Fifth Atlantic Colloquium, Stationary Office, Dublin, 107–25.

Jackson, J. 1980. 'Bronze Age copper mining in counties Cork and Kerry', in Craddock, P. T. (ed.) *Scientific Studies in Early Mining and Extractive Metallurgy*. British Museum Press, London, 8–29.

Jackson, J. 1984. 'Copper mining at Mount Gabriel, Co. Cork: Bronze Age bonanza or post-Famine fiasco? A reply', *Proceedings of the Prehistoric Society* 50, 375–7.

Jacobsen, J. 1975. 'Copper deposits in time and space', *Mineral Science Engineering* 7, 337–71.

James, D. 1988. 'Prehistoric copper mining on the Great Ormes Head, Llandudno, Gwynedd', in Ellis-Jones, J. (ed.) *Aspects of Ancient Mining and Metallurgy*. British School at Athens Centenary Conference, University of North Wales, Bangor, 115–21.

James, S. 2008. 'The economic, social, and environmental implications of faunal remains from the Bronze Age copper mines at Great Orme, North Wales', in Sykes, N., Newton, C., and Cool, H. (eds) *Food and Drink in Archaeology 2*, University of Nottingham postgraduate conference, Nottingham, 57–63.

James, S. 2011. The Economical, Social, and Environmental Implications of the Faunal Remains from the Bronze Age Copper Mines at the Great Orme, North Wales. Unpublished PhD thesis, University of Liverpool.

Janković, S. 1980. 'Ore-deposit types and major metallogenic units in Europe', in Janković, J. and Sillitoe, R. (eds) *European Copper Deposits*. Society for Geology applied to Mineral Deposits, Special publication 1, Belgrade, 9–25.

Janzon, G. O. 1984. *Stenredskap med Skaftränna—Indikation på Tidig Metallurgi?* Jernkontorets Bergshistoriska utskott. H32, Stockholm.

Janzon, G. O. 1988. 'Early non-ferrous metallurgy in Sweden', in Maddin, R. (ed.) *The Beginning of the Use of Metals and Alloys*. Massachusetts Institute of Technology, Cambridge, 104–17.

Janzon, G. O. and Noréus, D. 1996. 'Metoder att spåra förhistorisk kopparhantering. Exempel från Helleroprojektet (Småland)', in Forshell, H. (ed.) *Icke-järnmetaller. Malmfyndigheter och Metallurgi*. Jernkontorets Bergshistoriske Utskott H64, Stockholm, 105–17.

Jenkins, D. 1995. 'Mynydd Parys copper mines', *Archaeology in Wales* 35, 35–7.

Jensen, M. and Bateman, A. 1981. *Economic Mineral Deposits* (third edition). John Wiley, London.

Jockenhövel, A. 1982. 'Zeugnisse der primären Metallurgie in Gräbern der Bronze- und Alteisenzeit Mitteleuropas', *Archeologia Polski* 27, 2, 293–301.

Jovanović, B. 1971. 'Early copper metallurgy of the Central Balkans', *Actes du VIII Congres International des Sciences Prehistorique et Protohistoriques I*, Belgrade, 131–8.

Jovanović, B. 1979. 'The technology of primary copper mining in south-east Europe', *Proceedings of the Prehistoric Society* 45, 103–10.

Jovanović, B. 1980a. 'The origins of copper mining in Europe', *Scientific American* 242, 5, 152–67.

Jovanović, B. 1980b. 'Primary copper mining and the production of copper', in Craddock, P. (ed.) *Scientific Studies in Early Mining and Extractive Metallurgy*. British Museum Occasional Paper 20, London, 31–40.

Jovanović, B. 1982. *Rudna Glava. Najstarije Rudarstvo Bakra Na Centralnom Balkanu*. Arheološki Institut, Belgrade, Bor.

Jovanović, B. 1983. 'Mali Sturac—ein neus prähistorisches Kupferbergwerk in Zentralserbien', *Der Anschnitt* 36, 177–9.

Jovanović, B. 1989. 'Die rolle des kupfers in der ökonomie der frühäneolithischen gruppen des Balkans', in Hauptmann, A., Pernicka, E., and Wagner, G. (eds) *Old World Archaeometallurgy*. Der Anschnitt, beiheft 7, Bochum, 13–17.

Jovanović, B. 1995. 'Continuity of the prehistoric mining in the central Balkans', in Petrović, P.and Djurdjekanović, S. (eds) *Ancient Mining and Metallurgy in South-east Europe*. Archaeological Institute, Museum of Mining and Metallurgy, Bor, 29–35.

Jovanović, B. 2003. 'Search for copper from Early Neolithic copper mine Rudna Glava (Majdanpek, Eastern Europe)', *Cu^+ Bulletin du Groupe de Travail International sur la Paléométallurgie des Cuivres et des Minerais Associés* 3, 10–12.

Jovanović, B. and Ottaway, B. 1976. 'Copper mining and metallurgy in the Vinča group', *Antiquity* 50, 104–13.

Junghans, S., Sangmeister, E., and Schroder, M. 1960, 1968, and 1974. *Studien zu den Anfangen der Metallurgie*, Gebr. Mann. Berlin.

Kammenthaler, E. and Beyrie, A. 2007. *Activites Proto-industrielles et Industrielles dans le haut Béarn. Etat des Recherches en Archéologie Miniere et Métallurgique*. Rapport de prospection, SRA Acquitane, 14–17.

Kasseroler, A. 1959. *Das Urnenfeld von Volders*. Schlern-Schriften 204, Wagner, Innsbruck.

Kassianidou, V. 1999. 'Bronze Age copper smelting technology in Cyprus: the evidence from Politico *Phorades*', in Young, S., Pollard, M., Budd, P., and Ixer, R. (eds) *Metals in Antiquity*, Proceedings of the 1995 Harvard Conference. British Archaeological Reports (international series) 792, Archaeopress, Oxford, 91–7.

Kassianidou, V. 2012a. 'Metallurgy and metalwork in Enkomi: the early phases', in Kassianidou, V. and Papasavvas, G. (eds) *Eastern Mediterranean Metallurgy and Metalwork in the Second Millennium* BC. Oxbow Books, Oxford, 94–106.

Kassianidou, V. 2012b. 'The origin and use of metals in Iron Age Cyprus', in Iacovou, M. (ed.) *Cyprus and the Aegean in the Early Iron Age. The Legacy of Nicolas Coldstream*. Cultural Foundation of the Bank of Cyprus, Nicosia, 229–59.

Kemp, J. 1909. 'What is an ore?', *The Western Argus*, Kalgoorlie (7 December 1909), 2.

Kienlin, T. 2012. 'Working copper in the Chalcolithic: a long-term perspective on the development of metallurgical knowledge in central Europe and the Carpathian Basin', in Allen, M., Gardiner, J., and Sheridan, A. (eds) *Is there a British Chalcolithic?* Prehistoric Society Research Paper 4. Oxbow books, Oxford, 126–43.

Kienlin, T. and Stollner, T. 2009. 'Singen copper, Alpine settlement and Early Bronze Age mining: is there a need for elites and strongholds?', in Kienlin, T. and Roberts, B. (eds) *Metals and Societies*. Verlag Dr Rudolf Habelt GMBH, Bonn, 67–104.

Klassen, L. 2000. *Frühes Kupfer im Norden*. Århus Archaeology Monograph, Århus.

Klassen, L. and Stürup, A. 2001. 'Decoding the Riesebusch-copper: lead-isotope analysis applied to Early Neolithic copper finds from South Scandinavia', *Praehistorische Zeitschrift* 76, Berlin, 55–73.

Klaunzer, M., Goldenberg, G., Hye, S., Staudt, M., Töchterle, U., and Tomedi, G. 2010. 'Prehistoric fahlore mining and metallurgy in the Mauken Valley, Radfeld/Brixlegg, North-Tyrol', in Anreiter, P. et al. (eds) *Mining in European History and its Impact on Environment and Human Societies*. Proceedings of the 2009 SFB-HiMAT conference, University of Innsbruck, Innsbruck, 345–9.

Klemm, S. 2003. *Montanarchäologie in den Eisenerzer Alpen, Steiermark. Archäologische und naturwissenschaftliche Untersuchungen zum prähistorischen Kupferbergbau in der Eisenerzer Ramsau*. Mitteilungen der Prähistorischen Kommission 50, Vienna.

Klemm, S. 2004. 'Der prähistorische Kupferschmelzplatz S1 in der Eisenerzer Ramsau (Steiermark)', in Weisgerber, G. and Goldenberg, G. (eds) *Alpenkupfer—Rame delle Alpi*. Der Anschnitt 17, Bochum, 189–98.

Klose, O. 1918. 'Die prähistorischen Funde vom Mitterberge bei Bischofshofen im städtischen Museum Carolino-Augusteum zu Salzburg und zwei prähistorische Schmelzöfen auf dem Mitterberge', *Österreichs Kunsttopographie* 17, Beitrag 2, 1–40. Wien.

Knapp, A. B. 1985. 'Alasiya, Caphtor/Keftiu, and eastern Mediterranean trade: recent studies in Cypriot archaeology and history', *Journal of Field Archaeology* 12, 231–50.

Knapp, A. B. 1998. 'Social approaches to the archaeology and anthropology of mining', in Knapp, A. B., Pigott, V. C., and Herbert, E. W. (eds) *Social Approaches to an Industrial Past: The Archaeology and Anthropology of Mining*. Routledge, London, 1–24.

Knapp, A. B. 1999. 'The archaeology of mining: fieldwork perspectives from the Sydney Cyprus Survey Project', in Young, S., Pollard, M., Budd, P., and Ixer, R. (eds) *Metals in Antiquity*, Proceedings of the 1995 Harvard Conference. British Archaeological Reports (international series) 792, Cambridge, MA, 98–109.

Knapp, A. B. 2012. 'Metallurgical production and trade on Bronze Age Cyprus: views and variations', in Kassianidou, V. and Papasavvas, G. (eds) *Eastern Mediterranean Metallurgy and Metalwork in the Second Millennium* BC. Oxbow Books, Oxford, 14–25.

Kneussl, W. 1969. 'Die älterbronzezeitlichen Funde aus der Tischoferhöhle', *Innsbrucker Beitrage zur Kulturwissenschaft*, Sonderheft 29, Innsbruck, 39–135.

Krajnović, D. and Janković, S. 1995. 'Copper mineralization as potential raw material source of ancient copper metallurgy in Serbia', in Petrović, P. and Djurdjekanović, S. (eds) *Ancient Mining and Metallurgy in Southeast Europe*. Archaeological Institute, Museum of Mining and Metallurgy, Bor.

Krause, R. 1991. 'Bergbauspuren westlich und östlich des Reicher- bzw. Toifenbaches', *Heimatbuch St Veit*, St Veit im Pongau, Gemeindeamt, 52–3.

Krause, R. 2002. 'The cultural and chronological context of the development of Early Bronze Age metallurgy in central Europe', in Ottaway, B. and Wager, E. (eds) *Metals and Society*. British Archaeological Reports S1061, Archaeopress, Oxford, 33–44.

Krause, R. 2007. 'The prehistoric settlement of the inneralpine valley of Montafon in Vorarlberg (Austria)', *Preistoria Alpina* 42, 119–36.

Krause, R. 2009. 'Bronze Age copper production in the Alps: organisation and social hierarchies in mining communities', in Kienlin, T. and Roberts, B. (eds) *Metals and Societies*. Verlag Dr Rudolf Habelt GMBH, Bonn, 47–66.

Krause, R., Bechter, D., Lutz, J., Oeggl, K., Pernicka, E., Schwarz, A.S., Tropper, P., and Würfel, F. 2011. 'Prähistorische siedlungen und mittelalterlicher bergbau im Montafon, Vorarlberg', in Goldenberg, G., Töchterle, U., Oeggl, K., and Krenn-Leeb, A. (eds) *Forschungsprogramm HiMAT: Neues zur Bergbaugeschichte der Ostalpen*. Archäologie Österreichs Spezial 4, Vienna, 147–66.

Kresten, P. 2005. *Analysis of LBA Celts from the Collections of the Museum of Nordic Antiquities, Uppsala University*. Activity Report 2000/2001, Geoarchaeological Laboratory, Department of Archaeological Excavations, Central Board of National Antiquities, Uppsala, 36–48.

Kristiansen, K. and Larsson, T. 2005. *The Rise of Bronze Age Society*. Cambridge University Press, Cambridge.

Kyrle, G. 1916. 'Der prähistorische Bergbaubetrieb in den Salzburger Alpen', *Österreichs Kunsttopographie* 17, Beitrag 1, Wien, 1–50.

Large, D. 2003. 'Base metal (Cu, Pb, Zn) metallogeny of Europe: an overview', in Kelly, J., Andrew, C., Ashton, J., Bland, M., Earls, G., Fusciardi, L., and Stanley, G. (eds) *Europe's Major Base Metal Deposits*. Irish Association for Economic Geology, Dublin, 1–29.

Larocca, F. 2005. *La Miniera Pre-Protostorica di Grotta della Monaca (Sant'Agata di Esaro—Cosenza)*. Leonardo Zaccaro, Roseto, Italy.

Larocca, F. 2010. 'Grotta della Monaca: a prehistoric copper and iron mine in the Calabria region (Italy)', in Anreiter, P. et al. (eds) *Mining in European History and its Impact on Environment and Human Societies*. Proceedings of the 2009 SFB-HiMAT conference, University of Innsbruck', Innsbruck, 267–70.

Le Blanc, M. 2005. 'Pollution minière chalcolithique dans l'estuaire du Rio Tinto (Huelva, Espagne)', in Ambert, P. and Vaquer, J. (eds) *La Première Métallurgie en France et dans les Pays Limitrophes*. Mémoire 37 de La Société Préhistorique Française, Carcassonne, 233–9.

Léchelon, B. 1974. *La Mine Antique de Bouche-Payrol (Sud-Aveyron): Essai d'Archéologie Minière*. Fayet, France.

Léchelon, B. 2001. 'Mines et métallurgies préhistoriques', in Gruat, P. (ed.) *Su Silex au Metal: Mines et métallurgie en Rouerque*. Catalogue d'exposition, Montrozier, Musée du Rouerque, 116–42.

Léchelon, B., Morasz, J-G. and Lopez, M. nd. 'Les mines de cuivre préhistoriques de l'est de la Montagne Noire (Aveyron, Herault): état de la recherche'. Online publication (no date): <http://races.univ-tlse2.fr/accueil-traces/equipes-de-recherche/equipe-5-histoire-et-archéologie-du-metal/les-mines-de-cuivre-prehistoriques-de-l-est-de-la-montagne-noire-aveyron-herault-etat-de-la-recherche-61371.kjsp>

Lehtinen, L. 1996. 'Tidig kopparbaserad metallurgi I Finland', in Forshell, H. (ed.) *Icke-järnmetaller. Malmfyndigheter och Metallurgi*. Jernkontorets Bergshistoriske Utskott H64, Stockholm, 17–27.

Leshtakov, P. 2010. 'Archaeometallurgical evidences for exploitation of copper deposits from the Medni Rid ridge in the fifth millennium BC', *Abstract*, Geosciences 2010 conference, Geological Society, Bulgaria.

Lewis, A. 1990. 'Underground exploration of the Great Orme copper mines', in Crew, P. and Crew, S. (eds) *Early Metal Mining in the British Isles*, Plas Tan y Bwlch Occasional Paper 1, Snowdonia Press, Porthmadog, 5–10.

Lewis, A. 1994. 'Bronze Age mines of the Great Orme', *Mining before Powder: Peak District Mines Historical Society Bulletin* 12, 3, 31–6.

Lewis, A. 1996. *Prehistoric Mining at the Great Orme: Criteria for the Identification of Early Mining*. MPhil thesis, University of Bangor. Published online at: <http://www.greatormemines.info/MPhil.htm>

Lewis, A. 1998. 'The Bronze Age mines of the Great Orme and other sites in the British Isles and Ireland', in Mordant, C., Pernot, M., and Rychner, V. (eds) *L'Atelier du Bronzier en Europe du XX au VIII siècle avant notre ère*. Actes du colloque international Bronze' 96 (vol. 2). Neuchâtel and Dijon, Paris, 45–58.

Lindgren, W. 1933. *Mineral Deposits* (second edition). McGraw-Hill, New York.

Ling, J., Hjärthner-Holdar, E., Grandin, L., Billström, K., and Persson, P. 2013. 'Moving metals or indigenous mining? Provenancing Scandinavian Bronze Age artefacts by lead isotopes and trace elements', *Journal of Archaeological Science* 40, 1, 291–304.

Lippert, A. 1992. *Der Götschenberg bei Bischofshofen*. Österreichische Akademie der Wissenschatten, Vienna.

Llull, B., Perelló, L., Hunt, M., Perelló, D., and Salvà, B. 2012. 'L'explotació prehistòrica dels recursos cuprífers d'Illa den Colom (Maó, Menorca)', in Riera, M. (ed.) *IV Jornades d'Arqueologia de les Illes Balears* (Eivissa, 1 i 2 d'octubre, 2010), Arqueologia i Cultura SL, Vessants, 11–17.

Lo Schiavo, F. 1988. 'Early metallurgy in Sardinia', in Maddin, R. (ed.) *The Beginnings of the Use of Metals and Alloys*. Massachusetts University Press, Cambridge, 92–103.

Lo Schiavo, F. 2012. 'Cyprus and Sardinia, beyond the oxhide ingots', in Kassianidou, V. and Papasavvas, G. (eds) *Eastern Mediterranean Metallurgy and Metalwork in the Second Millennium* BC. Oxbow Books, Oxford, 142–50.

Lo Schiavo, F., Muhly, J., Maddin, R., and Giumlia-Mohr, A. (eds) 2009. *Oxhide Ingots in the Central Mediterranean.* Biblioteca di Antichità Cipriote 8. Rome, A. G. Leventis Foundation, CNR Istituto di Studi sulle Civilta dell'Egeo e del Vicino Oriente.

Loredo, J., Álvarez, R., Ordóñez, A., and Bros, T. 2008. 'Mineralogy and geochemistry of the Texeo Cu-Co mine site (NW Spain): screening tools for environmental assessment', *Environmental Geology* 55, 1299–310.

Lorenzen, W. 1965. *Helgoland und das früheste kupfer des Nordens. Ein Beitrag zur Aufhellung der Anfänge der Metallurgie in Europa.* Niederelbe-Verlag, Otterndorf/ Niederelbe.

Lutz, J., Pernicka, E., Pils, R., Tomedi, G., and Vavtar, F. 2010. 'Geochemical characteristics of copper ores from the Greywacke Zone in the Austrian Alps and their relevance as a source of copper in prehistoric times', in Anreiter, P. et al. (eds) *Mining in European History and its Impact on Environment and Human Societies.* Proceedings of the 2009 SFB-HiMAT conference, University of Innsbruck', Innsbruck, 145–50.

Lutz, J. and Schwab, R. 2012. 'Iron Age copper production in the Eastern Alps', in Anreiter, P. et al. (eds) *2nd Mining in European History Conference.* Abstract, conference guide, Innsbruck University Press, Innsbruck, 104.

Maass, A. 2005. 'Quelques réflexions concernant les structures minières chalcolithiques de la region de Cabrieres', in Ambert, P. and Vaquer, J. (eds) *La Première Métallurgie en France et dans les Pays Limitrophes.* Mémoire 37 de La Société Préhistorique Française, Carcassonne, 63–70.

Maggi, R., Campana, N., and Negrino, F. 1995. 'Valle Lagorara (I 28): a quarry of radiolarite (jasper) exploited during the Copper and Early Bronze Ages', *Archaeologia Polona* 33, 187–208.

Maggi, R. and Del Lucchese, A. 1988. 'Aspects of the Copper Age in Liguria', *Rassegna di Archeologia* 7, 331–8.

Maggi, R. and Pearce, M. 2005. 'Mid fourth millennium copper mining in Liguria, north-west Italy: the earliest known copper mines in western Europe', *Antiquity* 79, 303, 66–77.

Maggi, R. and Pearce, M. 2010. 'Changing subsistence structures and the origins of mining in the Ligurian Apennine mountains', in Anreiter, P. et al. (eds) *Mining in European History and its Impact on Environment and Human Societies.* Proceedings of the 2009 SFB-HiMAT conference, University of Innsbruck', Innsbruck, 283–7.

Magnusson Staaf, B. 2002. 'Changes of use: transformations of comprehension', in Ottaway, B. and Wager, E. (eds) *Metals and Society.* British Archaeological Reports S1061, Archaeopress, Oxford, 133–44.

Mahr, A. 1937. 'New aspects and problems in Irish prehistory: Presidential Address for 1937', *Proceedings of the Prehistoric Society* 3, 261–436.

Mantenant, J., Beyrie, A., Fabre, J-M., Kammenthaler, E., Cortés, N., Munteanu, G., and Rico, C. 2012. 'The mine of 'Les Barrencs' (South France). Technological approach of a copper mine of the 2nd Iron Age and the Late Republic (4th–1st centuries BC)', in Anreiter, P. et al. (eds) *2nd Mining in European History Conference*. Abstract, conference guide, Innsbruck University Press, Innsbruck, 50.

Marcello, A., Pretti, S., and Salvadori, I. 1978. 'Le prospezioni geominerarie in Sardegna la pospezione geochimica strategica', *Italian Journal of Geosciences* 99, 277–310.

Marescotti, P., Carbone, C., and De Capitani, L. 2010. 'Mineralogical and geochemical spatial analysis of a waste-rock dump at the Libiola Fe–Cu sulphide mine (Eastern Liguria, Italy)', *Environmental Earth Science* 61, 187–99.

Marino, S. 2010. 'Stone tools with hafting groove and prehistoric mining exploitation in Calabria (Italy)', in Anreiter, P. et al. (eds) *Mining in European History and its Impact on Environment and Human Societies*. Proceedings of the 2009 SFB-HiMAT conference, University of Innsbruck', Innsbruck, 321–7.

Marshall, P., O'Hara, S., and Ottaway, B. 1999. 'Early copper metallurgy in Austria and methods of assessing its impact on the environment', in Hauptmann, A. et al. (eds) *The Beginnings of Metallurgy*, Der Anschnitt, beihaft 9, 255–64.

Martín Cólliga, A., Mata-Perelló, J., Gallart Fernández, J., and Rovira Hortalà, C. 2005. 'Indices miniers et métallurgie ancienne dans la Catalogne sub-pyrénéenne', in Ambert, P. and Vaquer, J. (eds) *La Première Métallurgie en France et dans les Pays Limitrophes*. Mémoire 37 de La Société Préhistorique Française, Carcassonne, 37–50.

Martinek, K. 1995. 'Archäometallurgische Untersuchungen zur frühbronzezeitlichen Kupferproduktion und verarbeitung auf dem Buchberg bei Wiesing, Tirol', *Fundberichte aus Österreich* 34, 575–84.

Martinek, K. and Grutsch, C. 2012. 'Copper mineralizations in western North Tyrol: in prehistoric times exploited resources?, in Anreiter, P. et al. (eds) *Conference Guide, 2nd Mining in European History Conference*, HiMAT, Innsbruck, 31.

Martinek, K. and Sydow, W. 2004. 'Frühbronzezeitliche kupfermetallurgie im Unterinntal (Nordtirol). Rohstoffbasis, archäologische und archäometallurgische befunde', in Goldenberg, G. and Weisgerber, G. (eds) *Alpenkupfer: Rame delli Alpi*, Der Anschnitt, Beiheft 17, Deutsches Bergbau-Museum, Bochum, 199–211.

Matías Rodriguez, R., Neira Campos, A., and Alonso Herrero, E. 2001. 'Un lugar en el olvido: los restos de la explotación prehistórica del yacimiento de cobre de la mina 'La Profunda', Cármenes, León', *Promonumenta* 5, 14–24.

Mauss, M. 1923. *The Gift: Forms and Functions of Exchange in Archaic Societies*. Routledge, London.

McCullagh, M. and Pearce, M. 2004. 'Surveying the prehistoric copper mine at Libiola (Sestre Levante-GE), Italy', in Balzaretti, R., Pearce, M., and Watkins, C. (eds) *Ligurian Landscapes: Studies in Archaeology, Geography and History*. Accordia Research Centre, London, 83–95.

McKeown, S. 1994. 'The analysis of wood remains from Mine 3, Mount Gabriel', in O'Brien, W. *Mount Gabriel: Bronze Age Mining in Ireland*. Bronze Age Studies 3, Galway University Press, Galway, 265–80.

Melheim, L. 2012. *Recycling Ideas: Bronze Age Metal Production in Southern Norway*. Doctoral thesis, Department of Archaeology, Conservation and History, University of Oslo.

Menke, M. 1978/9. 'Studien zu den Frühbronzezeitlichen Metalldepots Bayerns', *Jahresberichte der Bayerischen Bodendenkmalpflege* 19/20, 5–305.

Merideth, C. 1998. *An Archaeometallurgical Survey for Ancient Tin Mines and Smelting Sites in Spain and Portugal.* British Archaeological Reports, (international series) 714, Archaeopress, Oxford.

Merkel, J. 1985. 'Ore beneficiation during the late Bronze/early Iron Age at Timna, Israel', *Masca Journal* 3, 5, 164–9.

Merrillees, R. 1984. 'Ambelikou-Aletri: a preliminary report'. *Report of the Department of Antiquities, Cyprus* (unpublished), 1–13.

Mighall, T. 2003. 'Geochemical monitoring of heavy metal pollution and prehistoric mining: evidence from Copa Hill, Cwmystwyth, and Mount Gabriel, county Cork', in Craddock, P. and Lang, J. (eds) *Mining and Metal Production through the Ages.* British Museum Press, London, 43–51.

Mighall, T., Chambers, F., Lanting, J., and O'Brien, W. 2000. 'Prehistoric copper mining and its importance on vegetation: palaeoecological evidence from Mount Gabriel, Co. Cork, south-west Ireland', in Nicholson, R. A. and O' Connor, T. P. (eds) *People as an Agent of Environmental Change.* Oxbow, Oxford, 19–29.

Mighall, T. and Chambers, F. M. 1993. 'The environmental impact of prehistoric mining at Copa Hill, Cwmystwyth, Wales', *The Holocene* 3, 260–4.

Mighall, T., Timberlake, S., Chambers, F., and O'Brien, W. 2012. 'Characterising vegetation changes in former mining and metalworking areas during prehistoric and Roman times', *Notizie Archeologiche Bergomensi* 20, 117–30.

Mighall, T., Tizzoni, M., Cucini Tizzoni, C., and O'Brien, C. 2003. 'The impact of Iron Age copper mining on vegetation palaeoecological evidence from Campolungo, Bienno, Lombard Alps, Northern Italy', *Notizie Archeologiche Bergomensi* 11, 253–71.

Mille, B. and Carozza L. 2009. 'Moving into the Metal Ages: the social importance of metal at the end of the Neolithic in France', in Kienlin, T. and Roberts, B. (eds) *Metals and Societies.* Verlag Dr Rudolf Habelt GMBH, Bonn, 143–71.

Monna, F., Galop, D., Carozza, L., Tual, M., Beyrie, A., Marembert, F., Chateau, C., Dominik, J., and Grousset, F. 2004. 'Environmental impact of early Basque mining and smelting recorded in a high ash minerogenic deposit', *Science of the Total Environment* 327, 197–214.

Montelius, O. 1872. *Sveriges forntid. Försök till framställning af den svenska fornforskningens resultat.* Atlas, 1, Stenåldern och bronsåldern. Nordstedt, Stockholm.

Montelius, O. 1888. *The Civilisation of Sweden in Heathen Times.* Macmillan, London.

Montero Ruiz, I. 1993. 'Bronze Age metallurgy in southeast Spain', *Antiquity* 67, 46–57.

Montero Ruiz, I. 1994. *El Origen de la Metalurgia en el Sureste Peninsular*', instituto de Estudios Almerienses, Almería, Spain.

Montero Ruiz, I. 1999. 'Sureste', in Delibes de Castro, G. and Montero Ruiz, I. (eds) *Las Primeras Etapas Metalúrgicas en la Península Ibérica: II Estudios Regionales*', instituto Universitario Ortega y Gasset y Ministerio de Educación y Cultura, Madrid, 333–57.

Montero Ruiz, I. (ed.) 2010. *Manual de Arqueometalurgia.* Museo Arqueológico Regional, Madrid.

Montero Ruiz, I. and Esperanza, M. J. R. de la 2008. 'Un pequeño campamento minero de la edad del Bronce: La Loma de la Tejeria (Albarracín, Teruel)', *Trabajos de Prehistoria* 65, 1, 155–68.

Montero-Ruiz, I., Rafel, N., Hunt Ortiz, M., Mata, J. M., Odriozola, C., Soriano, I., and Murillo-Barroso, M. 2012. 'Minería prehistórica en el Priorato: caracterización arqueométrica de minas de Cornudella, Alforja y Ulldemolins', in Mata, J. (ed.) *El Patrimonio Minero y Metalúrgico a lo largo del a Historia*. Libro de Actas del VII Congreso Internacional sobre Minería y Metalurgia Históricas en el Sudoeste Europeo (Utrillas, May 2012), 131–40.

Montero Ruiz, I. and Ruiz Taboada, A. 1996. 'Enterramiento colectivo y metalurgia en el yacimiento neolitico de Cerro Virtud (Cuevas de Almanzora, Almería)', *Trabajos de Prehistoria* 53, 2, 55–75.

Moslein, S. 1999. 'Neue depotfunde der älteren Bronzezeit aus dem oberbayerischen Alpenvorland. Ein Vorbericht', *Archäologie Osterreichs*, Sonderausgabe 9/10, 69–77.

Much, M. 1878. 'Das vorgeschichtliche Kupferbergwerk auf dem Mitterberg bei Bischofshofen (Salzburg)', *Mitteilungen der K.K. Central- Kommission für Erforschung und Erhaltung der Kunst- und Historischen Denkmale* 4, Vienna, 146–52.

Much, M. 1879. 'Das vorgeschichtliche Kupferbergwerk auf dem Mitterberg bei Bischofshofen (Salzburg)', *Mitteilungen der K.K. Central- Kommission für Erforschung und Erhaltung der Kunst- und Historischen Denkmale* 5,Vienna, 18–36.

Much, M. 1893. *Die Kupferzeit in Europa und ihr Verhältnis zur Kultur der Indogermanen*. Costenoble, Jena.

Muckelroy, K. 1981. 'Middle Bronze Age trade between Britain and Europe: a maritime perspective', *Proceedings of the Prehistoric Society* 47, 275–97.

Muhly, J. D. 1988. 'The beginnings of metallurgy in the Old World', in Maddin, R. (ed.) *The Beginning of the Use of Metals and Alloys*, Proceedings of the Zheng-Zhou colloquium, October 1986, Harvard, 2–20.

Muhly, J. D. 1989. 'Cayönu Tepesi and the beginnngs of metallurgy in the ancient world', in Hauptmann, A., Pernicka, E., and Wagner, G. (eds) *Old World Archaeometallurgy*. Der Anschnitt, beiheft 7, Bochum, 1–11.

Müller, R. and Cardoso, J. 2008. 'The origin and use of copper at the Chalcolithic fortification of Leceia (Oeiras, Portugal)', *Madrider Mitteilungen* 49, 64–93.

Müller, R., Goldenberg, G., Bartelheim, M., Kunst, M., and Pernicka, E. 2007. 'Zambujal and the beginnngs of metallurgy in southern Portugal', in La Niece, S., Hook, D., and Craddock, P. (eds) *Metals and Mines: Studies in Archaeometallurgy*. Archaeotype, London, 15–26.

Müller, R., Rehren, T., and Rovira, S. 2004. 'Almizaraque and the early copper metallurgy of southeast Spain: new data', *Madrider Mitteilungen* 45, 33–56.

Müller, R. and Soares, A. 2008. 'Traces of early copper production at the Chalcolithic fortification of Vila Nova de São Pedro, Portugal', *Madrider Mitteilungen* 49, 94–114.

Murillo-Barroso, M. and Montero Ruiz, I. (2012). 'Copper ornaments in the Iberian Chalclithic: technology versus social demand', *Journal of Mediterranean Archaeology* 25, 1, 53–73.

Needham, S. P. 1998. 'Modelling the flow of metal in the Bronze Age', in Mordant, C. et al. (eds), *L'Atelier du Bronzier en Europe* (vol. 3). Actes du Colloque International 'Bronze '96', Neuchâtel and Dijon, Paris, 285–308.

Needham, S. P. 2002. 'Analytical implications for Beaker metallurgy in north-west Europe', in Bartelheim, M, Pernicka, E., and Krause, R. (eds) *Die Anfänge der*

Metallurgie in derr Alten Welt. Freiberger Forschungen zur Archäometrie und Kulturgesch 1, Rahden, 99–133.

Neuninger, H. and Pittioni, R. 1963. 'Frühmetallzeitlicher Kupferhandel im Alpenvorland: zur Frage der Herkunft des Ringbarrenkupfers', *Archaeologia Austriaca* beiheft 6, Wien.

Neuninger, H., Pittioni, R., and Preuschen, E. 1957. 'Urzeitlicher Bergbau auf Kupfererz und Spureanalyse. Beiträge zur Relation Lagerstätte-Fertigobjekt', *Archaeologia Austriaca* beiheft 5, Wien.

Niederschlag, E. and Pernicka, E. 2002. 'The provenance of Early Bronze Age copper from Northern Bohemia and Central Germany: first results from physico-chemical investigations', in Ottaway, B. and Wager, E. (eds) *Metals and Society.* British Archaeological Reports S1061, Archaeopress, Oxford, 51–5.

Niederschlag, E., Pernicka, E., Seifert, Th., and Bartelheim, M. 2003. 'The determination of lead isotope ratios by multiple collector ICP-MS: a case study of Early Bronze Age artefacts and their possible relation with ore deposits of the Erzgebirge', *Archaeometry* 45, 1, 61–100.

Nocete, F. 2006. 'The first specialized copper industry in the Iberian Peninsula: Cabezo Juré (2900–2200 BC)', *Antiquity* 80, 646–54.

Nocete, F., Alex, E., Nieto, J., Saez, R., and Bayona, M. 2005. 'An archaeological approach to regional environmental pollution in the south-western Iberian Peninsula related to third millennium BC mining and metallurgy', *Journal of Archaeological Science* 32, 1566–76.

Nocete, F., Orihuela, A., and Peña, M. 1993. 'Odiel 3000–1000 A.N.E. 'Un modelo de análisis histórico para la contrastación del proceso de jerarquización social', *Anuario Arqueológico de Andalucía* 1991, 2, 259–66.

Nocete, F., Queipo, G., Sáez, R., Nieto, J., Inácio, N., Bayona, M., Peramo, A., Vargas, J., Cruz-Auñon, R., Gil-Ibarguchi, J., and Santos, J. 2008. 'The smelting quarter of Valencina de la Concepción (Seville, Spain): the specialized copper industry in a political centre of the Guadalquivir Valley during the third millennium BC (2750–2500 BC)', *Journal of Archaeological Science* 35, 717–32.

Noréus, D. 2001. 'Modern kopparmetod tillämpad redan under bronsåldern?', *Populär Arkeologi* 19, 1, 16–17.

Northover, J. P. 1982. 'The exploration of the long-distance movement of bronze in Bronze and Early Iron Age Europe', *Bulletin of the Institute of Archaeology, London* 19, 45–71.

Northover, J. P. 1999. 'The earliest metalworking in southern Britain', in Hauptmann, A. et al. (eds) *The Beginnings of Metallurgy.* Der Anschnitt 9, Bochum, 211–25.

Northover, J., O'Brien, W., and Stos, S. 2001. 'Lead isotopes and metal circulation in Beaker/Early Bronze Age Ireland', *The Journal of Irish Archaeology* 10, 25–48.

O'Brien, W. 1987. 'The dating of the Mount Gabriel-type mines of West Cork', *Journal of the Cork Historical and Archaeological Society* 92, 50–70.

O'Brien, W. 1990. 'Prehistoric copper mining in south-west Ireland: the Mount Gabriel-type mines', *Proceedings of the Prehistoric Society* 56, 269–90.

O'Brien, W. 1994. *Mount Gabriel: Bronze Age Mining in Ireland.* Bronze Age Studies 3, Galway University Press, Galway.

O'Brien, W. 1995. 'Ross Island and the origins of Irish-British metallurgy', in Waddell, J. and Shee Twohig, E. (eds) *Ireland in the Bronze Age*, Stationary Office, Dublin, 38–48.

O'Brien, W. 1996. *Bronze Age Copper Mining in Britain and Ireland.* Shire Publications, Princes Risborough.

O'Brien, W. 1997. 'Mount Gabriel and metal sourcing in the Bronze Age', *Historical Metallurgy* 31, 1, 8–11.

O'Brien, W. 1999. 'Resource availability and metal supply in the insular Bronze Age', in Hauptmann, A. et al. (eds) *The Beginnings of Metallurgy*, Der Anschnitt, Beihaft 9, 227–35.

O'Brien, W. 2000. 'Megalithic tombs, metal resources and territory in prehistoric south-west Ireland', in Desmond, A. et al. (eds) *New Agendas in Irish Prehistory.* Wordwell, Dublin, 161–76.

O'Brien, W. 2003. 'The Bronze Age copper mines of the Goleen area, Co. Cork', *Proceedings of the Royal Irish Academy* 103C, 2, 13–59.

O'Brien, W. 2004. *Ross Island: Mining, Metal, and Society in Early Ireland.* Bronze Age Studies 6. National University of Ireland Galway, Galway.

O'Brien, W. 2007. 'Miners and farmers: local settlement contexts for Bronze Age mining', in Burgess, C. (ed.) *Beyond Stonehenge.* Oxbow, Oxford, 20–30.

O'Brien, W. 2009. *Local Worlds: Early Settlement Landscapes and Upland Farming in South-west Ireland.* Collins Press, Cork.

O'Brien, W. 2010. 'Copper axes, stone axes: production and exchange systems in the Chalcolithic of Britain and Ireland', in Anreiter, P. et al. (eds) *Mining in European History and its Impact on Environment and Human Societies.* Proceedings of the 2009 SFB-HiMAT conference, University of Innsbruck', Innsbruck, 255–60.

O'Brien, W. 2013. 'Copper mining in Ireland during the Later Bronze Age', in Anreiter, P. et al. (eds) *Mining in European History and its Impact on Environment and Human Societies.* Proceedings of the 2012 SFB-HiMAT conference, University of Innsbruck', Innsbruck, 191–7.

O'Brien, W. and Hogan, N. 2012. 'Derrycarhoon: Bronze Age copper mining in West Cork', *Archaeology Ireland* 26, 1, 12–15.

O'Brien, W., Northover, P., and Cameron, E. 1989/90. 'An Early Bronze Age metal hoard from a wedge tomb at Toormore, County Cork', *The Journal of Irish Archaeology* 5, 9–17.

Oldeberg, A. 1960. *Skälbyfyndet. En boplatslämning från den yngre bronsåldern*, Antikvariskt Arkiv 15. Kungliga vitterhets historie och antikvitets akademien, Stockholm.

Ottaway, B. 1973. 'The earliest copper ornaments in northern Europe', *Proceedings of the Prehistoric Society* 39, 294–331.

Ottaway, B. 1982. *Earliest Copper Artifacts of the Northalpine Region: their Analysis and Evaluation.* Schriften des Seminars für Ur- und Frühgeschichte der Universität Bern, Bern.

Ottaway, B. 2001 'Innovation, production, and specialization in early prehistoric copper metallurgy', *European Journal of Archaeology* 4, 1, 87–112.

Otto, H. and Witter, W. 1952. *Handbuch der Altesten Vorgeschichtlichen Metallurgie in Mitteleuropa.* Barth, Leipzig.

Paniagua, A., Loredo, J. and García Iglesias, J. 1987. 'Epithermal mineralization in the Aramo mine (Cantabrian Mountains, NW Spain): Correlation between paragenesis and fluid inclusions data', *Bulletin of Mineralogy*, 111, 383–91.

Pearce, M. 2000. 'Metals make the world go round: the copper supply for Frattesina', in Pare, C. (ed.) *Metals Make the World Go Round: Supply and Circulation of Metals in Bronze Age Europe*. Oxbow, Oxford. 56–69.

Pearce, M. 2007. *Bright Blades and Red Metal*. Accordia Specialist Studies on Italy 14, Accordia Research Institute, University of London.

Pearce, M. 2009. 'How much metal was there in circulation in Copper Age Italy?', in Kienlin, T. and Roberts, B. (eds) *Metals and Societies*. Verlag Dr Rudolf Habelt GMBH, Bonn, 277–84.

Pearce, S. 1983. *The Bronze Age Metalwork of South Western Britain*. British Archaeological Reports (British series) 120, Archaeopress, Oxford.

Pellicer, M. and Hurtado, V. 1980. *El Poblado Metalúrgico de Chinflon (Zalamacea la Real, Huelva)*. Publicaciones del Departmento de Prehistoria y Arqueología, Seville.

Penhallurick, R. 1986. *Tin in Antiquity*, London, Institute of Metals.

Perelló Mateo, L., Llull Estarellas, B., Salvà Simonet, B., and Hunt Ortiz, M. 2013. 'Sa Mitja Lluna: una mina de coure de l'edat del bronze', *Revista Àmbit* 29, 1–5.

Pérez Macías, J., 1998. *Las minas de Huelva en la Antigüedad*. Diputación de Huelva, Huelva.

Perini, R. 1992. 'Evidence of metallurgical activity in Trentino from Chalcolithic times to the end of the Bronze Age', in Sanpaolo, E. A. (ed.) *Archeometallurgia Richerche e Prospettive*. Clueb, Bologna, 53–76.

Perini, R. 2005. 'Testimonianze de attività metallurgica dall' eneolitico alle fasi finali dell' età del Bronzo nel Trentino', *Scritta di Archeologia* II, 1115–46.

Pernicka, E. 1998. 'Whither metal analysis in archaeology?', in Mordant, C. et al. (eds) *L'Atelier du Bronzier en Europe I*. Actes du Colloque International 'Bronze 96', Neuchâtel and Dijon, Paris, 259–67.

Pernicka, E. 1990. 'Gewinnung und verbreitung der metalle in prähistorischer zeit', *Jahrbuch des Römisch-Germanischen Zentralmuseums Mainz* 37, 21–129.

Pernicka, E. 1999. 'Trace element fingerprinting of ancient copper: a guide to technology or provenance?', in Young, S. et al. (eds) *Metals in Antiquity*, Proceedings of the 1995 Harvard Conference. British Archaeological Reports (international series) 792, Cambridge, MA, 163–71.

Pernicka, E. and Anthony, D. 2010. 'The invention of copper metallurgy and the Copper Age of Old Europe', in Anthony, D. and Chi, J. (eds) *The Lost World of Old Europe: the Danube Valley, 5000–3500 BC*. Princeton University Press, New York, 163–77.

Pernicka, E., Begemann, F., Schmitt-Strecker, S., Todorova, H., and Kuleff, I. 1997. 'Prehistoric copper in Bulgaria. Its composition and provenance', *Eurasia Antiqua* 3, 41–180.

Pernicka, E., Begemann, F., Schmitt-Strecker, S., and Wagner, G. A. 1993. 'Eneolithic and Early Bronze Age copper artefacts from the Balkans and their relation to Serbian copper ores', *Praehistorisches Zeitschrift* 68, 1, 1–54.

Pétrequin, P. and Pétrequin, A-M. 1993. *Écologie d'un outil: la hache de Pierre en Irian Jaya (Indonésie)*. CNRS, Paris.

Pichler, T., Nicolussi, K., and Goldenberg, G. 2009. 'Dendrochronological analysis and dating of wooden artefacts from the prehistoric copper mine Kelchalm/Kitzbühel (Austria)', *Dendrochronologia* 27, 87–94.

Pickin, J. 1990. 'Stone tools and early metal mining in England and Wales', in Crew, P. and Crew, S. (eds) *Early Mining in the British Isles*. Plas Tan y Bwlch Occasional Paper 1, Snowdonia Press, Porthmadog, 39–42.

Pickin, J. and Worthington, T. 1989. 'Prehistoric mining hammers from Bradda Head, Isle of Man', *Bulletin of the Peak District Mines Historical Society* 10, 5, 274–5.

Piñon Varela, F. 1986. 'La Edad del Cobre: primeros metalurgicos en Huelva', *Huelva y su Provincia* 2, 62–89.

Pittioni, R. 1948. 'Recent researches on ancient copper mining in Austria', *Man* 133, 120–2.

Pittioni, R. 1951. 'Prehistoric copper mining in Austria: problems and facts', *Seventh Annual Report*, Institute of Archaeology, London, 17–40.

Ploquin, A., Happ, J., Barge, H., and Bourhis, J. 1997. 'Scories archéologiques et reconstitution expérimentale de réduction de sulfure de cuivre (minerai de St Véran, Hautes-Alpes): prémices d'une approche pétrographique', *Archéologie en Languedoc* 21, 11–120.

Pollard A. M. 2009. 'What a long, strange trip it's been: lead isotopes and archaeology', in Shortland, A. J., Freestone, I. C., and Pollards, A. M. (eds) *From Mine to Microscope: Advances in the Study of Ancient Technology*. Oxbow Books, London, 181–9.

Pollard, M., Thomas, R., and Williams, P. 1990. 'Experimental smelting of arsenical copper ores: implications for Early Bronze Age copper production', in Crew, P. and Crew, S. (eds) *Early Metal Mining in the British Isles*, Plas Tan y Bwlch Occasional Paper 1, Wales, 72–4.

Prange, M. and Ambert, P. 2005. 'Caractérisation géochimique et isotopique des minerais et des métaux base cuivre de Cabrières (Hérault)', in Ambert, P. and Vaquer, J. (eds) *La Première Métallurgie en France et dans les Pays Limitrophes*. Mémoire 37 de La Société Préhistorique Française, Carcassonne, 71–81.

Prescott, C. 2006. 'Copper production in Bronze Age Norway?', in Glørstad, H., Skar, B., and Skre, D. (eds) *Historien i Forhistorien. Festskrift til Einar Østmo på 60-årsdagen*, Skrifter nr. 4. Kulturhistorisk Museum, Universitetet i Oslo, Oslo, 183–90.

Presslinger, H. and Eibner, C. 1989. 'Bronzezeitliche kupferverhüttung im Paltental', in Hauptmann, A., Pernicka, E., and Wagner, G. (eds) *Old World Archaeometallurgy*. Der Anschnitt beiheft 7, Bochum, 235–40.

Presslinger, H. and Eibner, C. 1993. 'Prähistorischer kupfererzbergbau und die verhüttung der erze', in Presslinger, H. and Köstler H. (eds) *Bergbau und Hüttenwesen im Bezirk Liezen (Steiermark)*. Kleine Schriften der Abteilung Schloss Trautenfels am Steiermärkkischen Landesmuseum Joanneum 24, Innesbruk, 25–36.

Presslinger, H. and Eibner, C. 1996. 'Montanwesen und siedlungen in der Bronzezeit im Paltental (Österreich)', *Der Anschnitt*, Zeitschrift für Kunst und Kultur im Bergbau 48, 5–6, 158–65.

Preuschen, E. 1962. 'Der urzeitliche Kupfererzbergbau von Vettiolo (Trentino)', *Der Anschnitt* 14, 3–7.

Preuschen, E. 1967. *Urzeitlicher Kupfererzbergbau in den österreichischen Alpen.* Leobener Grüne Hefte 104. Montan-Verl, Vienna.

Preuschen, E. 1968a. 'Kupfererzlagerstätten in der Steiermark', in *Der Bergmann. Der Huttenmann. Gestalter der Steiermark.* Katalog der 4. Landesausstellung, Graz, 185–8.

Preuschen, E. 1968b. 'Bronzezeitlicher Kupfererzbergbau im Trentino', *Der Anschnitt* 20, 1, 3.

Preuschen, E. 1973. 'Estrazione mineraria dell'età del bronzo nel Trentino', *Preistoria Alpina* 9, 113–50.

Preuschen, E. and Pittioni, R. 1937. *Untersuchungen im Bergbaugebiete Kelchalpe bei Kitzbühel, Tirol.* Mitteilungen der Prähistorischen Kommission der Österreichischen Akademie der Wissenschaften. 3, Wien.

Preuschen, E. and Pittioni, R. 1953. 'Osttiroler Bergbaufragen', in *Beitrage zur Älteren Europäischen Kulturgeschichte.* Festschrift für Rudolf Egger, Klagenfurt, 68–74.

Preuschen, E. and Pittioni, R. 1954. 'Untersuchungen im Bergbaugebiet Kelchalpe bei Kitzbühel, Tirol', *Archaeologia Austriaca* 15, 2–97.

Preuschen, E. and Pittioni, R. 1956. 'Das urzeitliche bergbaugebiet wirtsalm bei Viehhofen im Mitterpinzgau, Salzburg', *Archaeologia Austriaca* 19/20, 264–78.

Pulak, C. 1998. 'The Uluburun shipwreck: an overview', *The International Journal of Nautical Archaeology* 27, 3, 188–224.

Pulak, C. 2008. 'The Uluburun shipwreck and Late Bronze Age trade', in Aruz, A., Benzel, K., and Evans, J. (eds) *Beyond Babylon. Art, Trade, and Diplomacy in the Second Millennium* BC. Yale University Press, New Haven and London, 289–385.

Radivojević, M., Rehren, T., Pernicka, E., Sljivar, D., Brauns, M., and Borić, D. 2010. 'On the origins of extractive metallurgy: new evidence from Europe', *Journal of Archaeological Science* 37, 2775–87.

Ramis, D., Hauptmann, A., and Coll, J. 2005. 'Reduction du minerai de cuivre dans la prehistoire de Majorque', in Ambert, P. and Vaquer, J. (eds) *La Première Métallurgie en France et dans les Pays Limitrophes.* Mémoire 37 de La Société Préhistorique Française, Carcassonne, 217–24.

Rehren, T. 2009. 'From mine to microbe: the Neolithic copper melting crucibles from Switzerland', in Shortland, A., Freestone, I., and Rehren, T. (eds) *From Mine to Microscope: Advances in the Study of Ancient Technology.* Oxbow Books, Oxford.

Reinecke, P. 1930. *Die Bedeutung der Kupferbergwerke der Ostalpen für die Bronzezeit Mitteleuropas.* Schumacher-Festschrift, Mainz.

Renfrew, C. 1967. 'Colonialism and megalithismus', *Antiquity* 41, 276–88.

Renfrew, C. 1969. 'The autonomy of the south-east European Copper Age', *Proceedings of the Prehistoric Society* 35, 12–47.

Renfrew, C. 1972. *The Emergence of Civilization: The Cyclades and the Aegean in the Third Millennium* BC. Methuen, London.

Renfrew, C. 1973. *Before Civilization: the Radiocarbon Revolution and Prehistoric Europe.* Cambridge University Press, Cambridge.

Renfrew, C. 1976. 'Megaliths, territories, and populations', in Laet, S. de (ed.) *Acculturation and Continuity in Atlantic Europe.* De Tempel, Bruges, 198–220.

Reschreiter, H., Pany-Kucera, D., and Gröbner, D. 2012. 'Kindergarden 100 m below ground? New 'life scenes' for the Hallstatt salt mines'. Abstract in Anreiter, P. et al.

(eds) *Conference Guide, 2nd Mining in European History Conference*, HiMAT, Innsbruck, 48.

Robb, L. 2005. *Introduction to Ore-forming Processes.* Blackwell Publishing, Malden, Maryland.

Roberts, B. 2008. 'Creating traditions and shaping technologies: understanding the earliest metal objects and metal production in Western Europe', *World Archaeology* 40, 3, 354–72.

Roberts, B. 2009. 'Origins, transmission, and traditions: analyzing early metal in western Europe', in Kienlin, T. and Roberts, B. (eds) *Metals and Societies.* Verlag Dr Rudolf Habelt GMBH, Bonn, 129–42.

Roberts, B., Thornton, C., and Pigott, V. 2009. 'Development of metallurgy in Eurasia', *Antiquity* 83, 1012–22.

Rodríguez Casal, A. 1991. 'Megalithic culture and the beginnings of metal working in central and southern Portugal', in Mohen, J. P. and Eluère, C. (eds), *Découverte du Métal*, Amis du Musee des Antiquites Nationales, Picard, 317–24.

Roeder, C. 1901. 'Prehistoric and subsequent mining at Alderley Edge, with a sketch of the archaeological features of the neighbourhood', *Transactions of the Lancashire and Cheshire Antiquarian Society* 19, 77–118.

Roeder, C. and Graves, F. 1905. 'Recent archaeological discoveries at Alderley Edge', *Transactions of the Lancashire and Cheshire Antiquarian Society* 23, 17–29.

Rohl, B. and Needham, S. 1998. *The Circulation of Metal in the British Bronze Age: The Application of Lead Isotope Analysis.* British Museum Occasional Paper 102, London.

Rossi, M., Gattiglia, A., and Rostan, P. 1999. 'Aspects économiques de l'exploitation du cuivre et de la métallurgie du bronze dans les Alpes Franco-Piémontaises', in Della Casa, P. (ed.) *Prehistoric Alpine Environment, Society and Economy.* Proceedings of the international colloquium PAESE '97, Zurich, Universitätsforschungen zur Prähistorischen Archäologie 55, 257–66.

Rostan, P. 1994. 'Collobrières, Maraval et Guillaumes, Clue de Roua', *Bilan Sciéntifique de la région Provence-Alpes-Cote d'Azur*, Service régional de l' Archéologie, Région PACA, 224–5.

Rostan, P. and Mari, G. 2005. 'L'Exploitation protohistorique de cuivre natif de Roua (Daluis et Guillaumes, Alpes-Maritimes)', in Ambert, P. and Vaquer, J. (eds) *La Première Métallurgie en France et dans les Pays Limitrophes.* Mémoire 37 de La Société Préhistorique Française, Carcassonne, 139–49.

Rostan, P. and Rossi, M. 2002. 'Approche économique et industrielle du complexe minier et métallurgique de Saint-Véran (Hautes-Alpes) dans le contexte de l'Âge du Bronze des Alpes du Sud', *Bulletin d'Etudes Préhistoriques et Archéologiques Alpines* 13, 77–96.

Rothenberg, B. 1972. *Timna: Valley of the Biblical Copper Mines.* Thames and Hudson, London.

Rothenberg, B. and Blanco-Freijeiro, A. 1980. 'Ancient copper mining and smelting at Chinflon (Huelva), south-west Spain', in Craddock, P. (ed.) *Scientific Studies in Early Mining and Extractive Metallurgy*, British Museum Occasional Paper 20, London, 41–62.

Rothenberg, B. and Blanco-Freijeiro, A. 1981. *Studies in Ancient Mining and Metallurgy in South-West Spain*', institute of Archaeometallurgical Research, London.

Rothenberg, B., Carrión, F., Moreno, A., and Mérida, V. 1988. 'A prehistoric copper mine and smelter in south-east Andalusia (Spain): first discovery by IAMS survey team', *IAMS Newsletter* 13, 1–5', institute of Archaeometallurgical Research, London.

Roussot-Larroque, J. 1998. 'Premiers objets de cuivre dans le Sud-Ouest de la France', in Frere-Sautot, M. Ch. (ed.) *Paléométallurgie des Cuivres*, Actes du colloque de Bourg-en-Bresse et Beaune, Octobre 1997. Montagnac, 131–47.

Roussot-Larroque, J. 2005. 'Première métallurgie du Sud-Ouest Atlantique de la France', in Ambert, P. and Vaquer, J. (eds) *La Première Métallurgie en France et dans les Pays Limitrophes*. Mémoire 37 de La Société Préhistorique Française, Carcassonne, 159–74.

Rovira, S. 1999. 'Una propuesta metodológica para el estudio de la metalurgia Prehistórica: el caso de Gorny en la región de Kargaly (Orenburg, Russia)', *Trabajos de Prehistoria* 56, 85–113.

Rovira, S. 2002. 'Metallurgy and society in prehistoric Spain', in Ottaway, B. and Wager, E. (eds) *Metals and Society*. British Archaeological Reports S1061, Archaeopress, Oxford, 5–20.

Rovira, S. and Navarrete, M. 2005. 'Kargaly: esplendor minero el la edad de bronze', *Tierra y Tecnología* 27, 29–38.

Rowlands, M. 1971. 'The archaeological interpretation of metalworking', *World Archaeology* 3, 210–23.

Rowlands, M. 1976. *The Production and Distribution of Metalwork in the Middle Bronze Age in Southern Britain*. British Archaeological Reports (British series) 31, Archaeopress, Oxford.

Ruiz Taboada, A. and Montero Ruiz, I. 1999. 'The oldest metallurgy in western Europe', *Antiquity* 73, 897–903.

Sainter, J. 1878. *The Jottings of some Geological, Archaeological, Botanical, Ornithological, and Zoological Rambles around Macclesfield*. Swinnerton and Brown, Macclesfield.

Salvà, B., Llull, B., Perelló, L., and Rovira, S. 2010. 'Aproximación a la metalurgia prehistórica de las Baleares: estudio analítico de minerales cupríferos de Menorca', in Saiz, A. et al. (eds), *VIII Congreso Ibérico de Arqueometría. Actas*, Seminario de Arqueología y Etnología Turolense, Teruel, 183–96.

Satpaev, K. 1941. *Doistoricheskiye pamiatniki v Dzhezkazghanskom rayone*. Narodnoye khoziaystvo Kazakhstana 1, Alma Ata.

Schaer, A. 2003. 'Untersuchungen zum prähistorischen bergbau im Oberhalbstein (Kanton Grabünden)', *Jahrbuch der Schweizerischen Gesellschaft für Ur- und Frühgeschichte* 86, 7–54.

Schibler, J., Breitenlechner, E., Deschler-Erb, S., Goldenberg, G., Hanke, K., Hiebel, G., Hüster Plogmann, H., Nicolussi, K., Marti-Grädel, E., Pichler, S., Schmidl, A., Schwarz, S., Stopp, B., and Oeggl, K. 2011. 'Miners and mining in the Late Bronze Age: a multidisciplinary study from Austria', *Antiquity* 85, 1259–78.

Schwab, R., Ullén, I. and Wunderlich, C.-H. 2010. 'A sword from Vreta Kloster, and black patinated bronze in Early Bronze Age Europe', *Journal of Nordic Archaeological Science* 17, 27–35.

Scott, B. and Francis, P. 1981. 'Native copper in north-east Ulster: a contributory factor in the establishment of the earliest Irish metalworking?', in Ó Corráin, D. (ed.) *Irish Antiquity, Essays, and Studies presented to Professor M. J. O'Kelly*. Tower Books, Cork, 28–42.

Serra Vilaro, J. 1920. *Mina I fundicio d'aram del primer periode de l'Etat del Bronze de Riner*. Bulleti del Centre Excursionista de Catalunya. Anuari VI, Barcelona.

Shennan, S. 1995. *Bronze Age Copper Producers of the Eastern Alps: Excavations at St Veit-Klinglberg*. Universitätsforschungen zur Prähistorischen Archäologie, Band 27, Bonn.

Shennan, S. 1999. 'Cost, benefit, and value in the organization of early European copper production', *Antiquity* 73, 280, 352–63.

Shennan, S. 2010. 'Early mining and subsistence: some theoretical considerations', in Anreiter, P. et al. (eds) *Mining in European History and its Impact on Environment and Human Societies*. Proceedings of the 2009 SFB-HiMAT conference, University of Innsbruck', Innsbruck, 159–64.

Shepherd, R. 1980. *Prehistoric Mining and Allied Industries*. Academic Press, London.

Sheridan, A. 1983. 'A reconsideration of the origins of Irish metallurgy', *The Journal of Irish Archaeology* 1, 11–19.

Sherratt, A. 1976. 'Resources, technology, and trade: an essay on early European copper production', in Sieveking, G., Longworth, I., and Wilson, K. (eds) *Problems in Economic and Social Anthropology*. Duckworth, London, 557–81.

Sherratt, A. 1981. 'Plough and pastoralism: aspects of the secondary products revolution', in Hodder, I., Isaac, G., and Hammond, N. (eds) *Pattern of the Past; Studies in honour of David Clarke*. Cambridge University Press, Cambridge, 261–305.

Sherratt, S. 2000. 'Circulation of metals and the end of the Bronze Age in the eastern Mediterranean', in Pare, C. (ed.) *Metals Make the World Go Round: The Supply and Circulation of Metals in Bronze Age Europe*, Oxbow, Oxford, 82–98.

Siret, E. and Siret, L. 1890. *Las Primeras Edades del Metal en el Sudeste de España*. Barcelona.

Skeates, R. 1993. 'Early metal use in the central Mediterranean region', *Accordia Research Papers* 4, 5–48.

Slavchev, V. 2010. 'The Varna Eneolithic cemetery in the context of the Late Copper Age in the East Balkans', in Anthony, D. and Chi, J. (eds) *The Lost World of Old Europe: the Danube Valley, 5000–3500 BC*. Princeton University Press, New York, 192–210.

Sljivar, D. 2003. 'New data about the beginning of copper mining and metallurgy in the Vinča culture of the central Balkans', *Cu⁺ Bulletin du Groupe de Travail International sur la paléométallurgie des cuivres et des minerais associés* 3, 5–10.

Sljivar, D. 2006. 'The earliest copper metallurgy in the Balkans', *Metalurgija—Journal of Metallurgy* 12, 93–104.

Snodin, S. R. 1972. *Copper and other Mineralization in the Upper Palaeozoic Strata of south-west Cork*. Unpublished PhD thesis, University of Leicester.

Solecki, R. 1969. 'A copper mineral pendant from northern Iraq', *Antiquity* 43, 311–14.

Soriano Llopis, I. 2011. 'De tumbras de metalúrgico en el Nordeste peninsular. El Forat de la Tuta (Riner, Solsonès, Lleida)', *Revista d'Arqueologia de Ponent* 21, 37–56.

Srejovic, D 1972. *Europe's First Monumental Sculpture: New Discoveries at Lepenski Vir*. Thames and Hudson, London.

Steiniger, D. and Giardino, C. 2012. 'Prehistoric mining in central Italy: new evidence from the Monti della Tolfa (Latium)', in Anreiter, P. et al. (eds) *Second Mining in European History Conference. Conference abstracts*. University of Innsbruck, Innsbruck, 46.

Stevenson, M. 1982. 'Toward an understanding of site abandonment behaviour: evidence from historic mining camps in the south-west Yukon', *Journal of Anthropological Archaeology* 1, 237–65.

Stöllner, T. 2003a. 'Mining and economy: a discussion of spatial organisations and structures of early raw material exploitation', in Stöllner, T., Körlin, G., Steffens, G., and Cierny, J. (eds) *Man and Mining*. Der Anschnitt beiheft 16, Bergbau-Museum, Bochum, 414–46.

Stöllner, T. (with contributors) 2003b. 'The economy of Dürrnberg-Bei-Hallein: an Iron Age salt-mining centre in the Austrian Alps', *The Antiquaries Journal* 83, 123–94.

Stöllner, T. 2009. 'Die zeitliche Einordnung der prähistorischen Montanreviere in den Ost- und Südalpen: Anmerkungen zu einem Forschungsstand', in Oeggl, K. and Prast, M. (eds) *Die Geschichte des Bergbaus in Tirol und Seinen Angrenzenden Gebieten*. Innsbruck University Press, Innsbruck, 37–60.

Stöllner, T. 2010. 'Copper and salt: mining communities in the alpine Metal Ages', in Anreiter, P. et al. (eds) *Mining in European History and its Impact on Environment and Human Societies*. Proceedings of the 2009 SFB-HiMAT conference, University of Innsbruck', Innsbruck, 297–308.

Stöllner, T., Breitenlechner, E., Fritzsch, D., Gontscharov, A., Hanke, K., Kirchner, D., Kovacs, K., Moser, M., Nicolussi, K., Oeggl, K., Pichlerm T., Pils, R., Prange, M., Thiemeyer, H., and Thomas, P. 2010. 'Ein nassaufbereitungskasten vom Troiboden', interdisziplinäre enforschung des bronbzezeitlichen montanwesens am Mitterberg (land Salzburg, Österreich)', *Jahrbuch des Römisch-Germanischen Zentralmuseums Mainz* 57, 1–32.

Stöllner, T., Breitenlechner, E., Eibner, C., Herd, R., Kienlin, T., Lutz, J., Maass, A., Nicolussi, K., Pichler, T., Pils, R., Röttger, K., Song, B., Taube, N., Thomas, P., and Thurner, A. 2011. 'Der Mitterberg – Der Grossproduzent für kupfer im östlichen Alpenraum während der Bronzezeit', in Goldenberg, G., Töchterle, U., Oeggl, K., and Krenn-Leeb, A. (eds) *Forschungsprogramm HiMAT: Neues zur Bergbaugeschichte der Ostalpen*. Archäologie Österreichs Spezial 4, Vienna, 113–44.

Stos, S. 2009. 'Across the wine dark seas . . . sailor tinkers and royal cargoes in the Late Bronze Age', in Shortland, A. J., Freestone, I. C., and Pollard, A. M. (eds) *From Mine to Microscope: Advances in the Study of Ancient Technology*. Oxbow Books, London, 163–80.

Stos-Gale, S. 1989. 'Cycladic copper metallurgy', in Hauptmann, A., Pernicka, E., and Wagner, G. (eds) *Old World Archaeometallurgy*. Proceedings of the 1987 Heidelberg Symposium, Bochum, 279–91.

Stos-Gale, S. 2000. 'Trade in metals in the Bronze Age Mediterranean: an overview of lead isotope data for provenance studies', in Pare, C. (ed.) *Metals Make the World*

Go Round: The Supply and Circulation of Metals in Bronze Age Europe. Oxbow, Oxford, 56–69.

Stos-Gale, S., Maliotis, G., Gale, N., and Annetts, N. 1997. 'Lead isotope characteristics of the Cyprus copper ore deposits applied to provenance studies of copper oxhide ingots', *Archaeometry* 39, 1, 83–124.

Strahm, C. 1994. 'Die anfänge der metallurgie in Mitteleuropa', *Helvetia Archaeologica* 25, 2–39.

Strahm, C. 2005. 'L'introduction et al diffusion de la métallurgie en France', in Ambert, P. and Vaquer, J. (eds) *La Première Métallurgie en France et dans les Pays Limitrophes*. Mémoire 37 de La Société Préhistorique Française, Carcassonne, 27–36.

Strahm, C. and Hauptmann, A. 2009. 'The metallurgical developmental phases in the Old World', in Kienlin, T. and Roberts, B. (eds) *Metals and Societies*. Verlag Dr Rudolf Habelt GMBH, Bonn, 116–28.

Sykes, C. 1796. *Journal of a Tour in Wales*. MS 2258C, National Library of Wales.

Taylor, J. Du Plat 1952. 'A Late Bronze age settlement at Apliki, Cyprus', *The Antiquaries Journal* 32, 133–40.

Taylor, T. 1999. 'Envaluing metal: theorizing the Eneolithic "hiatus"', in Young, S., Pollard, M., Budd, P., and Ixer, R. (eds) *Metals in Antiquity*, Proceedings of the 1995 Harvard Conference. British Archaeological Reports (international series) 792, Harvard, Cambridge, MA, 22–32.

Theocharis, D. 1952. 'Anaskaphai en Araphini', *Praktika tis Archaiologikis Etaireais*, 129–51.

Theocharis, D. 1954. 'Anaskaphai en Araphini', *Praktika tis Archaiologikis Etaireais*, 104–13.

Thrane, H., 1975. 'Europæiske forbindelser. Bidrag til studiet af fremmede forbindelser i Danmarks yngre broncealder (periode IV-V)', in: *Arkeologisk-historisk række, Bind XVI*. Nationalmuseets Skrifter, København.

Timberlake, S. 1987. 'An archaeological investigation of early mine workings on Copa Hill, Cwmystwyth', *Archaeology in Wales* 27, 18–20.

Timberlake, S. 1988. 'Excavations at Parys Mountain and Nantyreira', *Archaeology in Wales* 28, 11–17.

Timberlake, S. 1990a. 'Excavations at Pary's Mountain and Nantyreira', in Crew, P. and Crew, S. (eds) *Early Mining in the British Isles*, Plas Tan y Bwlch Occasional Paper 1, Snowdonia Press, Porthmadog, 15–21.

Timberlake, S. 1990b. 'Excavations and fieldwork on Copa Hill, Cwmystwyth, 1989', in Crew, P. and Crew, S. (eds) *Early Mining in the British Isles*. Plas Tan y Bwlch Occasional Paper 1, Snowdonia Press, Porthmadog, 22–9.

Timberlake, S. 1990c. 'Review of the historical evidence for the use of firesetting', in Crew, P. and Crew, S. (eds) *Early Mining in the British Isles*. Plas Tan y Bwlch Occasional Paper 1, Snowdonia Press, Porthmadog, 49–52.

Timberlake, S. 1992. 'Prehistoric copper mining in Britain', *Cornish Archaeology* 31, 15–34.

Timberlake, S. 1994. 'Archaeological and circumstantial evidence for early mining in Wales', *Mining before Powder, Bulletin of the Peak District Mines Historical Society* 12, 3, 133–43.

Timberlake, S. 2002. 'Ancient prospection for metals and modern prospection for ancient mines: the evidence for Bronze Age mining within the British Isles', in Bartelheim, M., Pernicka, E., and Krause, R. (eds) *The Beginnings of Metallurgy in the Old World.* Verlag Marie Leidorf GmbH , Rahden, 327–57.

Timberlake, S. 2003a. 'Early mining research in Britain: the developments of the last ten years', in Craddock, P. and Lang, J. (eds) *Mining and Metal Production through the Ages.* London, British Museum, 21–42.

Timberlake, S. 2003b. *Excavations on Copa Hill, Cwmystwyth (1986–1999): An Early Bronze Age Copper Mine within the Uplands of Central Wales.* British Archaeological Reports (British series) 348. Archaeopress, Oxford.

Timberlake, S. 2006. 'Excavations of early mine workings at Twll y Mwyn (Cwm Darren) and Erglodd, Ceredigion', *Archaeology in Wales* 46, 79–86.

Timberlake, S. 2009. 'Copper mining and production at the beginning of the British Bronze Age', in Clark, P. (ed.) *Bronze Age Connections: Cultural Contact in Prehistoric Europe.* Oxbow, Oxford, 94–121.

Timberlake, S. 2010. 'Geological, mineralogical, and environmental controls on the extraction of copper ores in the British Bronze Age', in Anreiter, P. et al. (eds) *Mining in European History and its Impact on Environment and Human Societies.* Proceedings of the 2009 SFB-HiMAT conference, University of Innsbruck, Innsbruck, 289–95.

Timberlake, S. and Mason, J. 1997. 'Ogof Wyddon (Machynlleth Park copper mine)', *Archaeology in Wales* 37, 62–5.

Timberlake, S. and Marshall, P. 2013. 'Understanding the chronology of British Bronze Age mines: Bayesian modeling and theories of exploitation', in Anreiter, P. et al. (eds) *Mining in European History and the Impact on Environment and Human Societies.* Proceedings of the 2012 SFB-HiMAT conference, University of Innsbruck, 59–66.

Timberlake, S. and Prag. A. 2005. *The Archaeology of Alderley Edge: Survey, Excavation, and Experiment in an Ancient Mining Landscape.* John Hedges, Oxford.

Timberlake, S. and Switsur, R. 1988. 'An archaeological investigation of early mine workings on Copa Hill, Cwm Ystwyth: new evidence for prehistoric mining', *Proceedings of the Prehistoric Society* 54, 329–33.

Tite, M. 1996. 'In defence of lead isotope analysis', *Antiquity* 70, 959–62.

Tizzoni, M., Cucini Tizzoni, C., Morin, D., and Jacquemot, D. 2003. 'Campolungo (Bienno-Brescia)', Campagna 2001. *Notizie Archeologiche Bergomensi* 11, 237–52.

Töchterle, U., Goldenberg, G., Klaunzer, M., Hye, S., and Tomedi, G. 2009. 'Neue Ergebnisse der Grabungen 2007/2008 am Kiechlberg bei Thaur in Nordtitol', in Oeggl, K. and Prast, M. (eds) *Die Geschichte des Bergbaus in Tirol und Seinen Angrenzenden Gebieten.* Proceedings 3 HiMAT project, University of Innsbruck, Innsbruck, 291–8.

Tocik, A. and Zebrák, P. 1989. 'Ausgrabungen in Spania Dolina-Piesky', in Hauptmann, A., Pernicka,E., and Wagner, G. (eds) *Old World Archaeometallurgy.* De Anschnitt beiheft 7, Bochum: Deutsches Bergbau-Museum, 71–8.

Todorova, H. 1978. *The Eneolithic Period in Bulgaria in the Fifth Millennium* BC. British Archaeological Reports (international series) S49, Archaeopress, Oxford.

Torrence, R. 1984. 'Monopoly or direct access? Industrial organization at the Melos obsidian quarries', in Ericson, J. and Purdy, J. (eds) *Prehistoric Quarries and Lithic Production.* Cambridge University Press, Cambridge, 49–64.

Trebsche, P. 2012. 'Resources and nutrition in the Urnfield period mining site of Prigglitz-Gasteilin Lower Austria: preliminary report on the excavations in 2010 and 2011', in Anreiter, P. et al. (eds) *2nd Mining in European History Conference*. Abstract, Conference Guide, Innsbruck University Press, Innsbruck, 32.

Tykot, R. 1996. 'Obsidian procurement and distribution in the central and western Mediterranean', *Journal of Mediterranean Archaeology* 9, 1, 39–82.

Tylecote, R., Ghaznavi, H. A., and Boydell, P. J. 1977. 'Partitioning of trace elements between the ores, fluxes, slags, and metal during the smelting of copper', *Journal of Archaeological Science* 4, 305–33.

Vandkilde, H. 1996. *From Stone to Bronze. The Metalwork of the Late Neolithic and Earliest Bronze Age in Denmark*. Jutland Archaeological Society Publications 32, Aarhus.

Vasseur, G. 1911. 'Une mine de cuivre exploitée a l'age du bronze dans les garrigues de l'Hérault (environ de Cabrières)', *L'Anthropologie* 22, 413–20.

Vicent, J., Rodríguez Alcalde, A., López Sáez, J., De Zavala, I., López García, P., and Martínez Navarrete, M. 2000. 'Catástrofes ecólogicas en la estepa? Arqueología del paisaje en el complejo minero-metalúrgico de Kargaly (región de Orenburg, Rusia)', *Trabajos de Prehistoria* 57, 1, 29–74.

Viehweider, B. and Oeggl, K. 2012. 'Reconstruction of the vegetation in the region of Kitzbühel (Tyrol) and the impact of different mining phases since the Neolithic of Kitzbühel', in Anreiter, P. et al. (eds) *2nd Mining in European History Conference*. Abstract, Conference Guide, Innsbruck University Press, Innsbruck, 74.

Vilaseca, S. and Vilaseca, L. 1957. 'Una explotación minera prehistórica. La Solana del Bepo, de Ulldemolins (prov. de Tarragona)', *IV Congreso Nacional de Arqueología, Burgos 1955, Zaragoza*, Nacional de Arqueología, Zaragoza, 135–9.

Wager, E. 2002. 'Mining as social process: a case study from the Great Orme, North Wales', in Ottaway, B. and Wager, E. (eds) *Metals and Society*. British Archaeological Reports S1061, Archaeopress, Oxford, 45–9.

Wager, E. 2009. 'Mining ore and making people: re-thinking notions of gender and age in Bronze Age mining communities', in Kienlin, T. and Roberts, B. (eds) *Metals and Societies*. Verlag Dr Rudolf Habelt GMBH, Bonn, 105–15.

Wagner, G., Gentner, W., Gropengiesser, H., and Gale, N. 1980. 'Early Bronze Age lead-silver mining and metallurgy in the Aegean: the ancient workings on Siphnos', in Craddock, P. (ed.) *Scientific Studies in Early Mining and Extractive Metallurgy*. British Museum Occasional Paper 20, London, 63–80.

Waldren, W. 1998. *The Beaker Culture of the Balearic Islands*. British Archaeological Reports, (international series) S709, Archaeopress, Oxford.

Webb, J. M. and Frankel, D. 2013. *Ambelikou Aletri. Metallurgy and Pottery Production in Middle Bronze Age Cyprus*. Studies in Mediterranean Archaeology 138, Åströms Förlag, Uppsala.

Weisgerber, G. 1982. 'Towards a history of copper mining in Cyprus and the Near East: possibilities of mining archaeology', in Muhly, J., Maddin, R., and Karageorghis, V. (eds) *Early Metallurgy in Cyprus 4000–500* BC. Pierides Foundation, Nicosia, 25–30.

Weisgerber, G. 1987. 'The technological relationship between flint mining and early copper mining', in Sieveking, G. de G. and Newcomer, M. (eds) *The Human Uses of Flint and Chert*. Proceedings of the fourth international flint symposium held at Brighton Polytechnic 10–15 April 1983. Cambridge University Press, Cambridge, 131–5.

Weisgerber, G. 1988. 'Oman: a bronze producing centre during the 1st half of the 1st millennium BC', in Curtis, J. (ed.) *Bronze-working Centres of Western Asia, c.1000–539 BC*, Kegan Paul International and the British Museum, London, 285–95.

Weisgerber, G. 1989. 'Montanarchäologie: Grundzüge einer systematischen Bergbaukunde für Vor- und Fruhgeschichte und Antike. Teil 1', *Der Anschnitt* 41, 6, 190–204.

Weisgerber, G. 1990. 'Montanarchäologie: Grundzüge einer systematischen Bergbaukunde für Vor- und Fruhgeschichte und Antike. Teil 2', *Der Anschnitt* 42, 1, 2–18.

Weisgerber, G. and Pernicka, E. 1995. 'Ore mining in prehistoric Europe: an overview', in Morteani, G. and Northover, J. (eds) *Prehistoric Gold in Europe*. Kluwer Academic Publishers, The Netherlands, 159–82.

Weisgerber, G. and Willies, L. 2001. 'The use of fire in prehistoric and ancient mining: firesetting', *Paleorient* 26, 2, 131–49.

Wertime, T. A. 1973. 'The beginnings of metallurgy: a new look', *Science* 182, 875–86.

Wijngaarden–Bakker, L. van 2004. 'The animal remains', in O'Brien, W. *Ross Island*. Bronze Age Studies 6, National University of Ireland Galway, Galway, 367–85.

Willies, L. 1994. Firesetting technology. *Mining before Powder: Peak District Mines Historical Society Bulletin* 12, 3, 1–8.

Winghart, S. 2000. 'Mining, processing, and distribution of bronze: reflections on the organisation of metal supply between the northern Alps and the Danube region', in Pare, C. (ed.) *Metals Make the World Go Round: Supply and Circulation of Metals in Bronze Age Europe*. Oxbow, Oxford, 151–9.

Witter, W. 1938. *Die älteste Erzgewinnung im nordisch-germanische Lebenskreis* (2 vols). Mannus Bücherei 60 and 63, Leipzig.

Zachos, K. 2007. 'The Neolithic background: a reassessment', in Day, P. and Doonan, R. (eds) *Metallurgy in the Early Bronze Age Aegean*. Sheffield Studies in Aegean Archaeology 7, Sheffield, 168–206.

Zebrák, P. 1995. 'The traces of the primary mining of non-ferrous metals in Slovakia', in Jovanović, B. (ed.) *Ancient Mining and Metallurgy in Southeast Europe*. Archaeological Institute, Museum of Mining and Metallurgy, Gornji Milanovac, 13–19.

Zschocke, K. and Preuschen, E. 1932. *Das urzeitliche Berbaugebiet von Mühlbach-Bischofshofen*. Materialien zur Urgeschichte Österreichs 6. Anthropologische Gesellschaft in Wien, Vienna.

Zuffardi, P. 1989. 'Italy', in Dunning, F., Garrard, P., Haslam, H., and Ixer, R. (eds) *Mineral Deposits of Europe: Southwest and Eastern Europe, with Iceland* volume 4/5. The Institution of Mining and Metallurgy, London, 221–77.

Zwicker, U. 1986. 'Ancient metallurgical methods for copper production on Cyprus: part 1 (natural copper, oxide, sulphate, and silicate ore); part 2 (sulphide ore and copper-arsenic-alloy production)', *Cyprus Association of Geologists and Mining Engineers, Bulletin* 3, 79–91, 92–111.

Index

The manufacturer's authorised representative in the EU for product safety is Oxford University Press España S.A. of el Parque Empresarial San Fernando de Henares, Avenida de Castilla, 2 – 28830 Madrid (www.oup.es/en or product.safety@oup.com). OUP España S.A. also acts as importer into Spain of products made by the manufacturer.

www.ingramcontent.com/pod-product-compliance
Ingram Content Group UK Ltd.
Pitfield, Milton Keynes, MK11 3LW, UK
UKHW020849150926
475595UK00003B/97

9780199605651